Innovative Renewable Energy

Series editor
Ali Sayigh
World Renewable Energy Congress, Brighton, UK

The primary objective of this book series is to highlight the best-implemented worldwide policies, projects and research dealing with renewable energy and the environment. The books are developed in partnership with the World Renewable Energy Network (WREN). WREN is one of the most effective organizations in supporting and enhancing the utilisation and implementation of renewable energy sources that are both environmentally safe and economically sustainable. Contributors to books in this series come from a worldwide network of agencies, laboratories, institutions, companies and individuals, all working together towards an international diffusion of renewable energy technologies and applications. With contributions from most countries in the world, books in this series promote the communication and technical education of scientists, engineers, technicians and managers in this field and address the energy needs of both developing and developed countries.

Each book in the series contains contributions from WREN members and cover the most-up-to-date research developments, government policies, business models, best practices, and innovations from countries all over the globe. Additionally, the series publishes a collection of best papers presented during the annual and bi-annual World Renewable Energy Congress and Forum each year.

More information about this series at http://www.springer.com/series/15925

Ali Sayigh • David Milborrow
Editors

The Age of Wind Energy

Progress and Future Directions from a Global Perspective

Springer

Editors
Ali Sayigh
World Renewable Energy Congress
Brighton, UK

David Milborrow
Lewes, East Sussex, UK

ISSN 2522-8927 ISSN 2522-8935 (electronic)
Innovative Renewable Energy
ISBN 978-3-030-26445-1 ISBN 978-3-030-26446-8 (eBook)
https://doi.org/10.1007/978-3-030-26446-8

This Springer imprint is published by the registered company Springer Nature Switzerland AG
The registered company address is: Gewerbestrasse 11, 6330 Cham, Switzerland

Contents

About the Editors

Ali Sayigh BSc, AWP, DIC, PhD is a UK Citizen who graduated from Imperial College London in 1966. He is Fellow of the Institute of Energy and the Institution of Engineering and Technology, Chartered Engineer, and Chairman of the Iraq Energy Institute.

He taught in various countries like Iraq, Saudi Arabia, and Kuwait, and in universities like Reading University and University of Hertfordshire from 1966 to 2004. He was Head of the Energy Department at Kuwait Institute for Scientific Research (KISR) and Expert in renewable energy at AOPEC, Kuwait, from 1981 to 1985.

He started working in solar energy in September 1969. In 1972, he established with some colleagues the *Journal of Engineering Sciences* in Riyadh, Saudi Arabia, and in 1984 the *International Journal for Solar and Wind Technology*, him as an Editor in Chief. This has changed its name in 1990 to *Journal of Renewable Energy*. He is Editor of several international journal published in Morocco, Iran, Bangladesh, Nigeria, and India. He established the World Renewable Energy Network (WREN) and the World Renewable Energy Congress in 1990. Moreover, he is Member of various societies related to climate change and renewable energy and is Chairman of the Iraq Energy Institute since 2010.

He was Consultant to many national and international organizations, among them, the British Council, ISESCO, UNESCO, UNDP, ESCWA, UNIDO, and

UN. He organized conferences and seminars in 52 different countries and published more than 600 papers. He has edited, authored, and associated with more than 75 books and has supervised more than 80 MSc and 35 PhD students. He is Editor in Chief of the yearly *Renewable Energy Magazine* since 2000. He is the Founder of WREN journal *Renewable Energy* published by Elsevier and was the Editor in Chief for 30 years from 1984 to 2014.

He is the Editor in Chief of *Comprehensive Renewable Energy* coordinating 154 top scientists, engineers, and researchers' contribution in 8 volumes published in 2012 by Elsevier which won 2013 PROSE Award in the USA. In 2011, he founded Med Green Buildings and Renewable Energy Forum. In 2016, he established peer-reviewed international open-access journal called *Renewable Energy and Environmental Sustainability* (*REES*), which is published in English online by EDP Sciences publisher in Paris.

He is also Winner of the Best Clean Energy Implementation Support NPO, UK. In 2018, WREN was rated globally as one of the best organizations in the UK promoting renewable energy. In November 2018, he was elected Fellow of the Royal Society of Art (FRSA).

David Milborrow is an Energy Consultant with more than 40 years of experience in renewable energy, principally wind. He has been an Active Member of the World Renewable Energy Network during the last 25 years.

He carries out studies on the economics of renewable energy sources and comparisons with those of the thermal sources of electricity generation. He is also a Specialist on integration issues. He is, or has been, an Adviser to a number of bodies including the European Commission, the UK Department of Trade and Industry, the British Wind Energy Association (now RenewableUK), and the Engineering and Physical Sciences Research Council. He is Technical Adviser to the journal *Windpower Monthly* and compiles a monthly contribution on wind energy economics. He has lectured at five universities, has contributed to several books on wind energy, and has written over 100 articles.

From 1984 to 1992, he was a Principal Engineer with the Central Electricity Generating Board (CEGB) in the UK and was closely involved in the planning of the country's first wind farms. Prior to that, he carried out research on interactions between wind turbines in the laboratory and in the field. He obtained his BSc degree in Mechanical Engineering from the University of London (1961–1965).

Chapter 1
Introduction

Ali Sayigh

Nowadays, wind energy is at last reaching its potential as this book fully illustrates. It has developed from humble beginnings over 80 years ago when modern turbines were first developed to provide 10–20 kW power. These early wind turbines were mainly used for pumping water and in small-scale, mainly domestic, electricity production. Since then, the technology has grown exponentially and the only obstacle to their widespread use is political and not technical. Many countries utilise wind power as a major source of electricity generation, and with the greater realisation of the immense danger of climate change, the importance of wind energy to meet the growing demand for the energy with the lowest carbon footprint is now accepted. Such is the sophistication of the current generation of wind turbines which embody machines without gearboxes capable of producing more than 8 MW which is enough to provide electricity to a community of over 10,000 inhabitants.

Early wind turbines had blades 10–15 m in length and a tower height of 20–30 m, whereas current turbines can have blades of 160 m and a tower more than 300 m in height. Cost-wise, electricity generated from wind turbine is on a par with that generated by fossil fuels. Wind turbines have the lowest embedded carbon among various forms of renewable energy generation, and unlike nuclear energy they present no long-term danger to humanity and are considerably cheaper. Installation of wind turbines whether onshore or offshore is much speedier than that of other energy sources and the time scale of investment to production is much quicker. Their productive life also compares favourably to other energy sources, and the cost of decommissioning is significantly less and presents no hidden risks.

Each chapter of this volume outlines an important aspect of wind energy generation as experienced in a variety of nations and economics, taking into consideration both technology and policy. It is envisaged that this book will be useful to readers involved in the technological development of wind energy as well as those who are

A. Sayigh (✉)
World Renewable Energy Congress, Brighton, UK
e-mail: asayigh@wrenuk.co.uk

A. Sayigh, D. Milborrow (eds.), *The Age of Wind Energy*, Innovative Renewable Energy, https://doi.org/10.1007/978-3-030-26446-8_1

responsible for energy planning. During the last 80 years, wind turbines proved to be cost-effective in generating electricity; they require smaller number of people to run and a small space to operate. They are reliable and can sustain production of electricity, in most cases, without storage.

Chapter 2
Wind Energy Development

David Milborrow

2.1 Introduction

At the end of the year 2000, the amount of wind energy capacity in the world was 17.7 GW. By the end of 2010, the capacity was 191 GW and by the end of 2018 the total was around 561 GW. That represents a compound annual growth rate of 21%. Such a high growth rate is not sustainable in the long term and the growth rate in 2018 was about 9%, but that represented an increase of 46 GW—over double the capacity that was operational in the year 2000. There are strong indications that 2019 is likely to be a higher-growth year, with the extra installed capacity likely to be around 50 GW. Figure 2.1 shows how wind energy has developed since 2000.

The reasons for the strong growth of wind can be explored by comparing its attributes with those of the other renewable energy sources and this is the basis of Table 2.1.

The fact that wind energy is now cost-competitive with gas and coal and cheaper than most of the other renewable energy sources, clearly works in its favour. The short build time is also an asset. The principal difficulties that have delayed or frustrated some developments have been the visual impact, noise issues and interference with bird flight paths. Although the latter was frequently a problem with some of the early wind farms, developers take considerable care to research the possible environmental impacts, and there are now fewer problems reported. Now that offshore wind is becoming increasingly competitive, this opens up a potentially enormous resource, mostly free of environmental constraints, although ecological issues still need to be taken into account.

The capacity of solar photovoltaics is likely to overtake that of wind in the fairly near future, but its lower load factor means that energy generation is likely to lag behind, at least in the short to medium term. Photovoltaic arrays require considerable land which is effectively taken out of use, whereas in the case of wind energy, only a

D. Milborrow (✉)
Consultant, Lewes, East Sussex, UK

A. Sayigh, D. Milborrow (eds.), *The Age of Wind Energy*, Innovative Renewable Energy, https://doi.org/10.1007/978-3-030-26446-8_2

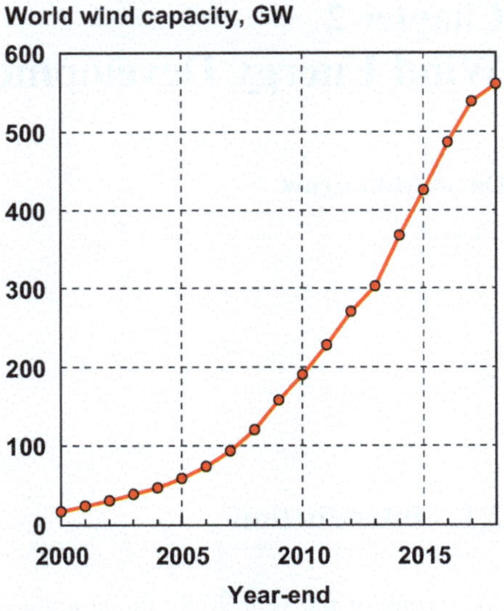

Fig. 2.1 Wind energy development since the year 2000. The figure for 2018 is provisional, and the final figure for the increase of capacity is likely to be similar to the 2017 figure

Table 2.1 Summary of the principal characteristics, advantages, and disadvantages of the renewable energy sources

Technology	Capacity (GWe)	Energy, 2017 (TWh)	Attributes (drawbacks in italics)	Locations
Hydro	1267	4185	Limited extra scope, due to need for reservoirs	Worldwide
Wind	540	1280	Short build time. Low-cost *Visual issues*	Worldwide
Solar PV	400	430	Modular, no moving parts *Land requirements*	Worldwide
Geothermal	14.3	86	Resource availability is limited to regions where high temperatures can be found at suitable depths	Pacific Rim, USA, some EU states
Tidal barrage of tidal stream	0.6		*Low load factor*, technology simple, but *cost of structures high*	Canada, UK
Wave	<1		Despite considerable R&D, no move yet to commercial viability	EU, Japan, USA
Energy crops	Not known		*Costs tend to be high; land use may be an issue*	Worldwide
Nuclear	392	2506	For comparison	

small area of land is sterilised by the tower bases and farming can continue around them, as shown in Fig. 2.2.

Although considerable research has taken place into the potential of tidal devices, very few have reached commercial viability. The same comments apply to wave

Fig. 2.2 Wind farm sited on a working farm, showing the minimal disturbance. (Photo: author)

energy, where research has been in progress for around 45 years, without yielding devices that are commercially viable. A possible reason is the complexity of the mechanical engineering required to convert the oscillatory motion of the waves into a rotary motion that is needed for the production of electricity. The prospects for energy crops, or bioenergy, are similarly being explored, worldwide, and the difficulty here is possibly the low energy density of the fuels. However, a very diverse range of options is being researched, which increases the likelihood of a breakthrough in the future.

The very rapid growth in the development of wind energy has been accompanied by an increase in the size of wind turbines and steadily falling generation costs. Of all the renewable energy sources, wind is now generally the cheapest, although there are instances where geothermal can be cheaper (where steam rises close to the surface) or where favourable geography enables hydro schemes to be developed economically. As noted in Table 2.1, at the end of 2018, there was around 600 GW of wind energy in the world, hydro approximately 1267 GW, and geothermal 11 GW, while solar photovoltaics—now developing very rapidly—accounted for 400 GW. According to the World Nuclear Association, there are now nuclear power plants with approximately 392 GW of nuclear power capacity that are operational, and so wind energy has overtaken nuclear energy—in capacity terms but not yet in energy terms.

The average load factor of wind energy, worldwide, is about 25% and so the energy-generating potential of the world's wind energy is around 1280 TWh. That is enough electricity to supply the whole of India or the Russian Federation. The success of wind energy can be put down to the plentiful resources, the fact that it is a proven technology and that it can be installed quickly. Average construction time of an onshore wind farm is generally less than a year and even offshore wind farms can take less than 2 years to construct.

2.2 Brief History

The Californian oil crisis of the late 1970s triggered growth in wind energy. In California, subsidies for wind energy led to the construction of several thousand small machines, with rotor diameters in the range 10–20 m and rated powers in the range 20–50 kW. Other countries followed suit, particularly Denmark, with similar results. In parallel with this activity, a number of governments initiated research and development programmes that aimed to develop megawatt size wind turbines. The thinking behind this approach was that it would cut down the number of machines required to produce quantities of electricity comparable with those from conventional power stations. Put another way, the aim was to produce the "jumbo jet" wind turbine more or less straight from the drawing board. By and large, however, few commercial machines emanated from the "large machine" research programmes, although they yielded considerable design insights. What happened instead was that commercial machine sizes gradually increased, so that 100 kW machines were available by the mid-1980s, 1 MW machines by the early nineties, and by the turn of the century, the largest machines had ratings around 3 MW and rotor diameters up to 70 m.

Figure 2.3 shows how the average rating of machines built in each year in the USA grew from 1980 onwards [1]. The rating reached 200 kW by 1990, 500 kW by 1997, 1 MW by 2002, and 2 MW shortly after 2010. A similar trend was followed in most of the other states that were building wind turbines.

The growth of offshore wind—with the first commercial farm being commissioned in 1991—accelerated the interest in large machines, to the point where the largest turbine now envisaged has a rating of 12 MW and a diameter of 220 m. Details of this and other very large machines now under development are shown in Table 2.2.

2.3 Design Options

There has been a wide variety of design options over the past 40 years. More recently, there has been some convergence, with the main design variations now being in the drive train. Table 2.3 summarises the main options for horizontal axis turbines, with an indication of the most common usage in 2018.

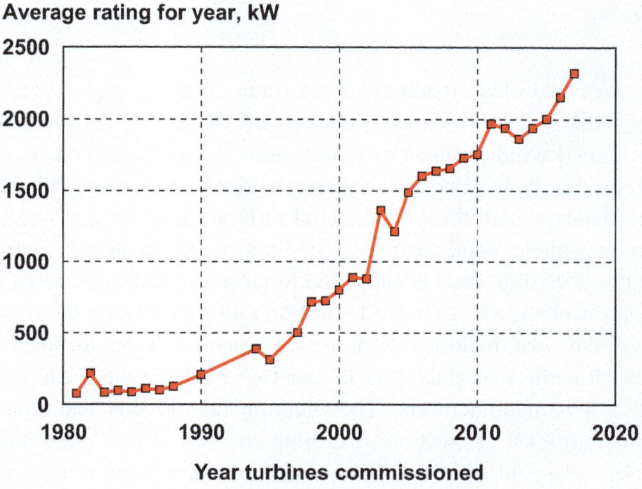

Fig. 2.3 Development of wind turbine power ratings in the USA

Table 2.2 The largest and most powerful machines in the world

Manufacturer	Rating (MW)	Diameter (m)	Comments/status
Adwen	8	180	Geared/prototype testing
GE wind	12	220	Under development
Vestas	10	164	Geared/9.5 MW prototypes testing
Siemens	10	193	Direct drive/prototype testing

Note that the two 10 MW machines have different diameters, which illustrates difference in rating philosophy, discussed in the text. In addition, it is reported that eight Chinese manufacturers are developing wind turbines with outputs of 10 MW and above [2]

Table 2.3 Summary of principal design options, with an indication of the most popular option (in bold) for new machines commissioned in 2018

Component			
Number of blades	1	2	3
Rotor orientation, relative to tower	**Upwind**	*Downwind*	
Speed	Fixed	Variable	*Multi-speed*
Control method	*Stall*	**Pitch**	*Partial-span pitch*
Drive train	Gearbox	**Direct drive**	Hydraulics
Blade type	Stiff	Teetered	Flexible
Tower type	Lattice	**Tubular**	Mixed

Where no indication is given, there is no clear "winner". Options in italics may have been popular in the early days, but have now fallen out of favour

2.3.1 Blades

In the early days two-blade machines were quite common and some of these ran "down wind", that is, the rotor ran downstream of the tower. Most of the large government-funded wind turbines that were built in the 1980s had two blades, but this concept has slowly dropped out of favour in preference to three blades. Contrary to what might be expected, three-blade wind turbine rotors are not necessarily 50% heavier than two-blade wind turbines [3]. One of the problems with two-blade machines is that the rotor shaft is subjected to large cyclic forces when the orientation of the rotor is changed—or yawed—to bring it into line with the wind direction.

Nowadays, the vast majority of large commercial wind turbines have three blades, although some manufacturers favour two blades. One-blade turbines were favoured by a few manufacturers. The thinking behind this was that the blades account for a significant proportion of turbine cost. However, one-blade machines are slightly less efficient than two-blade machines and rotor efficiency is further degraded by the necessity of having some form of counterweight. Although a number of machines were successfully commissioned, very few are now on the market. An example is shown in Fig. 2.4.

As machine sizes grew larger, there was speculation that the two-blade concept might come back into favour for offshore machines. For optimum performance, two-blade machines rotate faster than three-blade machines, which means they

Fig. 2.4 One-blade wind turbine at the Italian Alta Nurra test field in Sardinia. (Photo: author)

generate more noise, but this is rarely a problem offshore. One research project looked at the possibility in connection with a design study for 10 and 20 MW machines and found that there were difficulties due to the more pronounced vibrations encountered with two-blade machines. They did not rule out the concept, but focused more on the three-blade design concept.

A Dutch and a Chinese company are currently working on megawatt-size two-blade machines for offshore use, but examples of large machines using the concept are rare. Given the enormous amount of development work that the major manufacturers have put into optimising their three-blade designs, it is probably unlikely that there would be a significant trend back towards the use of the two-blade machines.

2.3.2 *Wind Turbine Size and Weight Trends*

The data shown in Fig. 2.3 suggest that the continuing upward trend in ratings is likely to continue. There have been corresponding increases in size, and Fig. 2.5 shows the relationship between power rating and rotor diameter. There is no universal relationship between size and rating and this is discussed later.

The weights of the rotors of current commercial machines are typically about half those of some of the early machines of similar size, designed as part of state-funded research and development programmes. This is a measure of how far the industry has advanced over the last 30 years.

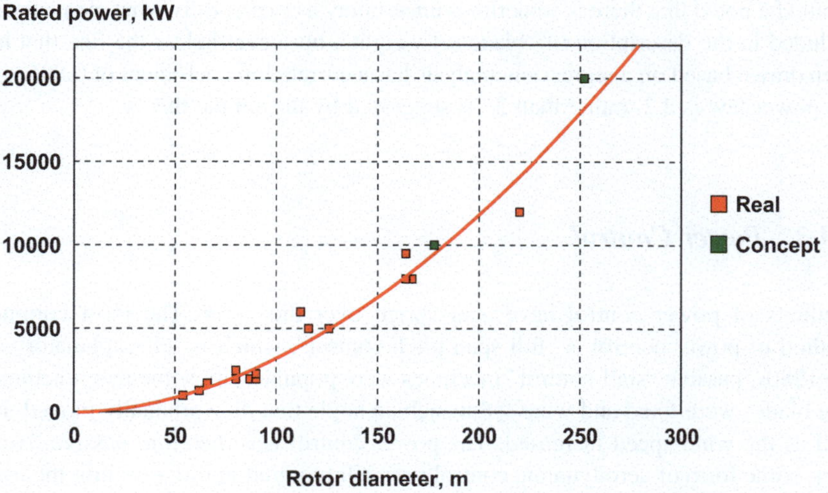

Fig. 2.5 The relationship between wind turbine rotor diameter and rated power. Two data points from design studies are included. The line linking the points comes from a regression analysis

Fig. 2.6 Blade set weights and rotor diameter

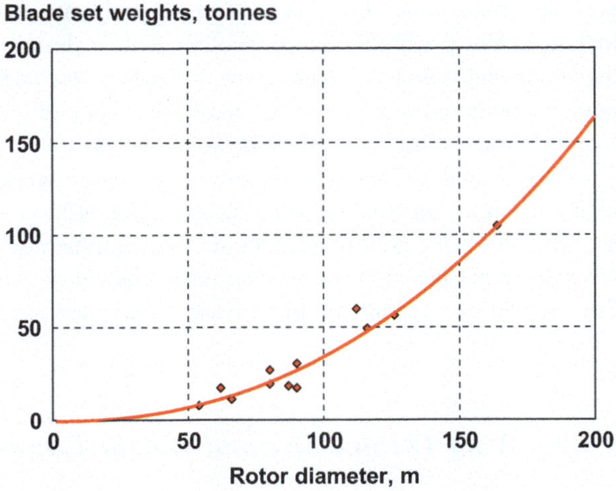

Although the blade set weight of the 164 m diameter machine is 105 tonnes, it is not eight times the weight of an 80 m machine, as simple theory suggests. A greater understanding of the dynamics and aerodynamics means that safety margins, which were possibly overgenerous in the early days, can now be calculated with a greater precision. Although weights (and sometimes costs) do sometimes increase slightly—in $/kW terms—with size, this may be acceptable, as there are considerable savings to be realised by using fewer, larger machines, in installation costs and internal cabling within a wind farm, for example.

The link between rotor diameter and rotor weight is illustrated in Fig. 2.6. It should be noted that there is sometimes uncertainty as to precisely what elements are included in the description of "blade set weight", but nevertheless the line that has been drawn based on a regression analysis has a correlation coefficient of 0.895, and the power law is 2.2, rather than 3, as suggested by the simple theory.

2.3.3 Power Control

Methods of power control have also varied over the years. The most common method of power control is "full span pitch control", which is self-explanatory. In the 1980s, passive "stall-control" machines were popular with some manufacturers. The blades were fixed and were set at such an angle that they gradually moved into stall as the wind speed increased. The power control was therefore passive. However, some form of aerodynamic control was still required in order to limit the rotor speeds should the machines become disconnected from the grid. The concept is now quite rare, except for some small machines. "Partial span pitch control" was employed by a number of manufacturers in the 1980s, as this meant that the duty required of the pitch control mechanism was less onerous than if the whole blade

Fig. 2.7 60 m diameter, 3 MW wind turbine on the island of Orkney, UK. The machine was funded by the UK Department of Energy and built by a consortium—the Wind Energy Group—that comprised Taylor Woodrow Construction, British Aerospace and GEC Energy Systems. It was completed in 1988, operated until 1997 and was demolished in 2000. (Photo: Author)

needed to be rotated. However, this concept also faded in popularity, as manufacturers sought the more positive control achievable with full span pitch control. The majority of wind turbines now have pitch control, as this is needed to enable them to meet increasingly stringent grid code requirements. Figure 2.7 illustrates a machine with partial-span pitch control. This was the UK's Department of Energy-funded 60 m diameter 3 MW machine, which was sited on the island of Orkney, where the higher wind speeds justified the high rating.

The relative popularity of fixed- and variable-speed concept has also changed. Fixed-speed operation was common in the early years, but variable speed has gradually become more established. It confers slight benefits with higher energy capture, but perhaps its principal advantage is that rotational speeds are low in low winds, which is when the noise from a wind turbine is likely to be most noticeable. In high wind speeds, the background noise of the wind itself is likely to mask the wind turbine noise.

2.3.4 Drive Train Possibilities

The other concept that has gradually become firmly established is direct drive. This dispenses with the gearbox, although direct drive generators tend to be quite heavy. Approximately 50% of the machines installed in Germany in 2013 were variable speed, and over 40% had direct drive (also variable speed) [4]. However, it may be noted that there is a well-established manufacturer of direct-drive machines based there.

The drive train is one feature of wind turbine design where the industry has not converged on a preferred solution. The "eighties" solution of an induction generator, driven by a step-up gearbox, is now becoming rare. Double-fed induction generators became popular from around 1998 onwards, and are still used, but permanent magnet generators have become more popular in the past 10 years. These are generally lighter than wound rotors and can be used in conjunction with a gearbox, or as part of a direct drive wind turbine. Because direct drive generators tend to be heavy, one intermediate option that is becoming popular is to step up the rotor speed to, say, 500 rpm (rather than 1000 or 1500 rpm). This, it is claimed, results in a compact gearbox, with a modest step-up ratio, a compact generator, and an economical solution [5].

2.3.5 Support Towers

Most of the early small wind turbines used lattice steel towers. A few of the large government-funded turbines used concrete towers, but both types were gradually superseded by tubular steel towers. This is the most common concept now in use. However, with tower heights steadily increasing, two difficulties arose. First, the size of the tower meant that transportation became an issue and secondly, providing sufficient stiffness demanded substantial wall thicknesses towards the base. To achieve the necessary rigidity at hub heights of over 100 m and to suppress resonance frequencies caused by turbine rotation, the lower part of the tower needs a diameter of over 4 m, according to manufacturer Nordex. By developing a concrete/steel hybrid tower, Nordex solved the logistics and resonance frequency problems arising with towers for turbines with a hub height of over 100 m [6].

They developed a hybrid tower comprising a concrete base structure with a height of around 60 m, mounted directly on the foundation at the location and then prestressed. It supports the three steel tower sections of the modular tower with a total height of a further 60 m. The concept is being tested in the German state of Mecklenburg-West Pomerania, with a 90 m diameter, 2500 kW turbine.

The advantage of this is that the concrete tube is produced on site, thus dispensing with the need for overland transportation, while the standard tower sections can be carried on conventional vehicles. At the same time, the overall system has an adequate resonance frequency as the diameter of the concrete element fitted in the lower part of the hybrid tower is adjustable. Manufacturer Enercon uses a concrete tower for its 7.5 MW, 126 m diameter turbine, and others are likely to follow suit as sizes increase.

2.4 Other Wind Turbine Concepts

2.4.1 Vertical Axis

Vertical axis wind turbines have a long history and formed part of research programmes in the 1980s, in Canada, the UK, and the USA. There are several types—Darrieus, with curved blades, H type, with straight blades, and V type, also with straight blades. The first two types are compared in Fig. 2.8. The principal advantages of the concept are twofold: firstly, it is not necessary to yaw the blade into wind in responses to changes in direction and, secondly, it is possible to locate the gearbox and generator at ground level or at the top of the support tower. However, the principal disadvantage is that the rotors are heavier, because the blades do not deliver power continuously as they rotate, but in a cyclic manner. As a consequence of this, even in a steady wind, they generate cyclic aerodynamic forces, both torque and thrust. These can be alleviated by using three-blade designs, but this concept is rare and would, in any case, increase rotor weights still further.

The H type rotor was comprehensively researched in the UK, although work terminated around 1990. It was concluded that "*in production they would be somewhat more expensive than contemporary horizontal axis wind turbines.*" [7] Research into the concept was restarted in 2009 when the Energy Technology Institute (a research body that had government support and included power generators and industrial bodies with an interest in wind energy) initiated a feasibility study on the technical and commercial viability of vertical axis turbines. The Nova project delivered a feasibility study which evaluated the technical and commercial

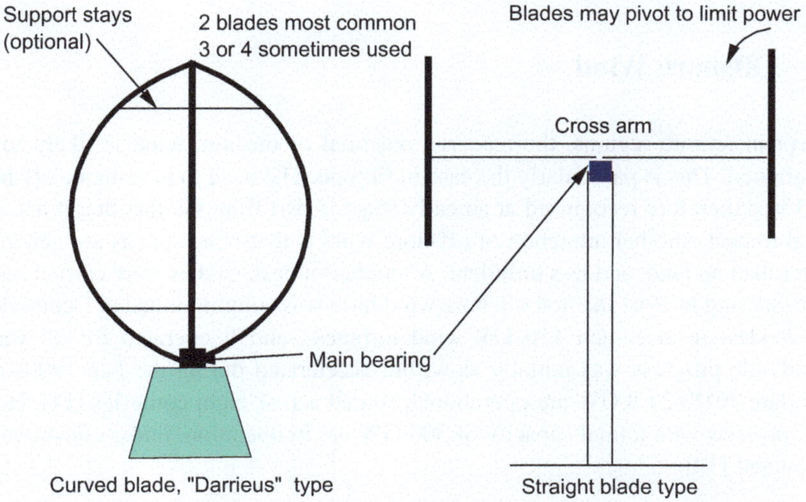

Fig. 2.8 Schematic impression of curved blade and straight blade—H type—vertical-axis wind turbines. Some Darrieus type turbines are mounted on a tower, rather than having the main bearing at or near ground level, as shown here

viability of a 5 and 10 MW vertical axis turbine, with a V-type rotor. It also evaluated specific design options for the rotor, drive train and foundations. "*The project showed that the Nova concept was commercially and technically feasible. Our further analysis suggested that horizontal axis wind turbines will evolve faster than vertical axis wind turbines and provide lower costs of energy in the short to medium term*" [8].

Research continues in the USA on vertical axis wind turbines and Sandia National laboratories recently published a report on a design study for a 5 MW turbine [9]. The projected near-term cost of energy from the design was $213/MWh, but the authors considered that this could be virtually halved by various measures, including a reduction in the weighted average cost of capital.

2.4.2 Multirotor Concept

Space precludes a review of all the alternative design concepts that have appeared over the years, but one recent project that was tested by a major manufacturer is worth mentioning. Vestas recently dismantled a four-rotor assembly after 2½ years of testing. In their words [10], "*we wanted to see whether there could be another way of increasing wind turbine size, apart from just extending the blades.*" Vestas used four 29 m diameter, 225 kW machines, mounted on a single tower, at two levels. Vestas found that there was a 1.5% increase in energy production, compared with a single 900 kW machine, but with no adverse effects on the loading of the structure.

2.5 Offshore Wind

Except in remote regions, the resource potential of onshore wind is likely to be constrained. This is particularly the case in Europe. The need to investigate offshore wind was therefore recognised at an early stage. Apart from the fact that it has less visual impact, another attraction of offshore wind is that wind speeds are generally higher than on land, and less turbulent. A number of desk studies were carried out in the 1980s and in 1991 the first offshore wind farm was commissioned in Denmark. It was modest in size—ten 450 kW wind turbines—and it operated for 25 years. Worldwide progress was initially slow, but accelerated during the late 1990s and now (late 2018) 21.8 GW are operational, spread across eight countries [11]. However, projects with a total capacity of 300 GW are in operation, under construction, or planned [12].

2.5.1 *Floating Wind Turbines*

It was recognised some time ago that the potential of floating wind turbines would need to be investigated. These would enable offshore wind to be exploited in regions where the depths increase rapidly close to land and, in addition, it may facilitate access to deep water zones far offshore, where wind speeds are generally higher. In the latter case, however, the increased costs associated with higher cabling and maintenance costs far offshore, plus the cost of a floating foundation, would need to be offset by the higher energy yield that would be realised. The depth at which floating wind turbines become economically preferable cannot be quantified with any precision but appears to be between 30 and 50 m. Potentially attractive deep-water areas close to the shore are found off California, New England, the Pacific Northwest, the Gulf of Mexico, the southwest tip of England, the Atlantic coast of Spain, and several areas in the Mediterranean (Fig. 2.9).

In September 2011 Japan's trade ministry announced a ¥10–20 billion ($130–260 million) project to install a 1 GW floating wind development in deep waters off its northern coast by 2020. In the same month the US Department of Energy announced offshore wind Power R&D Projects with funding totalling $43M. The funding allocated to floating wind turbines is $6.44M.

Statoil (now Equinor) commissioned a pilot project off Stavanger, in Norway, in 2009. It comprised one 82 m diameter 2.3 MW wind turbine. The same company

Fig. 2.9 Rampion wind farm, 13–20 km off the south coast of England, was completed in 2018. It comprises 116 wind turbines, each with a rated output of 3.45 MW, a rotor diameter of 110 m, and a hub height of 80 m. The projected output is 1.4 TWh/year. The photograph shows the substation where the power is marshalled. (Photo courtesy of Rampion Offshore Wind)

completed the first commercial wind farm off the coast of Scotland in 2017. The farm consists of five 6 MW turbines with a total installed capacity of 30 MW, and a rotor diameter of 154 m; the overall height is 253 m. Water depths vary between 95 and 129 m. The average wind speed in this area of the North Sea is around 10 m/s and the average wave height is 1.8 m. The export cable length to shore is 30 km. The total cost of the project was NOK 2 billion, which corresponds to around $8000/kW. The company has other offshore wind projects under development and expects the total market to reach over 3000 MW by 2025 and 13,000 MW by 2030. The total resource has been estimated to be around 7000 GW [13].

2.5.1.1 Performance Issues

It was noted in the caption to Table 2.2, that there is quite a large difference between the rotor diameters of the two 10 MW machines. Machines that are designed primarily for low wind speed sites tend to have low power ratings, simply because full power output would only be achieved for limited periods if the rating was too high. In this case the specific rating (rated power per unit rotor area) is typically around 300 W/m^2 or less. Machines designed for higher wind speed sites, where the maximum power can be utilised more often, may have ratings up to 600 W/m [2]. All wind turbine designers face the same dilemma—whether to have a high rating, which will extract as much energy as possible out of the air stream, but which will come with higher costs for the generator, gearbox and other drivetrain components, or to use a lower rating that will result in better utilisation of the equipment.

The data underlying the dilemma is illustrated in Fig. 2.10.

Up until the turn of the century, the tendency for many machines was to have ratings around 400–500 W/sq m [2], which usually corresponded to rated wind

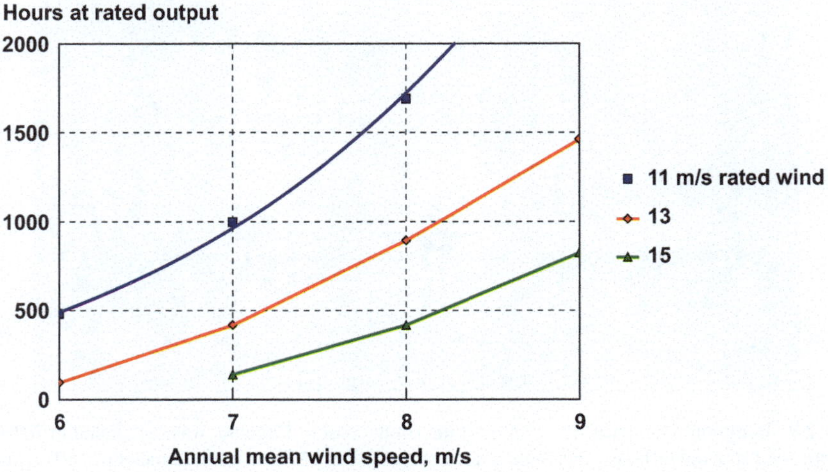

Fig. 2.10 The relationship between time spent at full rated output and annual mean wind speed, for different levels of rated wind speed

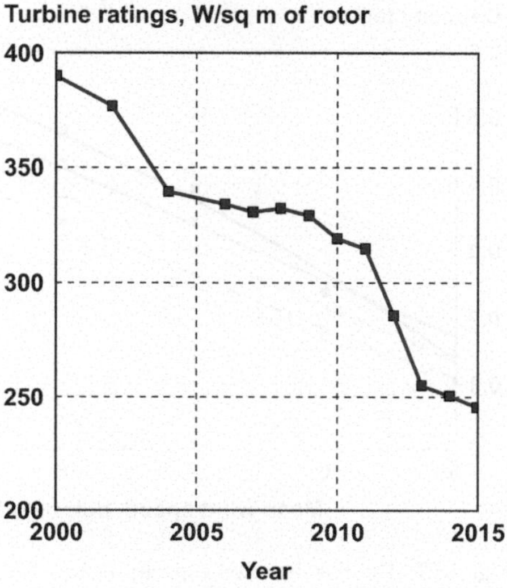

Fig. 2.11 Specific ratings of new machines installed in the USA from 2000 to 2015. (Source of data: Lawrence Berkeley National Laboratory, wind reports)

speeds around 13 m/s. However, several manufacturers did offer machines with different ratings for high or low wind speed sites. From around the turn of the century, however, there was a general trend towards using lower specific ratings and this is illustrated using data from the Lawrence Berkeley National Laboratory wind reports [14] in Fig. 2.11.

This change in rating philosophy has two "side-effects" both of which tend to mask the actual trends in cost and performance. As wind turbine and wind farm costs are traditionally quoted in this $/kW terms, machines with lower ratings will appear to be more expensive than those with high ratings. So the decline in costs has actually been more rapid than the headline figures suggest. The other "side-effect" of the lower specific ratings is that the performance of wind plant—measured by capacity factor—appears to have increased significantly. The effect is illustrated in Fig. 2.12.

The high capacity factors are sometimes cited in sales literature to imply that particular wind turbines are especially productive, whereas capacity factors have never been a valid parameter upon which to base comparisons between wind turbines. Notwithstanding despite the erroneous impression given by the recent trend towards lower specific ratings, overall energy productivity—measured in terms of kWh/m^2 of rotor area per year—has been increasing, driven partly by the use of taller turbine towers and partly by small improvements in blade efficiency.

2.5.2 Changes in Performance with Age

Offshore wind turbines operate in a hostile environment and hence some loss of output over a period of time may be expected, partly due to degraded aerodynamic

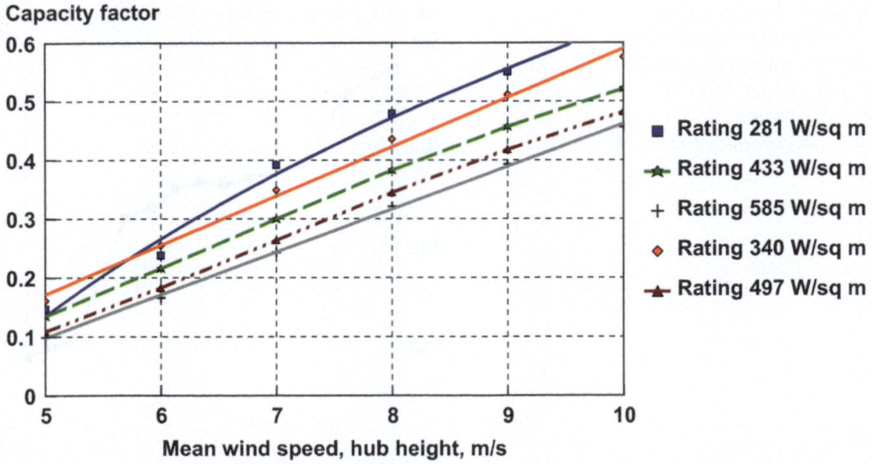

Fig. 2.12 The link between capacity factors and site mean wind speeds, for a range of specific ratings from 281 to 585 W/m², illustrating the higher capacity factors that are realised by the use of low specific ratings

performance associated with loss of finish on the blades and partly due to downtime to enable faults to be fixed. A number of studies have examined this topic and the latest concluded [15], "WTs constructed before 2007 lose around 0.15 capacity factor percentage points per year in absolute terms, corresponding to a life-time energy loss of 6%. A gradual increase of downtime accounts for around 1/3 of the decline and worsened efficiency for the rest." Other studies had suggested that the decline was more severe.

As Denmark has led the world in offshore wind, a scrutiny of data from there may shed light on this discrepancy. The first wind farm comprised ten 450 kW wind turbines, was commissioned in 1991 and taken out of service in 2017. During 27 years of operation it generated 22.1 GWh of electricity and so the average capacity factor was 20.8%. However, if the generation during the first and last 2 years was excluded, the generation was 21.6 GWh, giving a capacity factor of 22.8%. The projected capacity factor was 23.9%. This suggests there was a measurable decline. However, if the performance of the second wind farm, at Tuno Knob, is examined, the picture is somewhat different. This was commissioned in 1996 and is still operating. The average capacity factor over 23 years of operation, from 1996 to 2018, is 31%, which compares favourably with the projected value of 30.8%. However, when corrections are made to allow for the fact that some years are "good wind" years and other years have "bad wind" (relative to the long-term average), the Tuno capacity factor is 32.3–1.5% better than the projected value.

Denmark's third wind farm was commissioned in 2002 and comprised 80 m diameter, 2 MW machines. This had problems with the gearboxes during the first 2 years of operation and there was a significant loss of output as a result. Nevertheless the average capacity factor during the 16 years of operation from 2003 to 2018

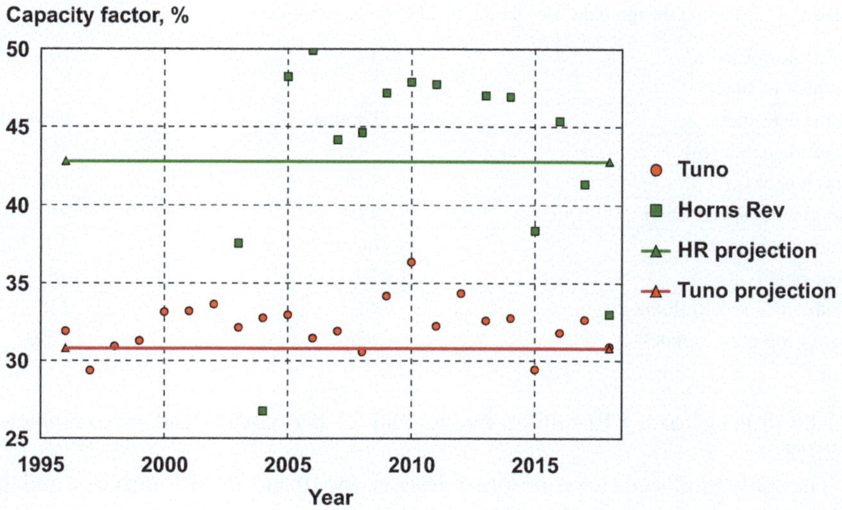

Fig. 2.13 Performance of the Tuno Knob and Horns Rev wind farms from the date of their commissioning until 2018. The data have been corrected to allow for the variability of the yearly winds using the Danish Wind Index calculated by EMD International [16]

was 41.5%, which compares with the projected value of 42.8%. When the data are corrected to allow for the variability of the wind, however, the deficit disappears and performance was slightly better than anticipated.

Figure 2.13 shows the performance of the Tuno Knob and Horns Rev wind farms from the time of their commissioning until 2018. The performance of the latter was significantly below the projected figure in 2018, possibly indicating that there were problems that needed rectifying which caused significant downtime.

2.5.3 Towards 20 MW and Beyond

Largely driven by offshore considerations, a number of feasibility studies for machines in the 10–20 MW range have been completed or are in progress. The "Up Wind" Project, supported by the European Commission, showed that a 20 MW, 152 m diameter wind turbine was a feasible proposition [17]. This was followed by the INNWIND project [18], which built on the success of UpWind, "*to accelerate the development of innovations that help realize the 20MW wind turbine. ... The overall objectives of the project are the high performance innovative design of a beyond-state-of-the-art 10–20 MW offshore wind turbine and hardware demonstrators of some of the critical components.*"

Table 2.4 Principal design data for 10 and 20 MW wind turbines

Rated power (MW)	10	20
Number of blades	3	3
Rotor placement	Upwind	Upwind
Rotor diameter (m)	178.3	252.2
Hub height (m)	119	167.9
Rated wind speed (m/s)	11.4	11.4
Rotor speed (rpm)	6–9.6	4.2–7.13
Gear ratio	50	48
Blade mass (each) (tonnes)	41.7	118
Tower top mass (tonnes)	676.7	1730

The project had a €19 million budget and 27 participants and was completed in 2017.

The study produced two conceptual designs, for 10 and 20 MW turbines, and the key features are summarized in Table 2.4 [19].

The design envisages using a "pseudo-magnetic direct drive", which is "a mechanical and magnetic integration of a non-contacting magnetic gear and a Permanent Magnet Generator (PMG). The high torque rotor blade loads are handled by the magnet-magnet coupling of the gear, which reduces the generator torque requirement, and increases the rotational speed of the excitation rotor. It is therefore analogous to a single stage PMG hybrid WT architecture, but with the benefits of eliminating a mechanical transmission."

In parallel with INNWIND, the European Energy Research Alliance (EERA) has initiated the AVATAR project. Its main goal is the development and validation of advanced aerodynamic models, to be used in integral design codes for the next generation of large-scale wind turbines (up to 20 MW).

In the USA, a team led by the University of Virginia is due to complete (in March 2019) a design for the world's largest wind turbine that uses a new downwind turbine concept called Segmented Ultralight Morphing Rotor (SUMR). It is claimed this will enable a large increase in power from today's largest turbines to a proposed 50 MW system. The SUMR concept allows blades to deflect in the wind, much like a palm tree, to accommodate a wide range of wind speeds (up to hurricane-wind speeds) with reduced blade load, thus reducing rotor mass and fatigue. The novel blades also use segmentation to reduce production, transportation, and installation costs. The new blades could be more easily and cost-effectively manufactured in segments, avoiding the unprecedented-scale equipment needed for transport and assembly of blades built as single units. The exascale turbines would be sited downwind, unlike conventional turbines that are configured with the rotor blades upwind of the tower. A sketch on the Sandia website shows a three-blade rotor, unlike most previous downwind machines which mostly had two blades. This innovative design, it is claimed, overcomes key challenges for extreme-scale turbines resulting in a cost-effective approach to advance the domestic wind energy market. The team working on the project includes experts at the National Renewable

Energy Laboratory (NREL) and Sandia National Laboratories in collaboration with the Colorado School of Mines, University of Colorado (Boulder), University of Illinois, and the University of Virginia.

2.6 Future Developments and Conclusions

Projections for the future are very positive. Figure 2.14 shows projections from the Global Wind Energy Council [20] and the International Energy Agency [21].

A number of other authorities have produced similar projections which, in the past, have often proved to be too conservative. Earlier projections from various sources suggested that 500 GW would be reached by 2020 (in practice, it was achieved during 2019). The moderate projection from the Global Wind Energy Council in Fig. 2.12 suggested that 1000 GW will be reached by 2025 or shortly after. As the industry is continuing to innovate, and generation costs are still falling, its low carbon credentials should ensure that it remains an attractive choice for electricity generation.

Now that the industry has moved closer to being able to operate free of subsidies, growth is likely to be strong. The basic design concepts are likely to remain unchanged, with three-blade upwind machines being the norm. There are likely to be further developments in drive train technology, and direct-drive machines may become more common. With sizes steadily increasing, 20 MW machines are likely to become established within a few years, but whether sizes will continue to increase beyond that point is more difficult to forecast.

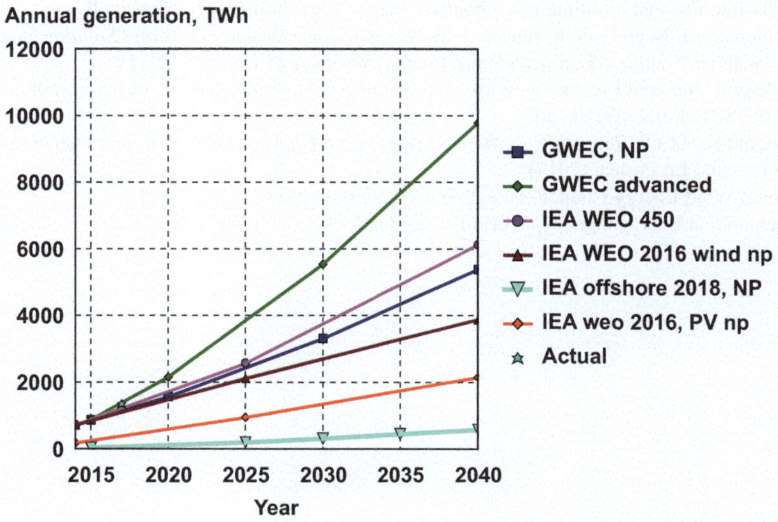

Fig. 2.14 Projections from the Global Wind Energy Council and the International Energy Agency's "World Energy Outlook", for the growth of wind. NP denotes "New Policies", and "Advanced" denotes Advanced Policies. To set the results in context, total world electricity generation in 2017 was 25,500 TWh

References

1. United States Geological Survey, Wind turbine database. https://www.usgs.gov/news/mapping-nations-wind-turbines
2. Y. Yu, Chinese wind turbine OEMs leap towards double digits. Recharge News (2018), https://www.rechargenews.com/wind/1605910/chinese-wind-turbine-oems-leap-towards-double-digits. Accessed 25 Jan 2019
3. D.J. Milborrow, Towards lighter wind turbines. *8th BWEA Wind Energy Conference, Cambridge*. MEP, London, 1986
4. Fraunhofer Institut fur Windenergie Und Energiesystemtechnik, Windenergie Report Deutschland 2013 (2014)
5. C. Halse, J. Keller, NREL next generation drive train. *American Wind Energy Association Conference*, Nevada, 5–8, May 2014
6. Nordex Acconia, Added yields in non-coastal regions Turbines installed on 120 meter high hybrid towers (2006), http://www.nordex-online.com/en/products-services/hybrid-towers.html
7. P. Musgrove, *Wind Power* (Cambridge University Press, England, 2010)
8. Energy Technology Institute [UK]. https://www.eti.co.uk/programmes/offshore-wind/nova. Accessed 22 Jan 2019
9. L. Brandon, et al., *System Levelized Cost of Energy Analysis for Floating Offshore Vertical-Axis Wind Turbines*. Sandia National Laboratories SAND2018-9131 (2018)
10. D. Weston, Vestas measures power gains in multi-rotor concept. Windpower Monthly **35**, 1 (2019)
11. Windpower Monthly, *Windicator Global Index* (Haymarket Media, London, 2019)
12. http://www.thewindpower.net/windfarms_offshore_en.php. Accessed 22 Jan 2019
13. Carbon Trust, *Floating Offshore Wind: Market and Technology Review*. Prepared for the Scottish Government (2015)
14. R. Wiser, M. Bolinger, *2017 Wind Technologies Market Report* (Lawrence Berkeley National Laboratory, US Department of Energy, 2018)
15. J. Olauson, et al., Wind turbine performance decline in Sweden. Energiforsk AB (2018), www.energiforsk.se
16. EMD International (continuously updated). Wind Index. http://www.vindstat.dk/
17. N. Fichaux, J. Beurskens, P. Jensen, J. Wilkes, *UpWind: Design Limits and Solutions for Very Large Wind Turbines* (European Wind Energy Association, Brussels, 2011)
18. P. Jensen, Innwind.Eu: overview of project and recent results (2014), http://orbit.dtu.dk/ws/files/87807100/INNWIND.pdf
19. Innwind.eu, *LCOE Reduction for the Next Generation Offshore Wind Turbines*. Outcomes from the Innwind.Eu Project (2017)
20. Global Wind Energy Council, Global Wind Energy Outlook 2016 (2017)
21. International Energy Agency, World Energy Outlook, 2016 (2017)

Chapter 3
What Is the Wind Energy Progress in Greece? Prospects and Problems

J. K. Kaldellis

3.1 Introduction: The First Steps of Wind Energy in Greece

The power of wind has impressed Greeks from the ancient times. Actually, according to the Greek mythology, wind was worshipped by the humans as god Aeolus, who along with his eight assisting gods, that is, the so-called Anemoi, each assigned with one wind direction (Boreas (N), Kaikias (NE), Eurus (E), Apeliotes (SE), Notus (S), Livas (SW), Zephyrus (W) and Skiron (NW)) (Fig. 3.1), was considered as the ruler of winds, underlining the importance of wind energy for the economic and production activities during even this early historic period [1]. In this context, wind energy was initially exploited mainly in maritime for the navigation of Greek sailing ships, while later, some relatively simple applications of wind energy exploitation for the production of mechanical work appeared.

In modern times the first commercial Greek wind park was created on Kithnos island in 1982. This park was a 5 × 20 kW pilot wind project, based on two-bladed MAN (Aeroman) first generation wind converters. These machines were replaced in 1990 by 5 × 33 kW wind turbines of the same manufacturer (Fig. 3.2). Between 1982 and 1990, no significant wind energy activity was encountered, excluding two ineffective installations on the islands of Mikonos (1 × 108 kW Micon) and Karpathos (1 × 175 kW HMZ). Both of the latter machines soon presented major failures that finally suspended their operation. At this point one should mention that until 1994, when the law 2244/94 was voted by the Greek Parliament, and actually until mid-1998 (when the entire legislative frame was integrated), the commercial exploitation of wind energy was the exclusive privilege of the State owned Public Power Corporation (PPC). Only some small wind turbines of private companies (the first private wind turbine was a Vestas V-55 installed in 1984), municipalities and

J. K. Kaldellis (✉)
Lab of Soft Energy Applications and Environmental Protection, Mechanical Engineering Department, University of West Attica, Athens, Greece
e-mail: jkald@uniwa.gr

A. Sayigh, D. Milborrow (eds.), *The Age of Wind Energy*, Innovative Renewable Energy, https://doi.org/10.1007/978-3-030-26446-8_3

Fig. 3.1 The tower of winds, situated in Athens (shown are Boreas and Skiron)

Fig. 3.2 One of oldest wind turbines (two-bladed Aeroman, 33 kW) in Kithnos island Greece

the Hellenic Telecommunications Organization (OTE SA) were installed (total installed power less than 3 MW) in the general frame of self-producers, i.e. the wind park owner should use the wind energy production only for self-consumption [2].

In 1994, after the new Renewables Law 2244/94 was voted by the Greek Parliament, the PPC domination in the local electricity market was challenged; while all wind energy exploitation activities were frozen up until the local national legislation frame was stabilized. During this perplexing period, only two isolated wind turbines were installed in the entire Greece from PPC, i.e. 1 × 500 kW Nordtank NTK-500 in Sitia of Crete (1995) and 1 × 500 kW Vestas V-39 in Kithnos hybrid station (1999). Finally, in October 1998 the first medium-sized private commercial application started its operation. This was at Modi-Sitia/Crete (17 × 600 kW NEG-Micon), within the new liberalized European electricity market status. Since then the local market has grown remarkably, taking into consideration the high wind potential of the country and more specifically of the Greek islands.

3.2 Wind Potential of Greece—Wind Energy Production

Greece, due to its position, possesses excellent wind potential in the islands and near the sea and relatively high wind potential in several areas of the mainland. According to the extended long-term measurements by PPC [3], the Hellenic Meteorological Agency [4], CRES [5] and private companies, one may easily conclude that the average wind speed in Greek islands of the Aegean Archipelago varies between 8 and 9.5 m/s, while the corresponding value in the mainland is approximately 6.5–8 m/s, see also Fig. 3.3.

As it is well known, the annual energy yield of a specific wind park mainly depends on the available wind potential, expressed usually via the corresponding probability density function ($f(V)$) [6]. For example the annual energy yield of a wind park "E_{wp}" including z similar wind turbines of rated power "P_R" is given by Eq. (3.1):

$$E_{\mathrm{wp}} = \mathrm{CF} \cdot (z \cdot P_{\mathrm{R}}) \cdot 8760 \tag{3.1}$$

where the capacity factor "CF" is the product of the mean power coefficient "ω" and the mean annual technical availability "Δ" of the installation [7, 8], that is,

$$\mathrm{CF} = \omega \cdot \Delta \tag{3.2}$$

where "ω" is defined as:

$$\omega = \int_0^{\infty} \frac{P_{\mathrm{ex}}(V)}{P_{\mathrm{R}}} \cdot f(V)\mathrm{d}V = \int_{V_{\mathrm{c}}}^{V_{\mathrm{F}}} \frac{P_{\mathrm{ex}}(V)}{P_{\mathrm{R}}} \cdot f(V)\mathrm{d}V \tag{3.3}$$

According to Eq. (3.3), the exact value of the mean power coefficient depends on both the local wind energy resource characteristics (normally the probability density "$f(V)$" curve of the local wind potential is used) and the operational, non-dimensional power curve "$P_{ex}(V)/P_R$" of the wind turbine each time examined (with "$P_{ex}(V)$" being the power output in relation to wind speed "V"), with both curves expressed as a function of the wind speed "V" at the machine hub height [8]. In this regard, it must be noted that the energy production of a wind turbine is

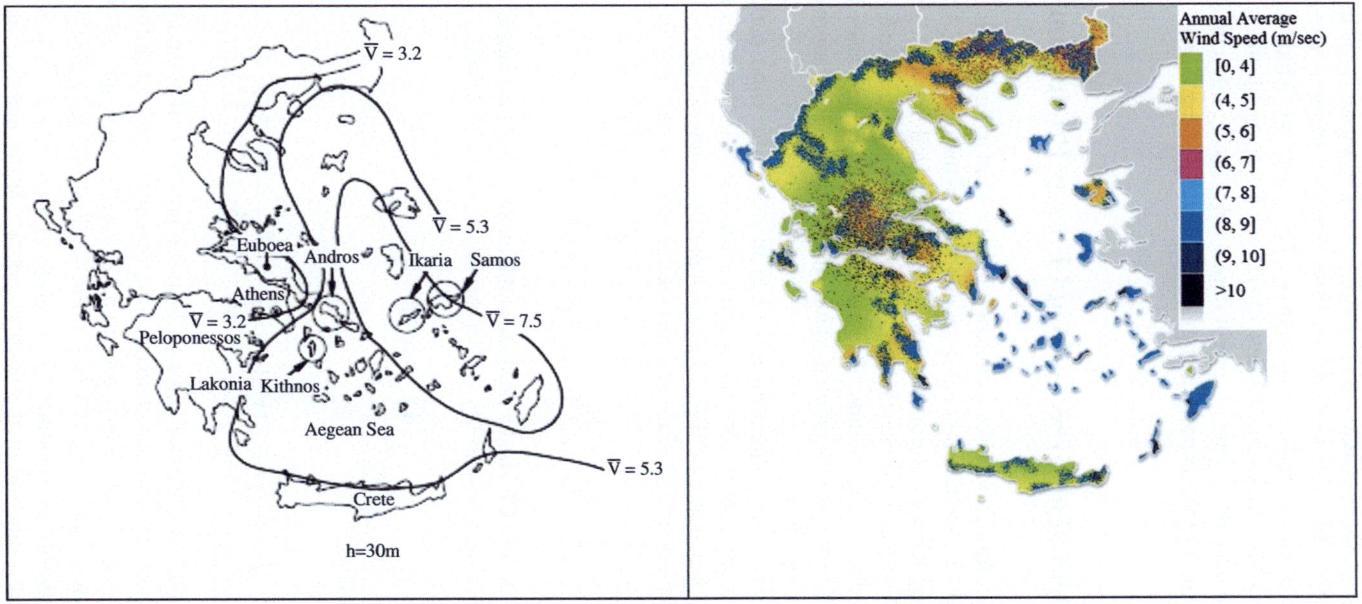

Fig. 3.3 Wind potential in Greece [5, 13]

limited within the range of wind speeds from the cut-in "V_c" to the cut-out "V_F" wind speed.

Considering the above, the importance of determining the mean technical availability in order to configure the energy yield of a wind turbine or a wind farm is reflected [9]. In this context, the technical availability of a wind turbine depends among others on the technological status, the age of the machine and the site of installation [10].

Hence, it is well established that the annual energy production can be fairly estimated on the basis of the local mean wind potential (mainly mean wind speed), [6–10], taking also into consideration the technical availability of each installation. Of course for detailed energy production calculations, the complete wind speed time-series data for every wind turbine at hub height is required, along with the corresponding ambient pressure and temperature [11]. In cases that all this information is not available, the expected annual energy production is based on the well-known Weibull parameters' long-term values.

Using the available information one may present the real wind energy production for the non-interconnected Greek islands during the last 3 years on a monthly basis, Fig. 3.4. As one may easily observe there is a remarkable wind energy production variation during the year, while the maximum wind energy production appears during the summer. At this point, it is important to underline the fact that although the wind potential is higher during the winter, the wind energy contribution is much higher during the summer due to the increased electricity demand (due to tourism) and the severe constraints imposed by the Island Networks Operator (HEDNO) concerning the wind power absorption, in order to safeguard the stability of the various weak autonomous island electrical networks [8, 12]. In this context, the annual mean capacity factor of the island wind parks is approximately 26–28%, which is quite lower than the theoretical expected value of 35–45%, mainly due to the wind power curtailments by the local network operator [13].

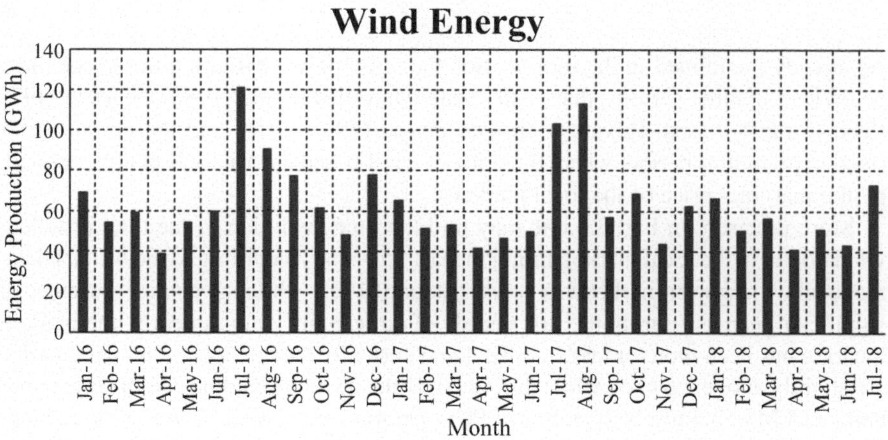

Fig. 3.4 Wind energy production in Greek islands (installed power 310 MW, end 2018)

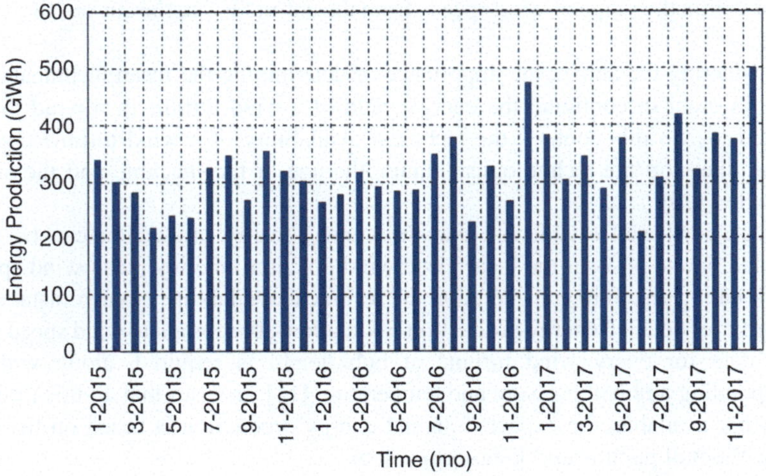

Fig. 3.5 Wind energy production in Greek mainland (installed power 2510 MW, end 2018)

An analogous situation appears also for the mainland wind parks. However in these cases there are no considerable wind power curtailments, since the produced wind energy (Fig. 3.5) is normally absorbed by the interconnected electrical network. On the other hand, the lower-quality wind potential of the mainland leads to lower wind energy capacity factor (i.e. CF = 23% for Greek mainland) in comparison with the corresponding island values. In any case, the contribution of the installed wind parks on covering the national electricity consumption during the last years approaches 10%.

3.3 Status of Wind Power Applications in Greece

As already mentioned in the introduction the wind power evolution in Greece may be divided in three periods, that is, the State-controlled period (1982–1998), the FIT (feed in tariff) based, Private-controlled period (1998–2018) and the current, FIP (feed in premium) period, where the price of wind-based electricity is mainly defined by the marginal price of the local market.

More precisely in Fig. 3.6 one may find the time evolution of the total installed wind power in the local market during the last 25 years. According to the official data the local wind power market was very weak until 1998, while the market presented a significant increase during the first decade of our century. Subsequently, the market increase has been more or less stabilized and currently the "in operation" wind parks are slightly above 2.8 GW, while during 2018, the decommissioning of the first wind parks (15 MW mainly belonging to PPC) operating since early 1990s started [14].

As far as the geographical location of the existing wind parks is concerned, the majority of them appears in Central Greece (900 MW) and Peloponnese (550 MW),

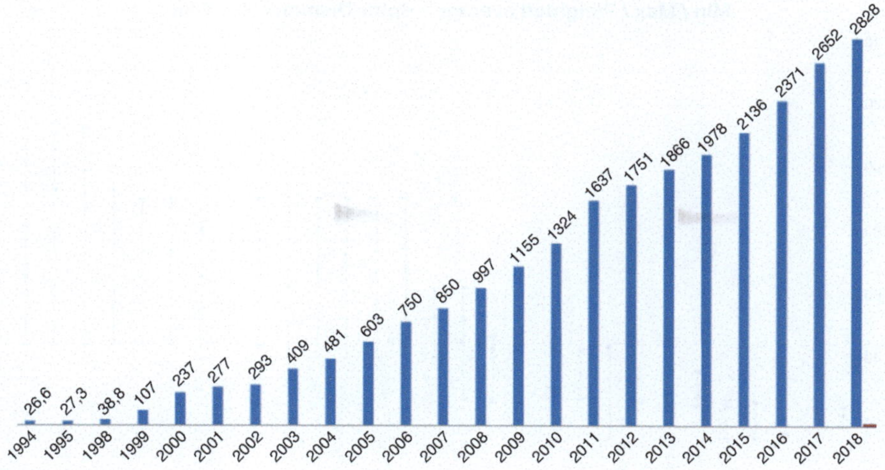

Fig. 3.6 Annual time evolution of the installed wind power in Greece [14]

due to the optimum combination of good infrastructure, access to electrical grid and fairly high wind potential. Unfortunately, the number of wind parks in North and South Aegean Sea remains modest for many years (approximately 120 MW), despite the excellent wind potential and the extremely high electricity generation cost in most islands [15]. On the other hand, Crete is really a special case, since the island has been the pioneer in the exploitation of wind energy since the 1990s [16, 17]. However, wind power in Crete has stagnated near 200 MW for the last 5 years, due to constraints imposed by the local grid stability [12].

Among the major local wind energy producers one may find [14] Terna Energy with almost 550 MW in operation, Anemos (Elaktor) slightly below 300 MW, Iberdrola-Rokas (250 MW), EDF EN Hellas (240 MW), EREN Group (210 MW) and ENEL Green Power (200 MW). Note that the PPC Renewables (being the first company involved in the development of wind parks in Greece) has less than 70 MW in operation.

Another interesting information [14] concerning the local market is the time evolution of the size of the new installed wind turbines, Fig. 3.7, on annual basis. Thus it is quite impressive to mention that the average wind turbine diameter was less than 20 m in 1987, while during the last 3 years wind turbines with rotor diameter between 80 and 100 m have been installed. According to available data, the biggest wind turbine in Greece at the end of 2018 has rated power of 3.6 MW.

Finally, the majority of the "in operation" wind turbines are based on the well-known Danish concept (geared machines with induction electrical generators and pitch control rotors), since Vestas has installed more than 50% of the "in operation" machines, while the gearless concept of Enercon has been adopted by 22% of the local market wind turbines. SGRE (Siemens-Gamesa Renewable Energy) group and Nordex are the other two main manufacturers appearing also in the local market

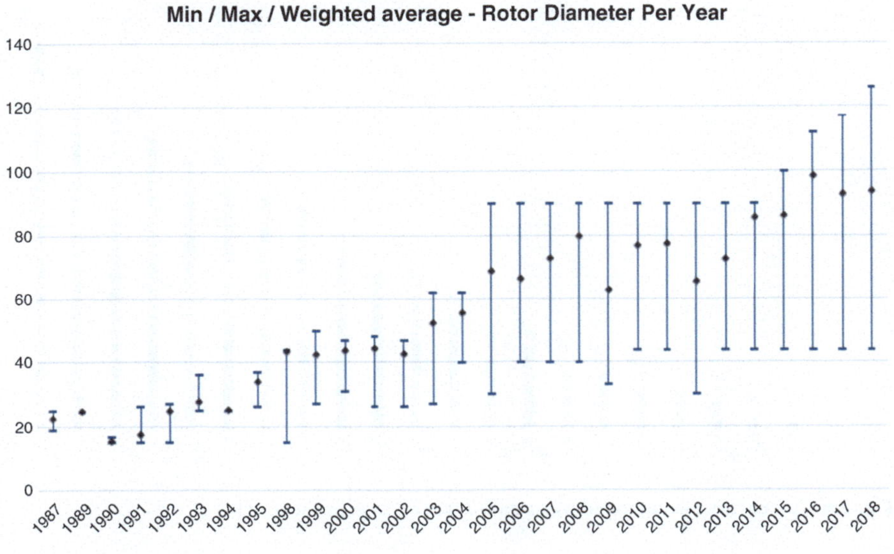

Fig. 3.7 Annual time evolution of installed wind turbines' size in Greece [14]

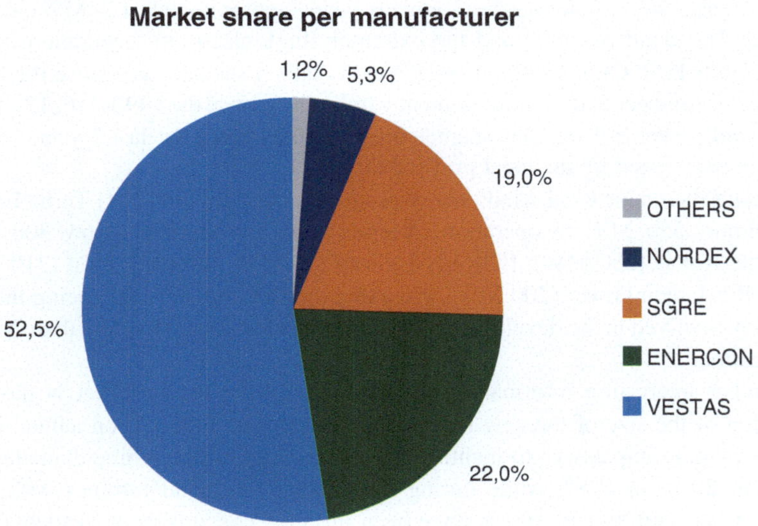

Fig. 3.8 Total market share per manufacturer [14]

(Fig. 3.8). Actually, from the total new installed capacity of 192 MW installed in 2018, 150 MW correspond to Vestas wind turbines, 30 MW to Enercon gearless machines and 12 MW belongs to the SGRE, see also Fig. 3.9.

At this point, it is important to mention that despite the long-term applications of wind power in Greece (since 1982), there is no serious manufacturing activity in the

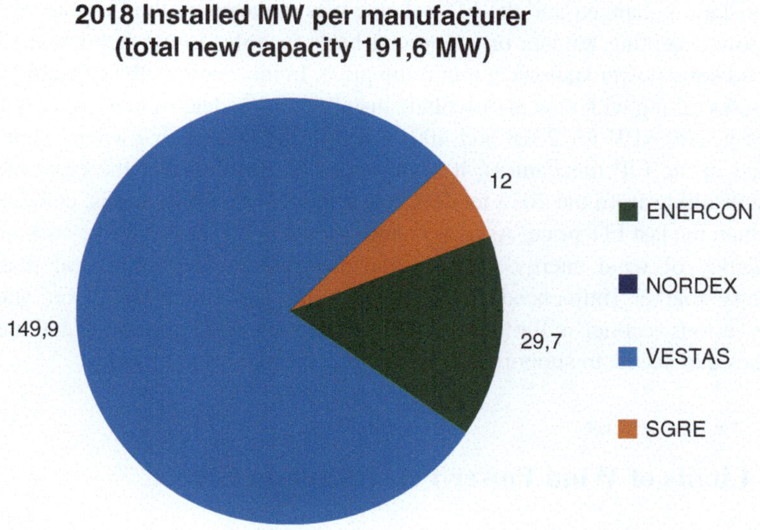

Fig. 3.9 2018 installed wind power per manufacturer [14]

country; hence, all the wind turbines installed have been imported. Since 1980 there are some inefficient attempts to produce commercial wind turbines in the country, which however for various reasons (local market size, international competitors, etc.) have not succeeded. The first unsuccessful experiment was the case of the Hellenic Airforce Industry (HAI) during the mid-1980s. In this case, the State controlled HAI finally decided not to manufacture 50 wind turbines (55 and 100 kW) for the State controlled PPC, in collaboration with the Danish company Windmatic (absorbed later by Vestas). Since then, no complete manufacturing activity has been realized concerning commercial wind turbines. For a remarkable period (until 2005) the towers of Bonus (a Danish manufacturer absorbed by Siemens) wind turbines were manufactured in Peloponnesus Greece. Finally, there are some small manufacturers developing very small wind turbines (around 1 kW), while recently one Greek company announced the local production of small wind turbines of 50 kW in Attica region.

Recently, an important change concerning the prospects of wind energy in Greece concerns the new legislative frame dominating the corresponding market. Actually, the first Law about the exploitation of wind energy in Greece was voted in 1985 (Law 1559/1985) giving permission only to the State controlled PPC to install commercial wind parks. All other individual actors had the permission to install single wind turbines only in order to cover their own electricity consumption (self-producers). Later in 1994 the Greek parliament voted (Law 2244/1994) in favour of private investors (ending the PPC monopoly), encouraging them to create their own wind parks (initially up to 50 MW) and sell the electricity to the national grid in a predefined price (FIT-feed in tariff). The corresponding prices recently approach 85 €/MWh for the mainland and almost 100 €/MWh for the islands. During 2018, this legislation changed and the FIT scheme was replaced by the FIP one, with the State now accepting tenders on an annual basis in order to cover predefined new wind and solar power stations at minimum price. In this context, the capacity of new wind parks (along with new photovoltaic installations) is determined annually by the State (e.g. 300 MW for 2018 and 400 MW for 2019), if their owners want to be included in the FIP mechanism. It is important to mention that the price of wind energy resulting from the 2018 tenders is less than 70 €/MWh, that is, considerably lower than the last FIT price. As it is obvious, the new legislative framework and the dependency of wind energy price on the marginal price of the local mainland electricity market (influenced by lignite, natural gas and hydropower stations) creates serious scepticism for the various investors, especially the small and medium size ones and seems to support mainly the large players of the market.

3.4 Limits of Wind Power Penetration in Greece

One of the main drawbacks concerning the significant wind energy contribution in covering the electrical power demand of an electrical network is the stochastic behaviour of the wind and the limited ability to match the wind power output with the corresponding load demand. The problem has been encountered during the first decade of the twenty-first century in most remote islands of Greece as the installed wind power becomes important [8, 12, 16]. During the last years, the same problem affects also the wind power contribution in the Greek mainland. More specifically the installed wind power in the mainland has exceeded 2500 MW, while the current peak load of the system is 9 GW and the mean annual load demand approximately 5.5 GW. In this context the existing wind power represents almost 30% of the peak power demand of the electrical network and almost 45% of the mean load demand. This is also the case in most islands, where the installed wind power represents 20–30% of the peak load demand of the island. Actually the problem is evident during high wind speed and load demand periods (e.g. winter), since, due to the local grids' stability constraints, wind power contribution is limited. As a result, zero new wind parks have been installed across the Aegean Archipelago during the last years, since no additional potential for wind energy absorption from the local electrical networks exists.

In most autonomous or partially interconnected grids the instantaneous wind power contribution is bounded by the optimum operation of the electrical grid thermal power units, their technical minima and the dynamic stability of the local network. Although not scientifically valid for every case, in most practical application cases the instantaneous maximum wind power penetration value is taken (as a rule of thumb) less or equal to 30%. This means that during a typical average day the mainland (interconnected) electrical network is not permitted to absorb more than 1650 MW of wind-based electricity (i.e. 30% of 5.5 GW) even in the case that due to high wind speed values the output of the wind turbines may approach their rated power of 2.5 GW.

Unfortunately this is also the case during the last 15 years concerning the wind power curtailments in all islands' wind parks. In Fig. 3.10 we present the monthly average wind speed as measured by the anemometer of each wind turbine from a small wind park in Kos island. In the next Fig. 3.11, one may compare the theoretical (maximum) wind energy production with the finally absorbed by the local electrical network, while one may also estimate the corresponding wind power curtailments. As a result, one may find in Fig. 3.12 the expected and the actual capacity factor monthly distribution for this typical island micro-grid. It is quite impressive to mention that the wind energy absorption by the local grid is rational only during the summer, when the real CF exceeds the value of 40% (due to high electricity consumption), while during the rest of the year the theoretically available capacity

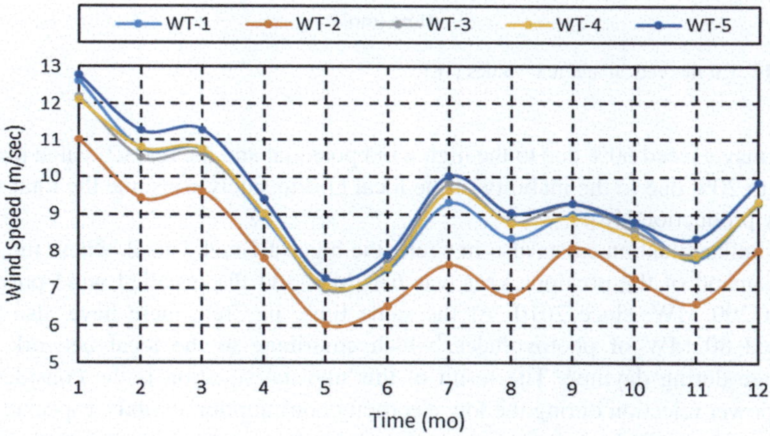

Fig. 3.10 Monthly mean wind speed values [18]

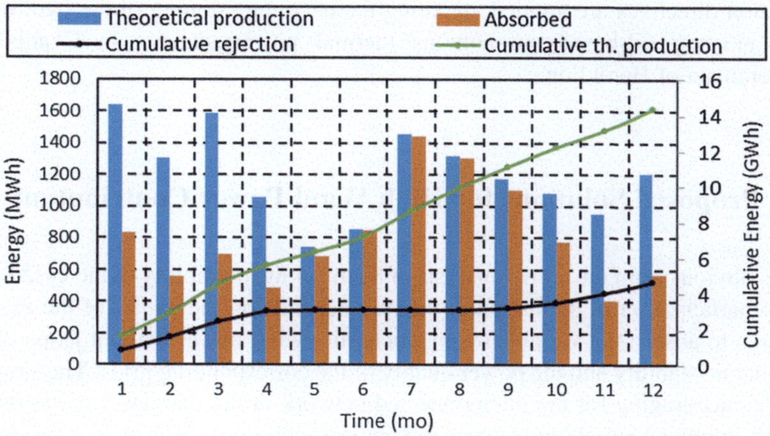

Fig. 3.11 Theoretical wind energy production vs. absorbed wind energy and wind energy rejections [18]

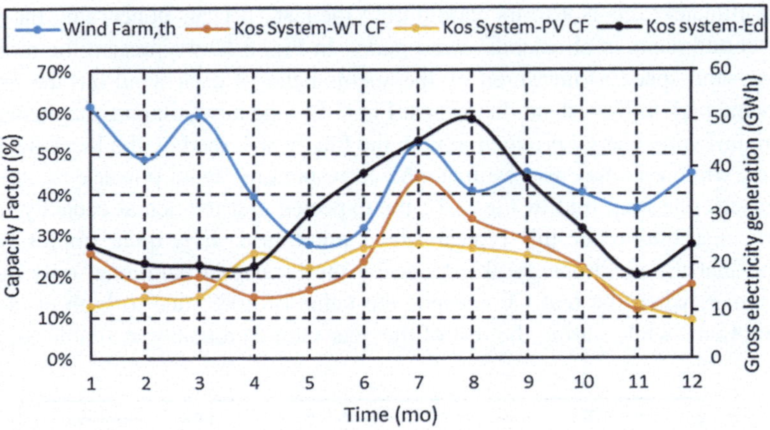

Fig. 3.12 CF theoretical and real values [18]

factor may exceed 60% due to the high wind potential and the real CF value may be down to 20% due to the inability of the local grid to safely consume the total wind energy production [18].

A similar situation exists also in Crete, the biggest Greek island, where the peak load demand of the previous year was 630 MW and the installed wind power is almost 200 MW, since 2010. At the same time, in Crete there have also been installed 80 MW of photovoltaics, which contribute to the local network load coverage during daytime. The result of this unusual situation is the considerable wind power rejection during the low electricity consumption periods, especially for the recently installed wind parks not protected by the corresponding power purchase agreement (PPA). On the other hand, Crete may face serious power demand coverage problems during the next 1 or 2 years in case that the EU environmental protection directives are applied, dictating the decommissioning of the two (out of three) existing oil-based autonomous thermal power stations in Chania and Linoperamata of Heraklion.

3.5 Proposed Solutions for High Wind Power Contribution

As it is obvious from the above analysis of Sect. 3.4 the possibility of increased wind energy participation in the national fuel mix depends on the ability of the existing networks to absorb the variable (even stochastic) wind power without jeopardizing the dynamic stability and the power quality of the corresponding grids. The situation is more encouraging for the interconnected network of the mainland due to its size and the interconnections with the neighbouring countries. However, even in this case, the idea of energy storage is under evaluation. More specifically, studies are under preparation concerning the transformation of the successive big hydroelectric power stations (e.g. Kremasta-Kastraki) to water pumping stations [19]. Actually, there are already two water pumping stations in Greece (in Aliakmon and Nestos rivers) operating for many years in order to save lignite-based excess energy during low demand periods (nights) and provide hydroelectric power during peak load demand hours. The same idea is now under investigation in order to store excessive wind power of the mainland grid.

The problem is much more pressing in the case of Aegean Sea islands, since the maximum penetration limit has been already surpassed by the existing wind parks, taking also into account the energy consumption decrease due to the economic recession of the last 10 years. According to the official data [20], in most islands the installed wind power is near or even higher than the upper limit (i.e. 30% of the island peak load demand of the previous year), thus the total in operation (permitted) wind power in the Aegean Sea islands, Crete included, is approximately 310 MW for the last 5 years. On the other hand, the Aegean Archipelago possesses excellent wind potential, while the operation cost of the existing oil-based thermal power stations is very high and the environmental impacts of the above mentioned thermal power stations quite important [15, 21].

In order to increase the exploitation of the available wind potential there are two complementary strategies. The first one is based on the interconnection of the main island groups with the mainland using HVDC undersea connections; see for example Fig. 3.13. During the last years the north and west part of Cyclades complex is under connection with the mainland, while two HV lines have been planned to connect Crete island either with Peloponnesus or with Attica. Frankly speaking there are several policy and social issues to be solved before the implementation of such connections [22]. In the meantime, a significant part of wind energy production is curtailed, while no additional wind power is to be installed. Moreover, another ambitious plan has been submitted in an attempt to exploit the high wind potential of the North Aegean Sea (i.e. islands of Limnos, Lesvos and Chios) and to connect all these islands either with North Greece (Thrace) or the Attica region. In all these cases, big-scale wind parks have been proposed by the investors which challenge the local population opposition, reacting in a similar way even against the small—already approved—wind parks.

Another important issue is the opportunity to create offshore wind parks in Greece [23]. Although there are some proposals for offshore wind parks in the NE Limnos and outside of Alexandroupolis (Thrace), the Greek sea topography is not in favour of similar installations. On the other hand, there are some interesting plans to create relatively big wind parks on several uninhabited small islands and sea rocks with very high wind potential. The first installation is already in operation (70 MW) on the Agios Georgios (Saint George) island outside of Lavrion-Attica. However, in order to develop a financially attractive installation in those remote islets, the major problem is the electrical interconnection cost; hence all these potentially attractive wind power investments strongly depend on the implementation of the above mentioned interconnections between the islands of Aegean Sea and the Greek mainland.

Fig. 3.13 Electricity interconnections (under revision) in Aegean Archipelago

The second strategy investigated is based on developing appropriate energy storage installations able to store the wind energy surplus during high energy production and low load demand periods and provide the necessary electricity during peak load demand periods. In this context, one may use a variety of energy storage technologies [24] depending on several parameters, like the network size, the storage period and the available land. In most cases the idea of developing a hybrid

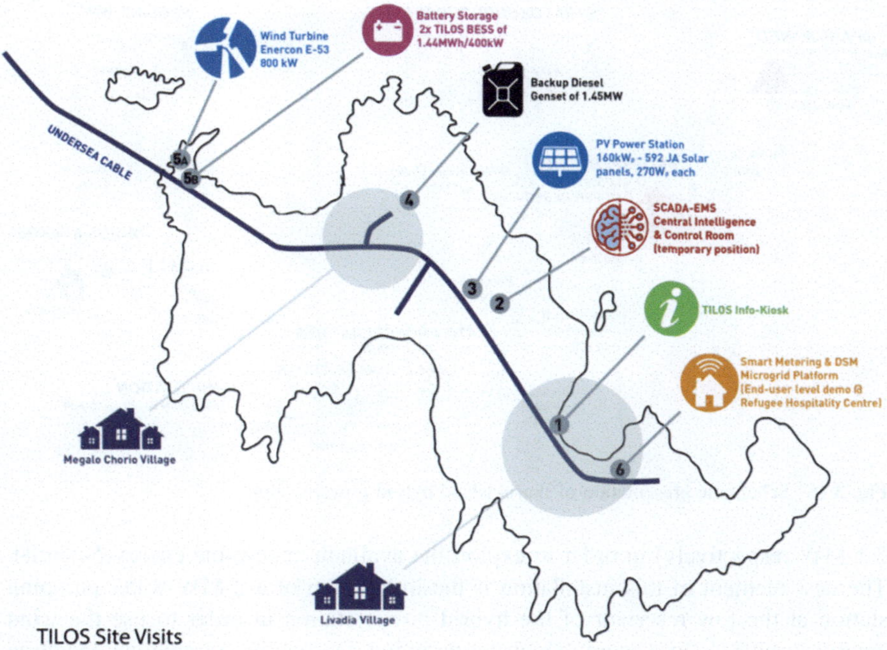

Fig. 3.14 Schematic presentation of Tilos island hybrid power station

power station taking advantage also of the high solar potential of the country is very challenging [25].

Until now there are two wind based hybrid power stations "in operation" in the Greek islands, using also different energy storage techniques. More precisely, the first hybrid power station is "in operation" in Tilos island (Dodecanese complex, NW of Rhodes island) [26]. Actually in Tilos island a medium-size wind turbine of 900 kW along with a small photovoltaic park of 160 kW_p have been installed in the frame of the TILOS Horizon 2020 research programme (Fig. 3.14). In the same island, innovative $NaNiCl_2$ batteries have been installed with energy storage capacity of almost 3 MWh [27]. According to the data provided by the corresponding research team one may cover 75% of the total annual energy consumption of the island on the basis of renewable energy resources, while the vast majority of the electricity provided (almost 65%) comes from the wind turbine of the hybrid power station.

Another interesting project is the one being developed in Ikaria island (N. Aegean Prefecture) by PPC Renewables since 2005 [28]. On the basis of the information given, the corresponding hybrid power station is almost ready to operate. In this case, one combines the high wind potential (average wind speed of the order of 9 m/s) and the hydro potential of the island. Note that in Ikaria there is a water reservoir of almost 1,000,000 m^3 at an elevation of 715 m [29]. Thus PPC Renewables has installed (Fig. 3.15) a wind park of 2.7 MW and two Pelton hydro turbines (1.05 and

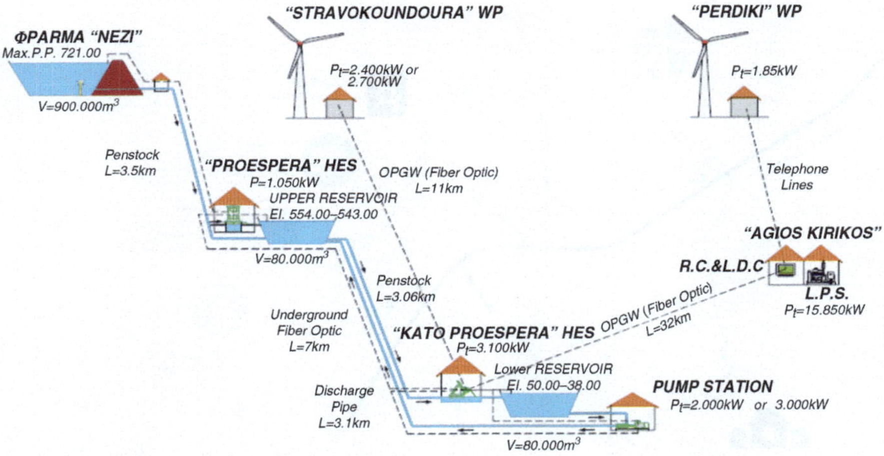

Fig. 3.15 Schematic presentation of Ikaria island hybrid power station

3.1 MW respectively) in order to exploit the available renewable energy potential. The new element of this installation is the installation of a 2 MW water pumping station at the low reservoir of the hybrid power station in order to use the wind energy surplus to store energy via water pumping. During the commercial operation of this hybrid power station, the contribution of the oil-based electricity in the island will be minimized in the range of 20–30%.

Recapitulating, both hybrid power stations presented describe two interesting integrated solutions that may significantly contribute in the limitless exploitation of the excellent wind potential of the Aegean Archipelago and maximize the contribution of clean-green energy in the remote islands fuel mix.

3.6 Conclusions

A brief historical evolution of Greek wind power stations since the early 1980s has been presented, including the time evolution of the installed power up to the end of 2018. Moreover, the existing wind parks are categorized on the basis of their location, the wind turbine manufacturer and the technology used, examining also the local manufacturing activity related to the wind energy sector. Subsequently, the wind potential in the mainland and in the Aegean Sea islands is examined, taking into consideration various long-term data from several official sources. Combining the available wind potential and the existing wind turbines, the wind energy yield has been presented and evaluated for the mainland and the islands of Greece.

Accordingly, the gradual changes of the corresponding legislative framework concerning the development of wind parks in Greece are described, starting from the first law of 1985 up to the current feed in premium status. Subsequently, the local wind energy related manufacturing activity has also been briefly discussed.

As far as the islands' wind power exploitation is concerned, the major problems related with achieving high wind power penetration in remote or interconnected islands are also analysed. Next, the application of wind based hybrid-energy storage solutions in selected Greek islands is also investigated, with special focus on the Ikaria island wind powered-water pumping solution, as well as on the Tilos island EU funded wind/PV-based hybrid power station, including advanced battery storage.

According to the analysis provided it is obvious that the wind energy applications in Greece are near their upper limit due to the existing technical barriers and the stochastic behaviour of the wind. However, taking into account the excellent wind potential of the country, especially in the islands, and the gradual adoption of several mature energy storage technologies, one may be optimistic that there are challenging wind energy prospects for Greece in view of the European-national electrical sector decarbonisation by 2050, since wind energy has a dominant role to play all around the globe.

References

1. J.K. Kaldellis, Wind energy—introduction, in *Comprehensible Renewable Energy Encyclopaedia*, vol. 2: Wind energy, ed. by A. Sayigh (Elsevier, 2012)
2. J. Kaldellis, D. Kodossakis, The present and the future of the Greek wind energy market. Presented at 1999 European Wind Energy Conference and Exhibition, Nice, France, 1999, pp. 687–691
3. Public Power Corporation, *Wind Speed Measurements for Greece, 1980-85* (PPC, Athens, 1985)
4. Hellenic Meteorological Agency, *Meteorological Data of Greek Meteorological Stations for the Period 1955-1998* (Department of Climatology, Athens, 2001)
5. Centre for Renewable Energy Sources (CRES-Greece), Wind potential map for Greece (2019), http://www.cres.gr/kape/
6. J.K. Kaldellis, G.T. Tzanes, C. Papapostolou, K. Kavadias, D. Zafirakis, Analyzing the limitations of vast wind energy contribution in remote island networks of the Aegean Sea Archipelagos, *9th International Conference on Applied Energy*, ICAE2017, Cardiff, 21–24 Aug 2017
7. D. Vlachou, G. Messaritakis, J. Kaldellis, Presentation and energy production analysis of commercial wind turbines. *Presented at 1999 European Wind Energy Conference and Exhibition*, Nice, France, 1999, pp. 476-480
8. J.K. Kaldellis, The wind potential impact on the maximum wind energy penetration in autonomous electrical grids. Renew. Energy **33**(7), 1665–1677 (2008)
9. J.K. Kaldellis, D. Zafirakis, The influence of technical availability on the energy performance of wind farms: overview of critical factors and development of a proxy prediction model. J. Wind Eng. Ind. Aerod. **115**, 65–81 (2013)
10. El. Kalafatis, Ph. Skittides, J.K. Kaldellis, Investigating the relation between the reliability and the technical availability of wind energy applications, *MedPower08 International Conference*, Thessalonica, Greece, Nov 2008
11. J.K. Kaldellis, D. Zafirakis, A.G. Paliatsos, Energy yield of contemporary wind turbines, in *Comprehensible Renewable Energy Encyclopaedia*, vol. 2: Wind energy ed. by A. Sayigh (Elsevier, 2012)
12. J.K. Kaldellis, Maximum wind potential exploitation in autonomous electrical networks on the basis of stochastic analysis. J. Wind Eng. Ind. Aerod. **96**(8–9), 1412–1424 (2008)

13. J.K. Kaldellis, D.S. Vlachou, A.G. Paliatsos, Twelve years energy production assessment of Greek state wind parks. Wind Eng. **27**(3), 215–226 (2003)
14. Hellenic Wind Energy Association (HWEA), Wind energy statistics 2018 (2019), http://eletaen.gr/wp-content/uploads/2019/01/2019-02-01-hwea-statistics-greece-2018f.pdf
15. J.K. Kaldellis, D. Zafirakis, Present situation and future prospects of electricity generation in Aegean Archipelago Islands. Energy Policy **35**(9), 4623–4639 (2007)
16. J.K. Kaldellis, K.A. Kavadias, D.E. Papantonis, G.S. Stavrakakis, Maximizing the contribution of wind energy in the electricity demand problem of Crete Island. Wind Engergy **30**(1), 73–92 (2006)
17. J.K. Kaldellis, Investigation of Greek wind energy market time-evolution. Energy Policy **32**(7), 865–879 (2004)
18. J.K. Kaldellis, D. Zafirakis, E. Kondili, Improved wind energy production prediction for Remote Aegean Sea Islands. *Conference Proceedings—17th IEEE International Conference on Environment and Electrical Engineering*, Milan, Italy, June 2017.
19. J.K. Kaldellis, Critical evaluation of the hydropower applications in Greece. J. Renew. Sustain. Energy Rev. **12**(1), 218–234 (2008)
20. Hellenic Electricity Distribution Network Operator, Issues of the operator for the non-interconnected islands, Athens, Greece (2018), https://www.deddie.gr
21. J.K. Kaldellis, Integrated electrification solution for autonomous electrical networks on the basis of RES and energy storage configurations. Energy Convers. Manage. **49**(12), 3708–3720 (2008)
22. Chr. Marouli, J.K. Kaldellis, Risk in the Greek electricity production sector. *7th International Conference on Environmental Science and Technology, Conference Proceedings*, vol. C (University of Aegean, Global-NEST, Syros, Greece, 2001), pp. 305–314
23. J.K. Kaldellis, M. Kapsali, Shifting towards offshore wind energy recent activity and future development. Energy Policy **53**, 136–148 (2013)
24. J.K. Kaldellis, D. Zafirakis, Optimum energy storage techniques for the improvement of renewable energy sources-based electricity generation economic efficiency. Energy **32**(12), 2295–2305 (2007)
25. J.K. Kaldellis, D. Zafirakis, Optimum sizing of stand-alone wind-photovoltaic hybrid systems for representative wind and solar potential cases of the Greek territory. J. Wind Eng. Ind. Aerod. **107–108**, 169–178 (2012)
26. Technology Innovation for the Local Scale Optimum Integration of Battery Energy Storage (TILOS), Horizon 2020 Program, No 646529 (2018), http://tiloshorizon.eu
27. J.K. Kaldellis, D. Zafirakis, E. Kondili, The Green Island Solution for Aegean Archipelago. The case of TILOS Project. *International Conference on Environmental Management, Engineering, Planning and Economics (CEMEPE) and SECOTOX Conference*, Thessaloniki, Greece, 26–29 June 2017
28. PPC Renewables, Hybrid systems (2019), https://www.ppcr.gr/en/hybrid-systems
29. P. Mastorakis, G. Betzios, J. Kaldellis, A proposal of installation of a combined wind–hydro station for the islands of Aegean Sea. *NTUA-RENES National Conference on the Application of Soft Energy Sources*, Athens, 1998

Chapter 4
Wind Energy Programme in Japan

Izumi Ushiyama

EIA: Japanese wind power market is still in a state of stupor due to the complicated and cumbersome EIA procedure which was applied for all wind farms over 10 MW since October 2012, and it needs about 4 years to finish this procedure. Only 203 MW with ten projects have won the EIA permission, and the other 6226 MW with 88 projects are still in the EIA process. The dawn light will shine in 2016, when most of these projects complete the EIA and start operation gradually. A lot of discussion and efforts are paid for cutting red tapes in Japan, especially for the EIA, grid access and land-use restrictions now.

Feed in Tariff: The feed-in-tariff (FIT) for onshore wind has been maintained at JPY 22/kWh (EUR 0.164/USD 0.185). And the Japanese Ministry of Economy, Trade and Industry (METI) had offered JPY 36/kWh (EUR 0.27/USD 0.30) for offshore wind. This tariff for offshore wind is set 1.64 times higher than onshore wind. The tariff is to be re-estimated every year according to the latest experience in Japan. On the other hand, the qualification of FIT can be gotten when the project almost finished the EIA procedure. The Japan Wind Power Association (JWPA) requests government to shift the qualification timing a little earlier so as to make Japanese wind power development bankable.

Deregulation: The Ministry of Environment (MOE) and the METI have tried to shorten the EIA process period to within 2 years. The re-estimation for the EIA contents was discussed. And the MOE started new subsidies supporting 50% cost for pre-EIA investigation and it was applied for about 20 sites in FY 2014. The strict rule for land use especially for farm was one of big red tape in Japan. The Ministry of Agriculture, Forest and Fisheries (MAFF) made a new law named "Act for the Promotion of Renewable Energy in Rural Districts (APRERD)" on 15 November in 2013, and APRERD went in operation on 1 May in 2014. It will help the change of

---

I. Ushiyama (✉)
Ashikaga University, Ashikaga, Tochigi, Japan
e-mail: ushiyama@ashitech.ac.jp

A. Sayigh, D. Milborrow (eds.), *The Age of Wind Energy*, Innovative Renewable Energy, https://doi.org/10.1007/978-3-030-26446-8_4

land use purpose from "farm/agriculture" to "wind power/industry". It means the big increase of potential land area for wind power in Japan.

Grid restriction: Solar power developers have rushed to propose FIT approval by March 2014, responding the announcement of FIT price reduction (from JPY 36/kWh to JPY 32/kWh) from April by the METI. Then, the METI had announced the new grid connection rule in October 2014. This new rule expands acceptable capacity by introducing the wider output control for renewable energies by power supply reasons. The new acceptable capacity for wind power at seven electric companies was 5.73 GW in total. 2.1 GW has been already in operation in these seven regions; therefore, 3.63 GW was available for new connection. The central three regions (Tokyo, Chubu, and Kansai) have huge electric demands and are free from this kind of restriction. If Japanese grid lines are operated more favourable for renewable energies, the above grid restriction shall be eased dramatically in future.

Most of Japanese wind resources exist in the northern rural regions called Hokkaido and Tohoku. Low population needs only small grid lines. The METI had started to build new grid lines for wind power in northern rural area in Japan (Hokkaido and Tohoku). The METI subsidizes about 50% of building cost (25 billion JPY for every year). The grid building consortium for Hokkaido (for about 3 GW) had decided in 2014, and new two consortiums for Tohoku (for 600 MW at Akita and for 900 MW at Aomori) have also decided in 2014. NEDO started new national project to build up nationwide wind power output forecast system for 5 years. The fund for first year was four billion JPY. The JWPA cooperated with this project.

Wind Roadmap 2050: One of the major issues in Japanese wind industry was a lack of long-term installation goal and roadmap. Then, JWPA proposed wind road map 2050 based on wind potential in Japan. The wind power target at 2050 was set as 50 GW so that it would produce 10% of national electricity demand by wind.

4.1 Brief History of Wind Power Generation in Japan

Today's low rate of wind power development in Japan could be accounted for because of the abundant rivers (30,000 in number) and annual precipitation of 1800 mm that have alternatively promoted hydropower. Thus, historically Japan has made little use of wind power compared to European countries. Furthermore, there have been several unfavourable conditions against wind power: seasonal gales due to the Asian monsoonal climate; visits of tropical typhoons; strong winter lightning on the side of the Sea of Japan due to Siberian cold air masses; and turbulences due to the unique Japanese terrain.

To overcome these severe environmental conditions, Japanese manufacturers of MW-class wind turbines such as Mitsubishi Heavy Industries Ltd., Fuji Heavy Industries Ltd., and The Japan Steel Works Ltd. have developed one of the most robust wind turbines in the world. These wind turbines meet Japanese guidelines of

the turbine design set up by the New Energy and Industrial Development Organisation (NEDO). A branch-like form of the national power grid that weakens towards tail ends rather than a ring-shaped grid as in Europe also sometimes thwarts wind power installation.

The biggest problem, however, has been the policies towards renewable energy. A national scheme named "Sunshine Project" initiated just after the oil crisis in the 1970s concentrated R&D support in solar energy and the support for wind power was delayed and limited. Moreover, the Government and electric power companies have given the priority to nuclear power, leaving renewable energy behind. As the Sunshine Project was focused on solar energy in the beginning, wind energy research started in 1978.

The first large-scale wind turbine in Japan, a two-bladed machine of 29 m diameter and 100 kW output, installed in Miyake Island started operation at the beginning of 1983 (Fig. 4.1). The output of the machine was connected to the diesel-powered network of the island. Then, a 500 kW wind turbine was developed in the beginning of the 1990s and installed at Cape of Tappi in Aomori prefecture (Fig. 4.2).

Demonstrative projects of wind power plants for isolated islands were conducted in the end of 1990s and two units of 100 kW were installed at Izena Island in Okinawa. Private companies such as Mitsubishi Heavy Industry, Fuji Heavy Industry, Japan Steel Works, and Komai-Haltech have developed large-scale wind turbines and wind farms.

4.1.1 National Strategy

At the UN Climate Change Conference in Kyoto in December 1997, the Japanese government agreed to reduce the output of greenhouse gases by 6% from the 1990 level by 2010 (in the period 2008–2012). To attain this target, the government decided to develop "new energy" (nearly equal to renewable energy) by 3% of the primary energy supply by 2010 in the Primary Energy Supply Plan. The contribution of wind power generation was 7% within the renewable energy, which made the target of wind power 3000 MW with 0.21% of primary energy supply by 2010.

In April 2003, the Japanese government passed legislation for a Renewables Portfolio Standard (RPS) in order to realise the national target for renewables by 2010. Under the RPS, Japan's utilities are obligated to source 12.2 TWh (1.35%) of total electricity supply from renewable energy by 2010. RPS targets were to be reviewed every 4 years. The RPS target established in 2007 was aiming at 16.0 TWh (equivalent to 1.63%) by 2014. During the fiscal year 2008, wind plants supplied 2865 TWh/year, which was one-third of the total supply from renewable energy.

Fig. 4.1 NEDO 100 kW wind turbine in Miyake Island

Fig. 4.2 NEDO 500 kW wind turbine at Cape of Tappi in Aomori

To counteract natural and social obstacles to wind power development, the government has been running the following investigations or research programmes:

1. Wind Energy Business Support Programme
2. R&D of Advanced Wind Energy Technology
3. Investigation on Offshore Wind Technology
4. Field Test Programme
5. Demonstration of Grid Stabilisation
6. Demonstration of Battery-Backup Wind Farms
7. Subsidies for Grid-Connected Systems
8. Wind Technology Standards (JIS/IEC)
9. Effect of Lightning Attack on Wind Turbines
10. Effect of Wind Turbine Noise on the Surrounding People

4.1.2 Installed Capacity

Japan's cumulative wind power capacity was 2450 MW with 1742 units at the end of December 2010. It is about 0.8% of total electric power plant capacity (about 275 GW). Wind power accounts for about 0.4% of total electric power supply in Japan (Fig. 4.3).

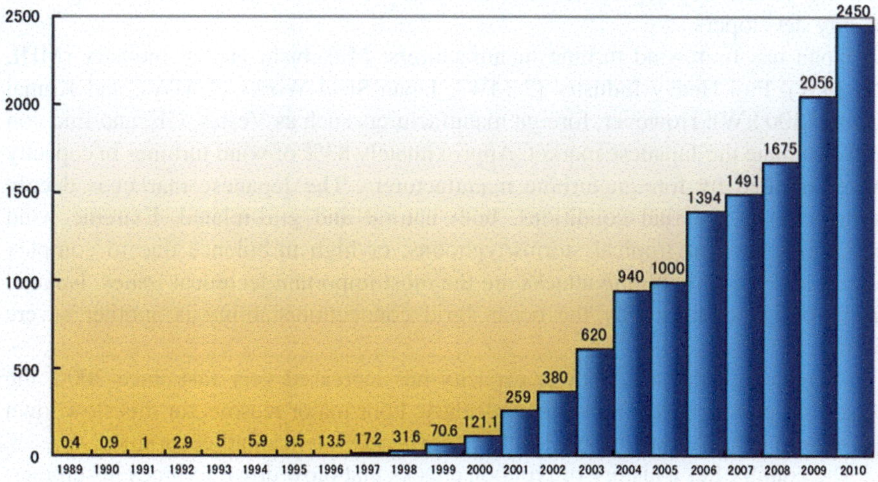

Fig. 4.3 History of wind power development in Japan

4.1.3 Benefits to National Economy

The wind turbine industry in Japan has been growing in recent years. Mitsubishi and other manufacturers like Japan Steel Works, Ltd. and Fuji Heavy Industries, Ltd. have started mass-production of 2 MW-class wind turbines (WTGs). Several bearing companies are expanding their factories and many electric companies are exporting power devices, which corresponds to the large amount of orders from wind turbine manufacturers throughout the world. The growth of the wind turbine industry brings the so-called "green money" and "green jobs" for Japan. The production of wind turbines and their components has reached around JPY 300 billion per year (ca. EUR 3 billion), according to the research of New Energy Foundation. More than 1000 people are working directly for wind turbine mass production, and as a global estimation, 4000–15,000 jobs have been created among parts and device companies.

4.1.4 Market Characteristics

The wind power market in Japan has rapidly progressed in the past 15 years. As a result, large wind farms have been developed (Fig. 4.4). The largest wind power plant, which was built in December 2008 in Izumo City in Shimane Prefecture, has 78 MW and consists of 26 units of Vestas 3 MW turbines. Its operation has started in April 2009. Today there are five large wind farms that exceed 50 MW. Many entities are developing wind power: citizen groups, NPOs, third sectors, local governments, and big private developers. Most of the large wind farms are owned by big wind energy developers.

Japan has four wind turbine manufacturers: Mitsubishi Heavy Industry (MHI, 2.4 MW), Fuji Heavy Industry (2 MW), Japan Steel Works (2 MW), and Komai Tekko (300 kW). However, foreign manufacturers such as Vestas, GE, and Enercon still dominate the Japanese market. Approximately 83% of wind turbines in capacity were supplied by foreign turbine manufacturers. The Japanese market is deeply influenced by external conditions, both natural and grid-related. Extreme wind conditions such as tropical storms/typhoons, or high turbulence due to complex terrain and heavy lightning attacks are the most important technical issues. Isolated from foreign countries by the ocean, grid connection/stability is another severe barrier.

Even though the wind power capacity has increased very fast since 2000, the sector has experienced a slowdown recently. Four major reasons for this slowdown are natural external conditions, legal system, grid systems, and economics.

The country has a history of typhoon attacks that blow down turbines in summer. Lightning strikes in winter and summer, as well as strong gusts and high turbulences in complex terrains are also technical hazards. The weather defines external conditions in Japan. Therefore in 2008, NEDO developed a safety guideline designed for

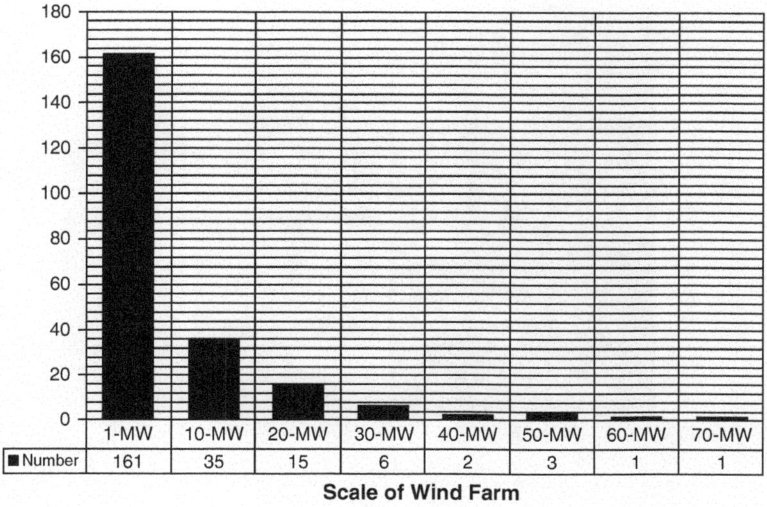

Fig. 4.4 Scale distribution of wind farms larger than 1 MW

Japanese meteorological and geographical conditions to provide technical measures against typhoons and lightning strikes and to help future wind turbine developments. As a result, some promising sites became carefully evaluated for safer development (Fig. 4.5).

According to the new Japanese Building Code that became effective in June 2007, a wind turbine of 60 m height or more shall be considered as a kind of building, and its height is defined as the top height of a blade from the ground level. Under this revised code, the installation of wind turbines needs the minister's sanction. The application procedure for planning permission is very complicated, time consuming and expensive. The first project authorised under the new code was in July 2008, which means absolutely no new projects started between June 2007 and July 2008. However, the permission process has become rather standardised and in recent years many projects are being authorised.

The third issue is the grid system in Japan, which has foiled new wind farm developments to keep the stability and security of electricity supply. Geographically, the most potential sites for wind power development are Tohoku and Hokkaido in the north of the country and Kyushu in the south. Unfortunately, these areas have relatively small grid capacity and are far away from the centre of Japan such as the Tokyo, Osaka, Nagoya areas, where the electricity demands are the greatest. As a social system, this regional monopolistic grid system with strong limitation of grid access makes wind farm developments even more challenging.

The relative high price of wind turbines due to the depreciation of Japanese Yen against the Euro was also the reason of slowdown, considered that more than 80% of wind turbines were imported. But the price of wind turbines started decreasing due to appreciation of Japanese Yen against the Euro from the second half of 2008. A number of developers are trying to manage shelved projects and get them

Fig. 4.5 Wind turbine destroyed by strong typhoon in Miyako Island

resurrected. However, the number of annually installed turbines did not recover until 2009 due to long lead-time for delivery of wind turbines.

4.1.5 Industrial Development and Operational Experience

There are several wind turbine manufacturers who produce their own turbines in Japan. The present main commercial turbines are listed in Table 4.1. The manufacturers have developed new turbines that are more suitable for Japanese external conditions, such as extreme gusting. Some of the turbines are designed to endure gust above 70 m/s over 50-years recurrence period (Fig. 4.6).

A stable wind farm development can be found in the efforts of the local government of Tomamae Town. The local government has been developing a small wind farm since December 1998. The wind power plant consists of two 600 kW turbines and one 1000 kW turbine. From Fig. 4.7, we may expect high capacity factors above 30% when the mean wind speed is 7 m/s or 40% when 8 m/s. Tomamae Town,

Table 4.1 Main commercial wind turbines

Manufacturer	Wind turbine	Technical characteristics
Mitsubishi Heavy Industries, Ltd. (MHI)	MWT92/2.4	$P = 2.4$ MW, $D = 92$ m, three-bladed, Upwind, SmartYaw control
	MWT-1000A	$P = 1.0$ MW, $D = 61.4$ m, three-bladed, Upwind, SmartYaw control
Fuji Heavy Industries Ltd. (FHI)	SUBARU 80/2.0	$P = 2$ MW, $D = 80$ m, three-bladed, Downwind
The Japan Steel Works, Ltd. (JSW)	J70/J82	$P = 2$ MW, $D = 70.65/82.6$ m, three-bladed, Upwind, Direct-drive
Komai Tekko Inc.	KWT300	$P = 300$ kW, $D = 33$ m, three-bladed, Upwind
		Iref = 20% Iref = 20%
Zephyr Corporation	Airdolphin	$P = 1$ kW, $D = 1.8$ m, three-bladed, Upwind
		Cut-in = 2.5m/s without Cut-out, Full carbon fibre blade

Fig. 4.6 (**a**) MWT 92/2.4 of Mitsubishi Heavy Industry (left); (**b**) SUBARU 80/2.0 of Fuji Heavy Industry

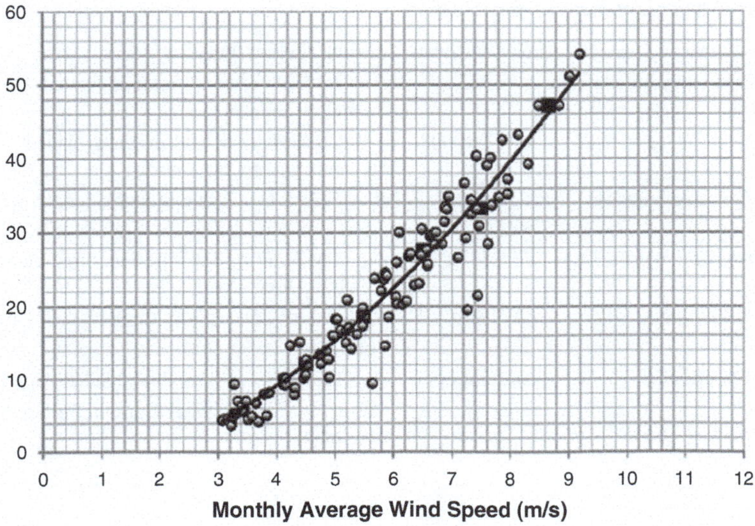

Fig. 4.7 Operation data of Tomamae Yuhigaoka wind farm owned by Tomamae Town: monthly capacity factors vs. monthly mean average wind speeds with bin-averaged data plotted with black squares and a fitting curve

Fig. 4.8 A view of Tomamae wind farm region

located in Hokkaido, has good winds and has also one of the biggest wind farms in Japan (Fig. 4.8). A steady effort to record and report the operation of a wind farm has encouraged many developers.

4.1.6 Economic Details

In general, the cost of a wind power plant in Japan is higher than in EU countries, where wider grid systems are well developed. In Japan, additional cost for grid connection/stability is particularly required including battery-backup plants.

During a couple of years before 2007, it was reported in a national committee that the average cost of energy (COE) for a 25 MW wind farm was 10.2 JPY/kWh with the aid of subsidy. Generally COE was from 9.00 to 11.0 JPY/kWh for medium-sized wind turbines (unit capacities between 500 and 1000 kW). For large-scale wind farms comprising wind turbines with capacities of more than 1000 kW, COE is in the range of 7.0–9.0 JPY/kWh. The average wind turbine cost was approximately 100,000 JPY/kW and the average initial cost was estimated at 190,000 JPY/kW in 2003. However, wind turbine cost sprang up by approximately 80% in 2007, where 50% was caused by worldwide tendency and 30% was due to currency exchange rate between the Euro, US Dollar and Yen reflecting that more than 80% of wind turbines have been imported from Europe and the United States. As of 2008, initial plant cost was around 300,000 JPY/kW on average and the electricity purchase price was 10.4 JPY/kWh. To find the value of COE for 2008 is quite difficult due to social/natural hazards described previously as well as the global economic crisis occurred in 2008.

4.1.7 National Incentive Programmes

National wind energy R&D programmes were once closed in fiscal year 2002 and the main stream of governmental incentive programmes on wind was shifted to subsidies for wind plant developments and investigations on grid issues. The grid issues and external conditions such as extreme wind, high gust and lightning were still found technically important, and investigations and demonstrations on these issues have been continued and unsolved during several years. NEDO established J-Class Wind Turbine Guideline in 2008 for the purpose of technical safety of wind turbines under Japanese conditions. Investigation programmes on offshore wind power have started as well.

After the accident of Fukushima nuclear power plant, wind power is gaining unprecedented recognition in Japan. Onshore and offshore wind power is now being introduced as effective and attractive choices for Japan, which has the sixth largest exclusive economic zone (EEZ) in the world.

NEDO started to develop offshore wind power generation from 2006. This project is focused on fixed foundation type. The rated output of the prototype is 2 MW and the test site is Chosi area, not far from Tokyo. From 2010, the Ministry of Environment initiated a demonstration programme to develop floating-type offshore wind power generation. A 100 kW model plant was installed in January 2012 then down-wind type 2000 kW wind turbine will be tested in Kamigoto Island in southwestern part of Japan.

Offshore wind turbines are being developed combining the leading shipbuilding technology. Mitsubishi Heavy Industries Ltd. has signed a memorandum of understanding with the UK government in February 2010 for the development of 5–7 MW-scale offshore wind turbines with the support of grants up to GBP 30 million. The development of offshore wind power plant is also taking place under Japanese national programmes and the installations of full-scale turbines were planned in the year 2011 for fixed foundations and in 2013 for floating structures respectively.

4.1.8 National R&D Efforts

Concerning R&D wind technologies, three programmes are running; it is worthy of special mention that the basic research revived, although the focus is still in the field of applied technology. Advanced Wind Technology is a comprehensive programme which includes basic research. Some keywords include the following: remote sensing for advanced wind measurements, lightning measurements for extended development of protection measures and J-class wind models also for future safety standard. The projects aim to seek international cooperation with IEA WIND R&D and IEC Standards.

Under the Offshore Wind Project, several candidate sites for offshore projects were selected for feasibility studies and project designs, based on wind/wave/soil measurements, offshore wind performance predictions, and detailed designs. Field Test Programme is a cooperative research project with NEDO under which by wind measurements at high altitudes at several promising sites, useful wind database will be developed.

Concerning demonstrations on grid issues, the Grid Stabilisation Programme was closed in 2008. One of the main technical targets was to investigate the performance of wind generation prediction technique using Computational Fluid Dynamics (CFD) models. The research was well appreciated but did not introduce any powerful tool to solve the grid issues.

Similar situation is with the Battery Back-Up Demonstration Project, where the technical experience and database will contribute to the future wind development with higher penetration. Standardisation of wind turbine technology in relation to IEC and JIS has been conducted by Ministry of Economy, Trade and Technology (METI), The New Energy and Industrial Technology Development Organisation (NEDO), Japan Electrical Manufacturers' Association (JEMA), and National Institute of Advanced Industrial Science and Technology (AIST). It is very important for reliability and safety since Japan has to face some severe external conditions onshore as well as offshore.

4.1.9 Collaborative Research

Japan has been the member of IEA R&D WIND since 1978. AIST is expanding the collaborative researches through TASKs from 2008 onwards. Since 1988, Japan has been involved in IEC activities aimed at establishing international standards for wind turbine technology. The Global Wind Energy Council Japan (GWEC Japan), consisting of Japanese Wind Energy Association (JWEA) and Japanese Wind Power Association (JWPA), has been cooperating as a member of the Global Wind Energy Council (GWEC) since March 2005.

4.2 Present Status of Wind Power Generation in Japan

Wind power new installation in Japan has increased gradually. 177 MW has installed in 2017 and it is 10% smaller than 196 MW in 2016. Several large wind farms have finished long EIA process and started operation (Fig. 4.9). The cumulative installation has reached 3400 MW at the end of 2017. It produced about 5000 GWh/year which is about 0.56% of electricity supply in Japan in 2017. More 6100 MW of projects have already gotten FIT approval (including grid connection right), which can start operation within a couple of years. Much more projects are in the EIA (Environmental Impact Assessment) process now, but most of them are facing grid connection problems which force us long time and huge money to solve them. Twenty-nine offshore wind turbines with 64.6 MW capacity are in operation. One Hitachi's 5 MW floating turbine started operation in May 2017 in Fukushima FORWARD project.

Japanese policies on wind power generation are confused and inconsistent. A new law is to be issued for promoting offshore wind in May 2018, but FIT may change into price based auction system for the offshore wind at the same time.

Wind power development process in Japan proceeds; wind investigation > Environmental Impact Assessment (EIA) > Securing grid connection > FIT approval. Some improvements (saving time) at EIA has been achieved. Then, grid restrictions bring many troubles for wind power business in Japan now.

Fig. 4.9 Shin-Aoyama-Kogen Wind Farm, in Mie pref., 80 MW (the largest in Japan), Hitachi 2 MW × 40 units, start operation in February 2017. (Courtesy of Aoyama Kogen wind farm)

4.2.1 Future Target

Japanese Ministry of Economy, Trade and Industry (METI) fixed wind power target as 10 GW (including 820 MW offshore wind power) by 2030, at their future energy plan called "Long-term Energy Supply and Demand Outlook (Energy Mix Plan)" in July 2015. It is re-estimated every 3 years. JWPA request more aggressive target for wind power. But unfortunately, METI won't change 10 GW target at next version in 2018, they only remove the internal limit for offshore wind power (up to 10 GW).

The increase of wind power at Energy Mix Plan means the shrinkage of nuclear and coal fired fossil share. It is unacceptable for Japanese establishment people. METI's Energy Mix Plan is made with the principle of balancing 3E&S, Energy Security, Economy, and Environmental Conservation and Safety. And there is an unofficial order; economy holds top priority, energy security second, followed by environment and safety. The principle comes from our past experience of the oil crisis in the 1980s. METI thinks nuclear power and cheap Australian (free from Middle East confliction) coal are good for Japan's economy and energy security. METI and Japanese huge electric power companies have spent lots of money and put in lots of efforts in these two technologies and business. Therefore, the switching cost from nuclear and coal to renewable energies is very huge in Japan both economically and politically, even after the Fukushima accident and the realisation of climate change. It is the so-called "Concorde fallacy" (Many people cannot make a right decision because they have strong misgivings about wasting resources/sunk cost.)".

4.2.2 Feed In Tariff

Both good and bad changes come in 2017. Japanese FIT law gets amended in April 2017. The FIT approval timelines for wind moves 2 years earlier, from the end of EIA to the middle of the EIA. But grid connection right is still necessary to get FIT approval. And FIT-approved wind power plants must start operation within 4 years (Table 4.2).

METI changes mind to terminate FIT in late 2017 because of shortage of funds.

At the start of FIT introduction in 2012, METI fixed upper cap of citizen's levy burden for FIT as 3.1 trillion JPY/year at 2030 in their Energy Mix Plan, in which renewable energies has 22–24% share. In FY2016, renewable energy share increases from 10% (in FY2010) to 15%, mainly by about 35 GW of new solar power plants, and FIT levy reaches 1.8 trillion JPY/year. More 42 GW of solar power gets FIT approval and waits for operation. This big boom at solar is beyond METI's original target (64 GW solar at 2030); it consumes all funds for FIT levy, then, there are no funds for other renewable energies. Then METI makes strong efforts to keep the levy burden upper cap. They introduce an auction system for large-scale solar power since FY2017, eliminate small wind power category in FY2018, and expand the auction to other solar, liquid fuel biomass, and fixed bottom type offshore wind power now.

Table 4.2 FIT price for wind power, decided in April 2018 (JPY/kWh) FY: April to March

Type	FY2017	FY2018	FY2019	FY2020
Onshore (>20 kW)	22/21	20	19	18
Small onshore (>20 kW)	55	20	19	18
Onshore at replace	18	17	16	16
Offshore, fixed bottom	36	36	36	(Auction)
Offshore, floating	36	36	36	36

Ref: Purchase cost and assessment cost of wind energy was decided on 23 March 2018 by METI

4.2.3 Grid Restrictions

Nine local electric power companies (huge utilities) divide and govern Japanese electric power system. It is old fashioned like before 1990s in Europe. They dominate power generation, transmission, and distribution at each region. The unbundling of power generation and transmission (Electric Power Reform) is to be introduced in FY2020, but until March 2020, huge utilities can deny new grid connection if they find technical insufficient problems. And who pays for grid reinforcement for new renewable energies is under discussion. "Causer (Generator/Polluter) pays principal (PPP)" and "Early comer has prior grid using right principle" are commonly applied in Japan. Therefore, existing power plants (nuclear and fossil) and new early planned coal fired power plants fill grid line capacity, before late coming new wind farms now. They are quite different from European rules, "Beneficiary Pays Principle (BPP)" and "Merit order". JWPA requests BPP for Japanese government. But BPP brings levy on electric fee, it is not easy to accept for METI who intends to keep electricity fee cheap and to restrict citizen's burden at lower level.

The location mismatch of wind resources and electricity demand in Japan causes grid connection problems. The northern area (Hokkaido and Tohoku) have most of wind resources in Japan, but their population are small, and their grid infrastructure are restricted for wind power. Tohoku Electric Power Co. announced to stop accepting new requests for grid connection at northern three prefectures (Aomori, Iwate, Akita) in October 2016. Hokkaido Electric Power Co. asks wind power developers to pay for large scale battery to stabilize the fluctuation of output. Tohoku Electric Power Co. has made grid reinforcement plan which enables more 2.8 GW and requested auction for grid connection rights with its cost share. This auction process closed in April 2017. 344 projects with 15.45 GW (including 4.46 GW onshore wind and 7.86 GW offshore wind) were bid, which is about six times larger than the targeted capacity.

The grid connection auction in Japan has several problems. The first one is the high cost burden. Ninety billion JPY was requested for a case in Niigata pref., and sixty billion JPY was requested for another case in Aomori pref. If the auction winner cannot afford it, they retire and the electric power company proposes re-bidding. It takes several months; if it delays to the next fiscal year, FIT price goes down 1 JPY/kWh, and the project profitability is severely damaged.

On the other hand, Professor Dr. Yoh Yasuda in the Kyoto University has revealed in October 2017 the following:

1. The average using rate of the trunk grid lines in Tohoku Electric Power Co. is only 2.0–18.2%
   http://www.econ.kyoto-u.ac.jp/renewable_energy/occasionalpapers/occasionalpapersno45
2. Tohoku Electric Power Co. announces 67% of their trunk lines are already full, but actual average using rate of them is only 9.5% (lesser than whole average 12.0%).
3. There might be some additional room for renewables, if we change the grid operation rule.

This exposure reported widely by TVs and newspapers and it caused a big argument in Japan, because it suggests more cost saving solution than spend huge money and long time for grid reinforcement. The reason of the low using rate was explained that N-1 safety control against power failure just in case set 50% upper limit (Japanese trunk grid line is tree shape. Loop shape grid like in Europe is not common in Japan.). And existing (sleeping by safety check) nuclear power plants and early planned new coal fired fossil plants (note: EIA process for coal fired fossil plant is shorter than for wind farm. It sounds strange, but is true in Japan) reserve grid line capacity because they have the prior grid connection rights against the late coming wind farm requests. It means nuclear and coal do not have to pay the grid reinforcement cost. But many people think it is not a fair deal against wind power. Then METI start discussion to amend the grid operation rule from past manner to the new so-called "Connect and Manage" solution. The details are under discussion now.

Another problem is "Unlimited curtailment by grid issue" without compensation. Current Japanese Electric Power system rule allows Electric power companies to do it when renewable energy rate reaches a certain ratio. Hokkaido and Tohoku Electric Power Co. announced it in 2016, and Kyushu and Shikoku Electric Power Co. announced it in Mar. 2017. The curtailment rates shall be estimated as lesser than 30% at Hokkaido and Tohoku, and 10% at Kyushu and Shikoku. (Formerly, the curtailment was limited 30 days/year, and if it exceeds the limit, the electric power companies should pay compensation money for it.)

This unlimited curtailment means profitability uncertainty at wind power business, so it becomes the new hurdle for getting development fund from investors.

4.2.4 Offshore Wind Power Development

By the end of 2017 Japan has 64.6 MW with 29 turbines of offshore wind power, including 16 MW with four floating wind turbines (Table 4.3).

Table 4.3 Offshore wind power experience in Japan (in March 2018)

Type	Location		Distance (km)	Depth (m)	Rated (MW)	No. of WTG	Total (MW)	Start operation
Fixed	Hokkaido	Setana Port	0.7	13	0.6	2	1.2	December 2003
	Akita	Akita Port	0.1	–	3.0	1	3.0	February 2015
	Yamagata	Sakata port	0.05	4	2.0	5	10.0	January 2004
	Ibaraki	Kamisu	0.04	4	2.0	7	14.0	February 2010
		Kamisu	~0.05	4	2.0	8	16.0	February 2013
	Chiba	Choshi[a]	3.1	12	2.4	1	2.4	March 2013
	Fukuoka	KitaKyushu[a]	1.4	14	2.0	1	2.0	June 2013
Floating	Nagasaki	Fukuejima	5.0	100	2.0	1	2.0	April 2016
	Fukushima	Iwaki city	20	120	2.0	1	2.0	December 2013
		Naraha[a]			7.0	1	7.0	March 2016
					5.0	1	5.0	May 2017
		Total				29	64.6	

[a]National projects

Fig. 4.10 Hitachi 5 MW floating offshore wind turbine. (Courtesy of Fukushima FORWARD)

Hitachi's 5 MW downwind turbine named "Hamakaze" on the advanced spar type floater started operation at Fukushima FORWARD project as the third turbine in May 2017 (Fig. 4.10).

The next one is to start operation in summer 2018 at Kitakyushu by NEDO (New Energy and Industrial technology Development Organization)'s new national project. 3.5 MW two bladed turbine designed by German Aerodyn on the French IDEOL designed moonpool type floater made by Hitz is applied for this project.

There are further 12 GW of offshore wind power projects currently under planning (Table 4.4). 5079 MW with 22 projects are well prepared. The first commercial project may start operation in 2021. The other 7 GW projects are the so-called "proposal rush" for Tohoku Electric Power Co.'s grid connection right auction. They are at pre-EIA stage now.

Sea area in Japan is categorized into two areas which are "Port associated area" and "General common sea area". However, The Japanese Ministry of land, Infrastructure, Transport and Tourism (MLIT) has amended "Port and Harbor Law" for promoting offshore wind power development at port associated area in May 2016 to allow 20 years occupation of the designated water zone in the port area for developers (which is repeatable) and to settle the bidding system of offshore wind power development in the port area.

As for general common sea area, the new is officially approved by the Japanese Cabinet on 9 March 2018. It will be discussed at the Diet and will be issued in around May 2018. The new legislation potentially takes effect 4 months after its promulgation.

The new bill is called the "Bill on promotion of use of territorial waters for offshore renewable energy generation facilities (Kaiyo saisei kanou enerugi hatsuden setsubi no seibi nikansuru kaiiki no riyo no sokushin nikansuru houritsuan)" (the "New Offshore wind Bill"). The bill allows 30 years occupation right for bidding winners.

Table 4.4 Offshore wind power under planning in Japan (in February 2018)

Type	Location		Area	WTG (MW)	No. of WTGs	Total (MW)	Start operation
Fixed	Hokkaido	Wakkanai port	Port			10	
		Ishikari new port	Port	4.0 MW	24	96	2020~
	Aomori	Mutsuogawara port	Port	2.0 MW	40	80	2019~
		Mutsu	Gen			800	
		Yokohama	Gen			80	
		Tsugaru	Gen			1000	
		Tsugaru East	Gen			480	
	Akita	Noshiro port	Port	3.3–6.0	20	100	2021
		Happo Noshiro	Gen			180	
		Akita port	Port	3.3–6.0	14	70	2022
		Akita North	Gen.	3.3–5.0	120	455	2023
		Yurihonjo	Gen.			1000	
	Yamagata	Sakata port	Port			15	
	Toyama	Toyama	Gen			7.5	
	Ibaraki	Kashima port	Port	5.2 MW	36	187	2021
	Yamaguchi	Yasuoka, Shimonoseki	Gen.	4.0 MW	15	60	
	Fukuoka	Kitakyushu port	Port			220	
		Kitakyushu	Gen.			300[a]	
	Nagasaki	Saikai Enoshima	Gen			240	
Floating	Fukuoka	Kitakyushu[b]	Gen.		1	3.5	2018
	Nagasaki	Fukuejima	Gen.	2.0–5.0	10	22	
Test Field	Niigata	Awashima	Gen.				
	Nagasaki	Kabashima	Gen.				
Planned projects (location defined)						5079	
Grid connection request (locations are disclosed)						7000	
						12,000	

[a]Estimated by JWPA
[b]National projects

In this alert we outline the general scheme for offshore wind development proposed under the New Bill, the specific provisions of the New Bill, the number of zones to be released and key issues we see with the New Bill. At least five sea areas will be nominated for the first round auction. A new organization will be made for harmonizing adjusting stakeholders.

This new bill reduces business risk. But grid connection and EIA process remain uncertain in this new bill. So some conflicts may happen when the bidding winners are different at offshore sea area occupation and grid connection. JWPA requests Japanese government to apply the so-called "Central system" like in Netherland to eliminate business uncertainties. And METI intends to introduce price base auction system for fixed bottom type offshore wind power development at general common sea area after this law starts operation. It brings another type of business risk, PPA price uncertainty. JWPA also request to keep FIT for offshore wind now.

4.3 Potential and Future Prospects of Wind Power in Japan

Available onshore wind energy in Japan was estimated to be 6400 MW in March 2000. However, there was the need to revise the available wind energy in Japan including offshore, based on the latest analytical methods, recent developments in wind turbine technology, and the latest national land map. The latest study re-estimated the Japan's onshore and offshore wind energy resources as well as the available wind energy in consideration of social and economic constraints. Consequently, the potential estimation found 6,434,830 MW for the wind energy resources and 782,220 MW for the available wind energy in onshore and offshore of Japan at an annual mean wind speed of 6.5 m/s for onshore and 7.5 m/s for offshore at 80 m height. Furthermore, the roadmap and annual installation capacity were estimated to achieve the long-term installation goal, "the amount of electricity produced by wind turbines exceeds 10% of the Japan's national electricity demand by the year 2050". As a result, the wind turbine installation targets were found as follows: 3000 MW in 2010; 11,310 MW in 2020; 27,000 MW in 2030; 44,300 MW in 2040; and 50,000 MW in 2050. The annual production capacity including replacement of turbines after the year 2030 continues to exceed 2500 MW every year; therefore, the sustainable development of the wind power generation industry is expected.

4.3.1 *Wind Turbine Installation Roadmap*

The roadmap to achieve the long-term installation goal (50,000 MW by the year 2050) as shown in Table 4.5, is estimated from a cubic polynomial of X as elapsed years to obtain Y as annual installation capacity. The obtained roadmap is shown in Fig. 4.11. Its premises are achieving the national installation goal of 3000 MW by the year 2010 and the introduction of fixed and floating structure offshore wind turbines starting in the year 2015 and 2020 respectively.

Table 4.5 Wind turbines installation roadmap

Year	Long-term installation goal (MW) (supply more than 10% of electricity demand in 2050)			
	Onshore	Offshore		Total
		Fixed	Floating	
2008	1854	0	0	1845
2010	3000	0	0	3000
2015	6500	10	0	6510
2020	11,100	200	10	11,310
2025	16,300	1200	600	18,100
2030	21,200	2900	2900	27,000
2035	24,500	5100	7100	36,700
2040	25,000	7000	12,300	44,300
2045	25,000	7500	16,600	49,100
2050	25,000	7500	17,500	50,000

Onshore: reaches 25,000 MW in the year 2038
Fixed Foundation Offshore: reaches 7500 MW in the year 2044
Floating Structure Offshore: reaches 17,500 MW in the year 2048

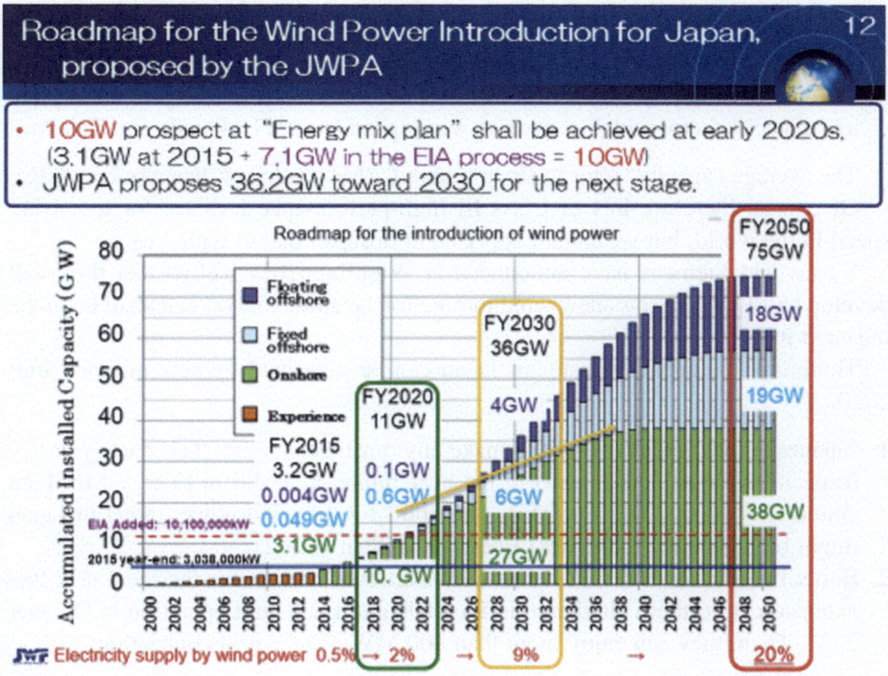

Fig. 4.11 Wind turbine installation roadmap

4.4 Conclusion

Installed capacity of wind power generation at the end of 2018: 3653 MW, 2310 units Revised on 1 February. JWPA announces the installed capacity of wind power generation in Japan as of the end of December 2018.

JWPA promotes wind power development in Japan and makes efforts to achieve our target 36.2 GW, including 10 GW of offshore wind towards 2030.

We think the difficulties in Japan shall get lower around 2020 in several fields.

- The unbundling of electric power generation, transmission, and distribution shall be applied for Japanese electric power system in FY2020.

The new grid operation company might be able to use Japanese grid system interregionally than now.

It enables improvement at wind power grid connection, especially at the good wind-resource regions (Hokkaido and Tohoku).

- The construction cost (materials, machines, and labour) in Japan is very high due to the big profitable works (demand) for Olympic Games preparation in 2020.

Once the Olympic Games in 2020 are over, construction companies will become hungry and will offer more moderate price for wind farm construction.

- The low profitability in Japanese wind power business is partially caused by the low capacity factor with low average wind speed (except Hokkaido and Tohoku).

The average capacity factor in japan is about 20%, it is lower than world average.

Of course, there are lots of Class III high-performance turbines for low wind speed in the world, but we cannot apply them in Japan due to typhoons.

Vestas and Siemens have announced at WindExpo2017 Tokyo that they will develop anti-typhoon class new wind turbine for Japanese market and shall be in the business in 2019.

Therefore, the capacity factor in Japan can be improved by new turbines after 2020.

1. Japanese government will not to make any drastic change in their energy policy (especially for nuclear power) in the near future, as it did in Energy Mix Plan announcement in 2018. Therefore, the uplift for the wind power target in Japan might be postponed to the next Energy Mix Plan in 2021.
2. Eurus Energy has announced that they start construction for the local grid line extension at northern Hokkaido in FY2018 and it will start operation in October 2021. Then, they can enjoy more than 600 MW of new grid connection.

Chapter 5
Wind Power Generation in Jordan: Current Situation and Future Plans

Ali Hamzeh and Mahmoud Awad

5.1 Introduction

Jordan lacks local energy resources, making energy the biggest challenge in Jordan. Jordan imports about 97% of its energy requirements, which includes mainly crude oil, oil derivatives, and natural gas. Local sources cover the remaining 3% of requirements with renewable energy contributing only a small proportion to this mixture [1]. In 2016, total primary energy consumed in Jordan was about 11.3 million tons of oil equivalent. The high annual growth rates of energy demand (5.5% for primary energy and 6.4% for demand for electricity), are to cover population growth (2.2%) [2] and overall economic development [3]. The cost of imported energy in Jordan for 2016 stood at 10% of the gross domestic product (GDP) while in 2014 the ratio was around 18% [1]. The primary energy consumption per capita is 1250 kilograms oil equivalent (kgoe), and per capita share of electricity is 2230 kWh, which are higher than the figures in comparable developing countries. Energy consumption density was 20 kgoe/$1000 of GDP, which is high compared to developed economies (150 kgoe/$1000 at fixed prices).

To meet the energy demand and the challenges of the energy sector a comprehensive energy strategy was approved by the Cabinet in December 2004 revised in 2007—"Master Strategy of Energy Sector in Jordan." The Strategy is to provide a vision for development of the energy sector over the next 10 years. The Strategy studied all options and alternatives for securing all types of energy from the following points of views: (1) the optimal options to cope with the energy demands and its investment cost, (2) reforming and restructuring the energy sector to open the market for competition, and (3) expanding on renewable energy projects and implementing energy conservation programs. To this extent, the future goals of the strategy can be summarized as follows:

A. Hamzeh (✉) · M. Awad
Faculty of Engineering, Electrical Engineering Department, Al-Ahliyya Amman University, Amman, Jordan

A. Sayigh, D. Milborrow (eds.), *The Age of Wind Energy*, Innovative Renewable Energy, https://doi.org/10.1007/978-3-030-26446-8_5

- Reduce the dependence on foreign energy sources (energy independence);
- Security of supply with energy production based on a variety of sources;
- The target for 2015 is for domestic resources to cover 25% of demand reducing imports to 75%;
- The target for 2020 is for domestic resources to cover 39% of demand reducing imports to 61% and achieving energy production from additional energy sources; and
- Promote renewable energy sources to share to 7% in the primary energy mix in 2015, and 10% in 2020. This is to be met through 600–1000 MW from wind energy and 300–600 MW from solar energy [4]. The renewable energy strategy has been modified to target 20% of the total energy mix in 2020 [5].

Jordan is rich in renewable sources of energy, especially solar energy with the annual daily average insolation on horizontal surface of about 5.4 kWh/m^2·day, and wind energy with average wind speeds of 7–8.5 m/s in some regions.

To promote renewable energy sources and in order to open the way for private sector to effectively participate in the implementation of renewable energy project, the Renewable Energy and Energy Efficiency Law was issued and officially entered into force in April 2012 under Law No. 13. With this law, and for the first time in Jordan, investors had the opportunity to identify and develop renewable grid-connected electricity production through the Direct Proposal Submission [4].

The renewable energy sector is regulated and monitored by the Energy and Minerals Regulatory Authority by specifying the different responsibilities of the licensees and granting them the necessary licenses under the laws, regulations and regulations in force for practicing the various activities based on the balance between the interests of consumers and licensees, investors and any other related parties [5].

This chapter presents wind power generation program in Jordan since its inception to the present trends and developments as well as the future prospects.

5.2 Wind Potential in Jordan

Studies on Jordan's wind potential have been conducted over a period of years and have shown that the country has a rich wind energy resource. The average annual wind speed exceeds 7 m/s in some areas. The long term climatic data are available at the Jordan Meteorological Department (JMD). Some other institutions like the Ministry of Energy and Mineral Resources (MEMR) and the Royal Scientific Society (RSS) have some measurements of such data especially wind and solar data for the purpose of assessing the potential of these resources for power generation and other applications in Jordan [6].

In 1988, the MEMR, JMD and other local institutions by cooperation with RISO National Laboratory in Denmark, have conducted the Wind Atlas for Jordan. This atlas is the first of its kind in the region and is considered as a reference for determining and selecting the areas that have promising potential for electricity

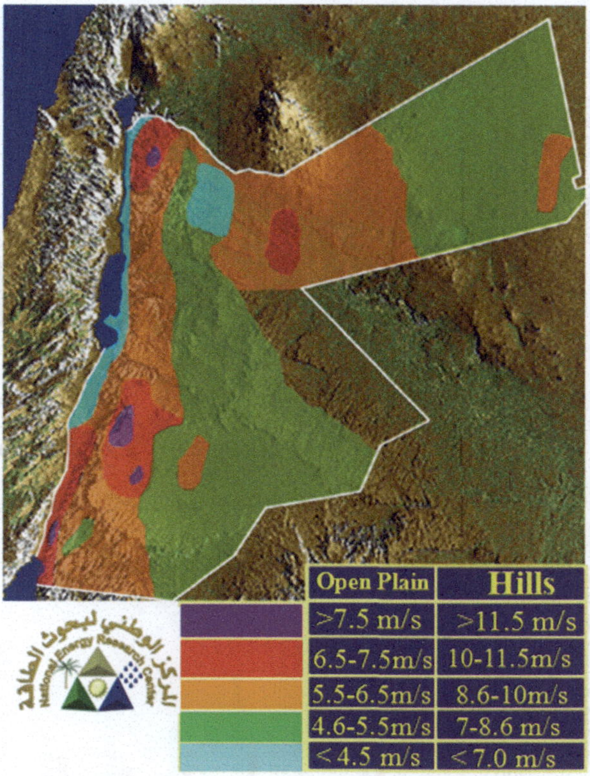

Fig. 5.1 Wind map of Jordan [7]

generation in Jordan. Figure 5.1 shows the Wind Map of Jordan in this Atlas which was recently updated by MEMR and the Royal Jordanian Geographic Centre depending on the new wind measurements conducted by MEMR for several years [6].

Figure 5.2 shows the wind speed at 10 m height above ground of all Jordan governorates all over the year. Figure 5.3 shows Mean power density of Jordan governorates at 50, 100 and 200 m heights. As shown in Fig. 5.3, the mean power density is increased with increasing of the elevation from 50 to 200 m. Aqaba governorate shows a maximum mean power density for the three different heights. The mean power density is increased toward the south governorates of Jordan until the maximum in Aqaba governorate the farthest south of Jordan. In the middle governorates of Jordan, the mean power density is within an average of 350 W/m^2. Concerning the wind rose (how many hours per year the wind blows from the indicated direction), the majority of the wind in Aqaba and Maan governorates which are the highest in the production of the mean power density in the NW direction. Ajloun, Jarash, Mafraq, Amman, Zarqa, Karak, Balqa, Madaba, and Tafilah wind distribution direction is toward W. Irbid governorate wind distribution is in the direction of WSW [8].

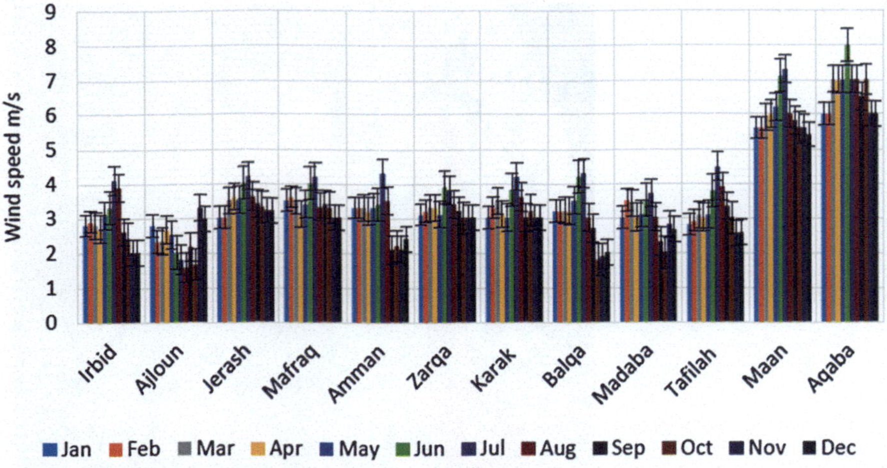

Fig. 5.2 Wind speed at 10 m height above ground of all Jordan governorates [8]

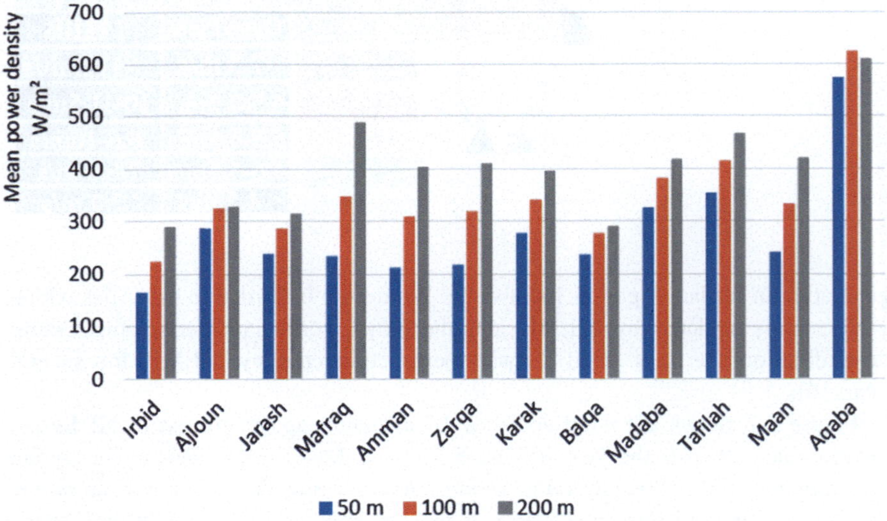

Fig. 5.3 Mean power density of Jordan Governorate at 50, 100 and 200 m heights [8]

The average wind speed for different promising locations in Jordan is about 6.5–7 m/s, as shown in Fig. 5.4. The highest average wind speed is in Tafila 2 location (altitude 1400 m above sea level) with a value of (8.54 m/s) followed by Al-Harir (8.4 m/s), which are both located in Tafila area.

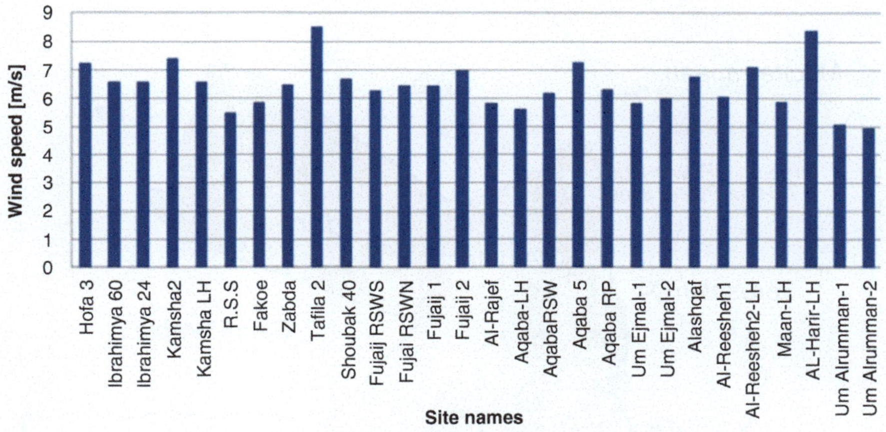

Fig. 5.4 Average wind speed comparison for different promising locations [9]

5.3 Wind Power Plants in Jordan

As of 2018, six operational wind farms with a total capacity of 373.645 MW have been built and connected to the grid in Jordan. These are Tafial wind farm (117 MW, 2015), Maan wind farm (80 MW, 2016), Fujeij wind farm (89.1 MW, 2018), Rajef wind farm (82 MW, 2018), Ibrahimyya wind farm (located approximately 80 km north of Amman, consists of four wind turbines with capacity 0.08 MW for each), and Hofa wind farm (located approximately 92 km north of Amman, consists of five wind turbines with capacity 0.225 MW for each). Moreover, there are three wind farms under construction with a total capacity of about 148 MW; these are Shobak wind farm (45 MW), Daehan wind farm near Tafila, 180 km southwest of Amman (51 MW) [10], and Abur wind farm near Tafila (51.75 MW). Figure 5.5 shows locations of some wind power plants in Jordan.

5.3.1 Tafila Wind Farm

The Tafila Wind Farm (Fig. 5.6) is the first commercial utility-scale wind power project in the Middle East. Located in the Hashemite Kingdom of Jordan, the 117 MW wind farm has increased the country's total power capacity by 3%. The U$287 million project became operational in September 2015. It was officially inaugurated in December the same year. The Tafila wind farm was developed by Jordan Wind Project Company, a co-development partnership between InfraMed (50%), Masdar (31%), and EP Global Energy (19%) [12, 13]. It is the first privately sponsored utility scale wind farm in the Middle East and the first successfully commissioned RE project onto the National Electric Power Company (NEPCO) grid in September 2015.

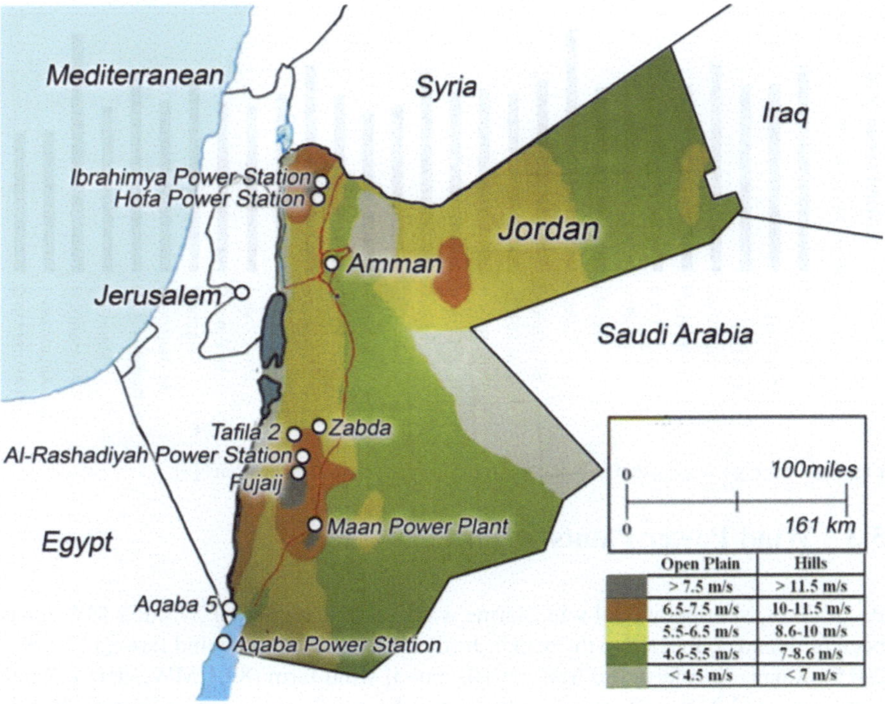

Fig. 5.5 Locations of some wind power plants in Jordan

Fig. 5.6 View from Tafila Wind Farm [11]

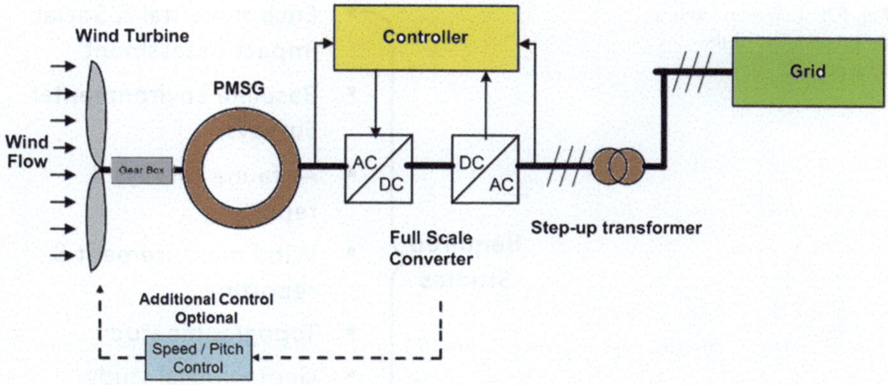

Fig. 5.7 Wind turbine system configuration (Type 4/D) in Tafila plant [14, 15]

The wind farm consists of 38 wind turbines (WTs) with the following technical specifications: Vestas V112, 3.075 MW rated capacity for each WT, 112 m rotor diameter, 94 m tower height, 3–25 m/s operating range, Fully automated. Countries of origin for the farm components are Spain (towers), Denmark (Nacelle and blades), Germany (Gearbox), and the USA (blades). The generator is a three-phase synchronous generator with a permanent magnet rotor that is connected to the grid through a full scale converter (Fig. 5.7). The converter is a full scale converter system controlling both the generator and the power quality delivered to the grid. The converter consists of four converter units operating in parallel with a common controller. The converter controls conversion of variable frequency power from the generator into fixed frequency AC power with desired active and reactive power levels (and other grid connection parameters) suitable for the grid. The converter is located in the nacelle and has a grid side voltage rating of 650 V. The generator side voltage rating is up to 710 V dependent on generator speed.

The Farm production is 400 GWh/year and its capacity factor (CF) is about 39% [16]. The average wind speed at hub height is 8.846 m/s. To verify these values, the following calculation is introduced.

$$\text{Capacity factor} = \text{CF} = 0.087V_{\text{avg}} - \frac{P_{\text{r}}}{D^2} = 0.087 \times 8.846 - \frac{3075}{90^2} = 0.389$$

$$\text{Annual farm production} = \text{CF} \times 38 \times 3075 \times 8760 \times 10^{-6} = 399.2 \text{ GWh}$$

The operation and maintenance activities are carried out by JWPC through weekly/monthly inspections of all farm issues, including electrical equipment, buildings and site roads, spare parts, wind turbines, monitoring performance, and environmental issues (bird monitoring), etc.

It is noteworthy that the project took about four and a half years from the first signature until the start of operation in September 2015. The first signing of MOU

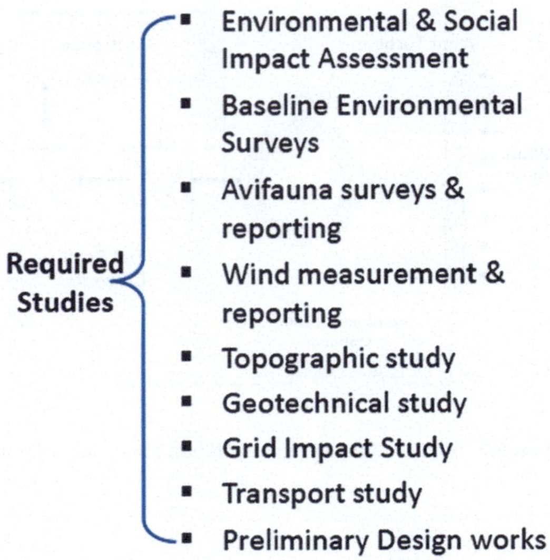

Fig. 5.8 Required studies for preparation of the project [13]

took place in Q 2 2011, the site works commenced in Q 2 2014 and took 21 months, whereas the turbines erection started in Q 1 2015 [13]. The development process of the project comprised numerous studies which are summarized in Fig. 5.8.

5.3.2 *Fujeij Wind Farm*

The project (Fig. 5.9) consists of a utility-scale wind farm with 27 wind turbines (Vestas V126-3.3 MW), each with an approximate height of about 110 m. The total operating capacity of 89.1 MW at a cost of $184 million. It is located in the Fujeij area near the southern city of Maan, 150 km south of Amman with annual wind speed of 6.7 m/s. The generator is a three phase asynchronous induction generator with cage rotor that is connected to the grid through a full scale converter. The converter is a full-scale converter system controlling both the generator and the power quality delivered to the grid. The converter consists of four converter units operating in parallel with a common controller. The converter controls conversion of variable frequency power from the generator into fixed frequency AC power with desired active and reactive power levels (and other grid connection parameters) suitable for the grid. The converter is located in the nacelle and has a grid side voltage rating of 650 V [17]. The generator side voltage rating is up to 750 V dependent on generator speed. The commercial operation started in the fourth quarter 2018 [18–20]. The turbine's 126 m rotor is equipped with light structural shell blades that ensure greater wind energy capture in areas with low to medium wind speeds, which are characteristic for Jordan. The capacity factor and annual yield are calculated as:

Fig. 5.9 View of Fujeij wind farm

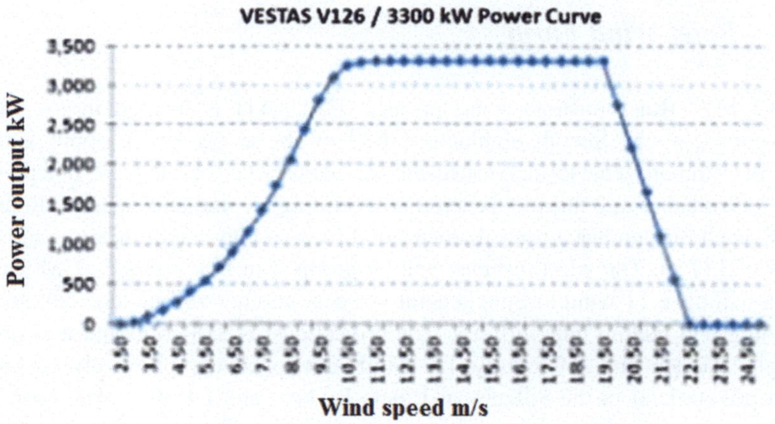

Fig. 5.10 The power curve of the wind turbine V126-3.3 MW [17]

$$\text{Capacity factor} = \text{CF} = 0.087V_{\text{avg}} - \frac{P_{\text{r}}}{D^2} = 0.087 \times 6.7 - \frac{3300}{126^2} = 0.385$$

$$\text{Annual farm production} = \text{CF} \times 27 \times 3300 \times 8760 \times 10^{-6} = 300.2 \text{ GWh}$$

The power curve of the wind turbine V126-3.3 MW is shown in Fig. 5.10.

Fig. 5.11 View of Rajef wind farm

5.3.3 Rajef Wind Farm

The 82 MW Rajef onshore wind project (Fig. 5.11) is located in the Ma'an Governorate on the Sherah highlands (1600 m above sea level), about 200 km south of Amman. The farm consists of 41 Gamesa G114-2.0 MW independent pitch-controlled wind turbine generators operating at variable speeds. Rated at 2 MW, each turbine has a rotor diameter of 114 m, a hub height of 80 m and a tip height of 137 m. The wind turbines will be grouped in four strings, of which two strings will have 11 wind turbine generators each, another will include ten, and the remaining will feature nine turbines. The project cost is $184.6 m and it is funded through 75% debt and 25% equity. The wind farm covers approximately 7.5 km^2 of leased private land in the villages of Rajef, Taybeh, and Dlaghah. The farm construction took about 22 months and operation started in October 2018 [21, 22].

The type of generator is doubly fed induction generator (DFIG) [23] and the wind turbine configuration is shown in Fig. 5.12.

The capacity factor and annual yield are calculated as:

$$\text{Capacity factor} = \text{CF} = 0.087V_{\text{avg}} - \frac{P_{\text{r}}}{D^2} = 0.087 \times 6.0 - \frac{2000}{114^2} = 0.365$$

$$\text{Annual farm production} = \text{CF} \times 41 \times 2000 \times 8760 \times 10^{-6} = 262 \text{ GWh}$$

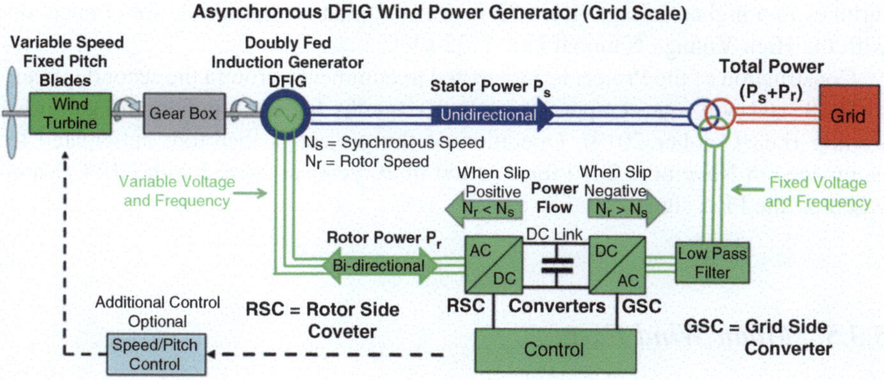

Fig. 5.12 Wind turbine configuration of type DFIG in Rajef wind plant [23]

Table 5.1 Summary of key components of Shobak wind farm [24]

Component	Description
Project generation capacity (MW)	44.85
Technology type	Wind power
Number of wind turbines	Vestas V136 3.45MW model (13 turbines)
Rated power per turbine (MW)	3.45
Rotor diameter (m)	136
Hub height (m)	112
Tip height (m)	180
Project area to be covered	14.5 km^2
Infrastructure and utilities	This includes: (1) internal road network; (2) underground cables; (3) warehouse and offices; (4) substation; and (5) associated facilities

5.3.4 *Shobak Wind Farm [4]*

The under construction 45 MW Project is located in Ma'an Governorate in the South of Jordan approximately 160 km south of the capital city of Amman. The Project is located within the District of Shobak within a hilly terrain area with altitudes ranging from around 1200 to about 1322 m above sea level. The Project area is around 14.5 km^2 with a maximum length of around 7.2 km and a maximum width of 1.5 km. Table 5.1 provides a summary of the key Project components for the Project, along with a detailed description of each of those components below.

Each turbine is equipped with a transformer that converts/steps up the output from the turbine to a higher voltage (from 1 to 33 kV) to meet a specific utility voltage distribution level that is appropriate for connection with a substation. The wind turbines will be connected through medium voltage cables (33 kV) to a substation located within the Project site, which collects and converts the output from the turbines to a higher voltage (from 33 to 132 kV) that is appropriate for connection with the High Voltage National Grid (132 kV).

Construction of the Project is anticipated to commence around the second quarter of 2018, and will require approximately 16 months for construction and commissioning (i.e., October 2019). Operation of the project is therefore anticipated to commence in November 2019 for a period of 20 years as agreed with NEPCO and based on the PPA signed.

5.3.5 *Maan Wind Farm*

The Maan wind farm (Fig. 5.13) has been constructed in 2 phases installing 40 Gamesa G97 wind turbines each with a capacity of 2 MW. The 80 MW wind farm is located in the city of Mann, Southern Jordan with a cost of $148 m, funded by the Kuwait Fund for Arab Economic Development (KFAED) [25].

The technical specifications of the wind turbine are: Gamesa G97, 2.0 MW rated capacity for each WT, 97 m rotor diameter, 90 m tower height, 2.5–25 m/s operating range, fully automated. The generator type is doubly fed induction generator (DFIG) with standard operating temperature range from −20 to 30 °C [28]. With average wind speed of 5.0 m/s, the capacity factor and annual farm production are calculated as:

Fig. 5.13 Maan win farm

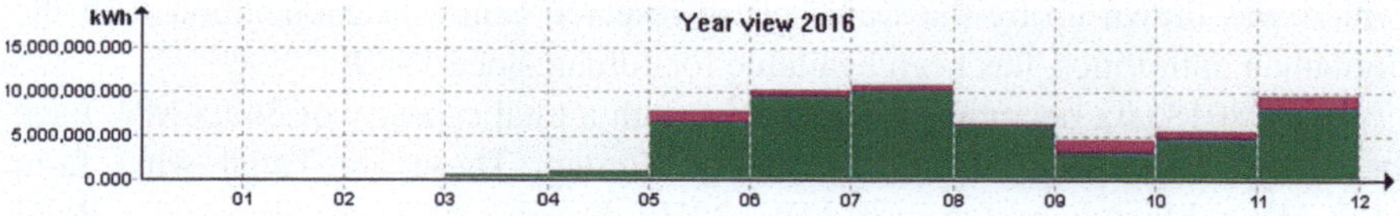

Fig. 5.14 Monthly yield in 2016 for Maan power plant [29]

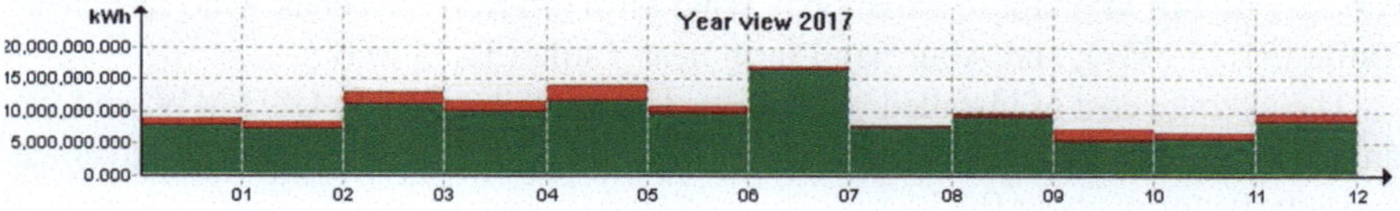

Fig. 5.15 Monthly yield in 2017 for Maan power plant [29]

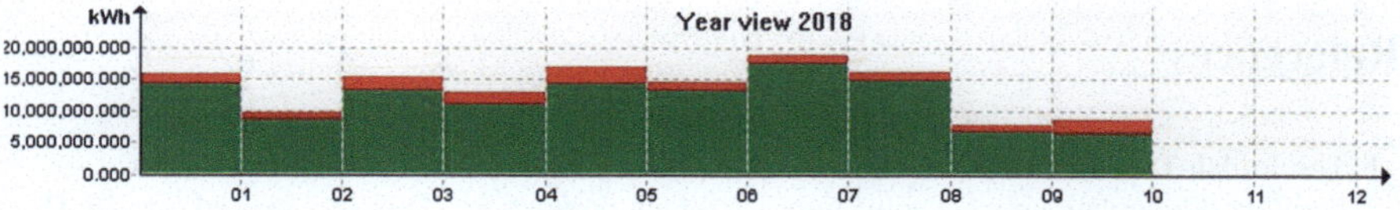

Fig. 5.16 Monthly yield in 2018 for Maan power plant [29]

$$\text{Capacity factor} = \text{CF} = 0.087V_{\text{avg}} - \frac{P_{\text{r}}}{D^2} = 0.087 \times 5.0 - \frac{2000}{97^2} = 0.214$$

$$\text{Annual farm production} = \text{CF} \times 40 \times 2000 \times 8760 \times 10^{-6} = 150 \text{ GWh}$$

For this wind farm, the energy production statistics is available from April 2016 (operation starting) to October 2018, as shown in Figs. 5.14, 5.15, and 5.16.

5.4 Conclusions

Jordan lacks local conventional energy resources, making energy the biggest challenge in Jordan, where fossil fuel imports constitute about 97% of energy requirements of the country. To meet the energy demand a comprehensive energy strategy was established. In the area of renewable energy (RE), the strategy goal is to promote RE sources to share to 7% in the primary energy mix in 2015, and 10% in 2020. This is to be met through 600–1000 MW from wind energy and 300–600 MW from solar energy [4]. The renewable energy strategy has been modified to target 20% of the total energy mix in 2020. Jordan has significant wind energy resources that could be potentially exploited for power generation where the annual average wind speed exceeds 7 m/s (at 10 m height) in some areas of the country. The regions with the greatest potential are located in the North and South of the country. A wind atlas, which was drawn up by the Danish Risø research centre in cooperation with the Jordanian authorities, has been available for Jordan since 1989.

As of 2018, six operational wind farms with a total capacity of 369.5 MW have been built and connected to the grid in Jordan. These are Tafial wind farm (117 MW), Maan wind farm (80 MW, 2016), Fujeij wind farm (89.1 MW), Rajef wind farm (82 MW, 2018), Ibrahimyya wind farm (0.32 MW), and Hofa wind farm (1.125 MW). Moreover, there are three wind farms under construction with a total capacity of 148 MW; these are Shobak wind farm (45 MW), Daehan wind farm near Tafila (51.75 MW), and Abur wind farm near Tafila (51.75 MW).

The average cost of installed wind farms is around 2.2 million USD/MW, and the annual average produced electrical energy is about 2.3 GWh/MW where the average capacity factor is about 0.27.

References

1. The Jordan Times, Energy cost ratio to GDP, 14 Jan 2017
2. http://worldpopulationreview.com/countries/jordan-population/, Website viewed on 26 Dec 2018
3. N. Abdul Rahim, N. Eid, Energy projects in Jordan, invest-export. Brussels/documents, October 2017
4. Energy & Minerals Regulatory Commission, Renewable Energy Sector, http://www.emrc.gov.jo/index.php/en/renewable-energy-sector, Website viewed on 29 Dec 2018
5. Z. Sabra, Wind Energy in Jordan—Use and Perspectives, DEWI Magazine No. 15, Aug 1999
6. Z. Sabra, Renewable Energy & Energy Efficiency Forum, 7 June 2010
7. S. Alrwashdeh, Map of Jordan governorates wind distribution and mean power density. Int. J. Eng. Technol. **7**(3), 1495–1500 (2018)
8. Y. El-Tous, S. Al-Battat, S. Abdel Hafith, Hybrid wind-PV grid connected power station case study: Al Tafila Jordan. Int. J. Energy Environ. **3**(4), 605–616 (2012)
9. Masdar., https://masdar.ae/en/masdar-clean-energy/projects/tafila-wind-farm, Website viewed on 07 Jan 2019
10. Jordan Wind Project Company (JWPC)., http://www.jordanwind.com/, Website viewed on 07 Jan 2019
11. A. Hamzeh, Integration of Distribution Generation into Electrical Distribution Generation, © 2010–2011, Damascus University Publications.
12. Ministry of Energy and Mineral Resources., www.Memr.gov.jo
13. M.F. Khan, M.R. Khan, Wind power generation in India: Evolution, trends and prospects. Int. J. Renew. Energy Dev. **2**(3), 175–186 (2013)
14. S.S. Judeh, Tafila Wind Farm (26 April 2017), http://www.eib.org/attachments/general/events
15. Vestas, General Specification V126-3.3 MW 50/60 Hz, Document no.: 0034-7616 V10, http://www.rettetdenbuschberg.at/wp-content/uploads/2016/11/10_V126-3.3-Allgemeine-Spezifikation.pdf
16. Wind Energy and Electric Vehicle Magazine (REVE)., https://www.evwind.es/2017
17. Renewables Now., https://renewablesnow.com/news
18. Jordan MEMR—Fujeij Wind IPP., https://www.zawya.com/mena/en/project/171209125536/jordan-memr-fujeij-wind-ipp/
19. Future Power Technology, Al-Rajef Wind Farm, https://www.power-technology.com/projects/al-rajef-wind-farm
20. https://en.wind-turbine-models.com/turbines/428-gamesa-g114-2.0mw

21. Eco Consult., Al-Rajef 82mw Wind Power Project (2016), http://alcazarenergy.com/wp-content/uploads/2016/10/ESIA_Rajef-Wind-Farm_Eng.pdf
22. A. Hamzeh, Lecture in course "Renewable Energy systems", Chapter 3, 2018–2019 Semester 1, Al-Ahliyya Amman University, Department of Electrical Engineering, https://vclass.ammanu.edu.jo/
23. Siemens Gamesa, G97-2.0 MW Technical specifications, https://www.siemensgamesa.com/en-int/-/media/siemensgamesa/downloads/en/products-and-services/onshore/brochures/legacy-gamesa/siemens-gamesa-onshore-wind-turbine-g97-en.pdf
24. ECO Consult, ESIA for Shobak Wind Power Project, 13 Aug 2017
25. khader al-janaideh, khader.al-janaideh@emrc.gov.jo, 21 Oct 2018

Chapter 6
Wind Energy in Australia

Jonathan Whale, Craig Carter, and Ali Arefi

This chapter is dedicated to the memory of Mr. Geoff Hill, who founded Westwind Turbines, and who was a pioneer and innovator in the age of wind energy in Australia.

6.1 Introduction

This book chapter reviews the use of wind energy in Australia including the wind potential, history, wind farms, and the corresponding technologies. Also, this chapter discusses the environmental and social aspects of wind energy and the related policies and regulations in Australia. In addition, the research conducted on wind energy and an outlook of this source of energy in Australia are presented in this chapter.

6.2 Wind Energy Potential in Australia

Australia has the greatest amount of wind energy resources in the world, considering the large coastal region areas, especially in south, southwest, southeast, and northeast as seen in Fig. 6.1. There are also very good resources of wind energy in and off the southern coastal area of Australia. These areas between latitudes 35° and 50° [1] have the highest wind energy potential in the world. In northern Australia, the generation of wind is mainly due to the monsoon and trade wind systems. Winds in the eastern regions of Australia are influenced by the large scale topography, such as the Great Dividing Range, which channels wind through valleys, deflecting or

J. Whale (✉) · C. Carter · A. Arefi
Discipline of Engineering and Energy, College of Science, Health, Engineering and Education, Murdoch University, Murdoch, WA, Australia
e-mail: J.Whale@murdoch.edu.au

A. Sayigh, D. Milborrow (eds.), *The Age of Wind Energy*, Innovative Renewable Energy, https://doi.org/10.1007/978-3-030-26446-8_6

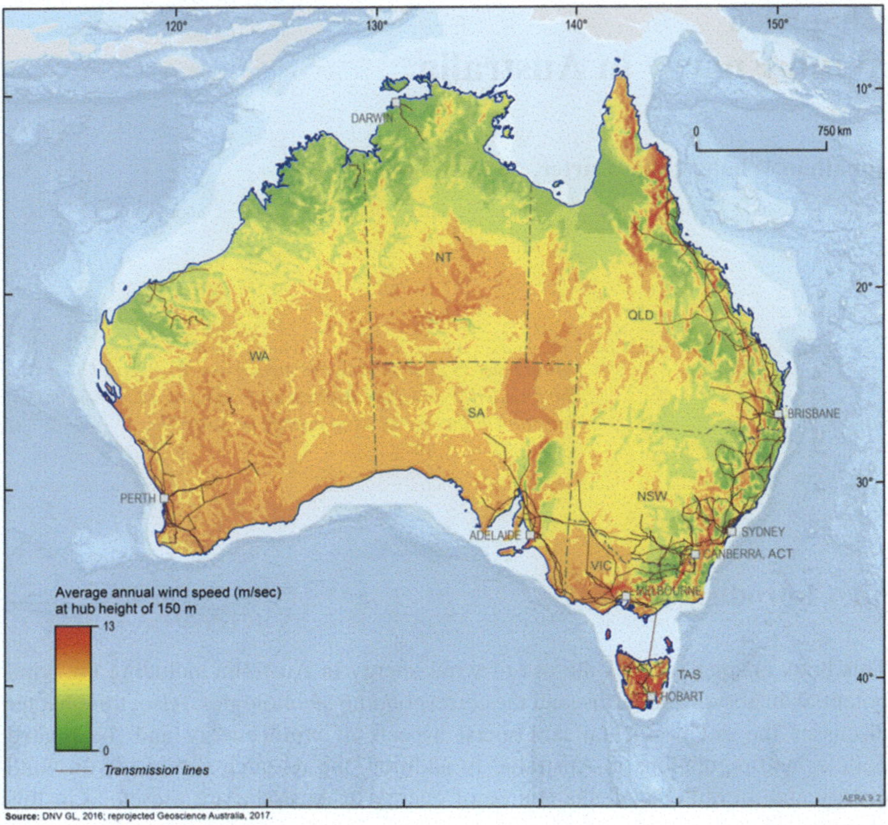

Fig. 6.1 Average annual wind speed at the height of 150 m in Australia (Source: Geoscience Australia 2017 [35])

blocking them from other areas [1]. As seen in Fig. 6.1, the existing power transmission network is relatively close to the areas with high potential for wind generation in Australia, which provides a good opportunity to invest in wind farms and readily transmit the produced energy to load centres.

The maximum installable generation capacity for the capture of wind energy resources in Australia is estimated at 880 and 660 GW for onshore and offshore, respectively. The maximum amount of energy that could be recoverable from this size of installation is 3100 TWh/year for each of onshore and offshore, totally 6200 TWh [2]. Western Australia (WA) has the greatest potential for wind generation but the electricity network in this state is not connected to the other states.

Based on the Australian Energy Statistics, the total wind energy production in 2018 was 16.3 TWh, which is 33.1% of total renewable energy generation and 6.2% of total electricity generation in Australia [3]. Figure 6.2 shows the wind energy generation (GWh) in Australia between 2015 and 2018 in the different states, which indicates a growth of 12.5% on average over this period. As depicted, South

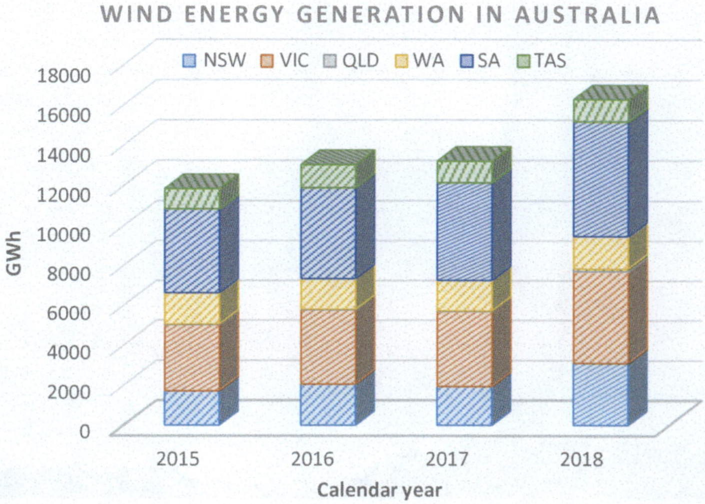

Fig. 6.2 Wind energy generation in GWh in the states of Australia from 2015 to 2018

Australia (SA) and Victoria (VIC) account for 35% and 28% of total wind energy production, respectively, in Australia in 2018. This figure is less in New South Wales (NSW), Western Australia (WA), Tasmania (TAS), and Queensland with 19%, 10%, 7%, and 1%, respectively in 2018. In Northern Territory (NA), there is currently no significant wind generation.

6.3 Wind Energy Harnessing History in Australia

WA was the first State in Australia to manufacture and install wind turbines. This was helped by the fact that WA had a local innovator, Geoff Hill, who was interested in wind turbines and developed 30 and 60 kW stall-regulated induction generator Type 1 wind turbines through his Westwind Company. The machines were tested by the State Electricity Commission of Western Australia (SECWA) at a coastal site in the suburb of Hamilton Hill in Perth, WA. This generated interest in turbines from other manufacturers and three types of turbines were installed and trialled on Rottnest Island in the 1980s, as there is a good wind resource on this island, which is located 18 km off the coast of Fremantle, WA. There was a 20 kW turbine horizontal axis variable pitch machine from MAN of Germany, which was installed and tested on the Island, before being moved to the Hamilton Hill site. A 50 kW Darrieus wind turbine supplied by DAF Indal of Canada and a horizontal axis fixed pitch 55 kW machine from Nordtank of Denmark were also installed and tested on Rottnest [4], as shown in Fig. 6.3. The Darrieus unit, which had a vertical axis "egg beater" type rotor went into uncontrolled overspeed during testing and flew apart,

Fig. 6.3 DAF Indal vertical machine and the Nordtank turbine behind in Rottnest Island, WA [4]

which brought its testing to an abrupt end. Horizontal axis type three blade d machines dominate, with two blades being used on some smaller machines.

A more advanced 60 kW Westwind turbine was installed and tested at Woodman's Point, south of Perth, which lead to the installation of the Salmon Beach Wind Farm, Australia's first wind farm, near Esperance, WA, in 1987. This wind farm of six 60 kW Westwind turbines remained operational on the local diesel power system until 2002, when it was decommissioned, mainly due to urban encroachment. Westwind went on to later develop small permanent magnet generator type wind turbines in the 2.5–20 kW range. A prototype 5 kW unit was installed at Murdoch University for testing in 1996. It was an up-wind machine with only two blades to reduce weight and was mounted on a double tapered 30 m steel tower supported by four guy wires. An "A" frame allowed it to be winched up and down for maintenance. The two-blade design caused tower vibrations due the constantly changing angular inertia of the two-blade rotor in combination with the passive yawing that followed the changing wind direction. The rotor blades had low angular inertia while yawing when the blades were in a vertical orientation and the opposite when they were in a horizontal orientation. Yawing to limit power and avoid rotor overspeed during high wind conditions was achieved by a trailing offset blade. Westwind changed the design to have three blades and a simpler non-tapered tubular steel tower. After developing and trialling its 20 kW model at Murdoch University, a Federal Govt grant was awarded to Western Power to install a small wind farm at Exmouth called the Advanced Mini Wind Farm comprising three 20 kW Westwind Turbines on guyed towers. Exmouth on the mid-west coast of WA, is in a cyclone prone region, so wind turbines, which can be lowered before cyclonic winds occur, were the only possible option. This project was completed in 2002. Westwind was later sold to an Irish business and went offshore.

While the wind farm at Exmouth (now removed) was not utility scale, it did lead to the concept of installing 275 kW Vergnet two-bladed downwind induction generator Type 1 wind turbines on guyed towers at Coral Bay in 2007, which is also a cyclone prone township south of Exmouth [5]. Due to the exceptionally high wind resource, Vergnet, a French company, derated the units to 200 kW but offered larger rotors to compensate. This was accepted, as the extra energy production from having a larger rotor in most wind conditions outweighed the energy loss of having a lower power limit, which often needed limiting in any case to maintain minimum diesel loadings. While the Vergnet Type 1 units could be power limited, the method of using blade pitching was too slow to avoid short-term power output excursions up to 50% during wind gusts. Hence an innovative flywheel type ±500 kW, 5 kWh flywheel energy storage system, called PowerStore by the manufacturer—PowerCorp in Darwin, NT, (now ABB) was included to smooth and limit the wind power variations, to manage the loading and load fluctuations on the 320 kW diesel generators, as described in this section later.

The market for small wind turbines has been adversely affected by the plummeting cost of solar PV, which for most locations now produces renewable energy at lower cost. Small wind turbines are now a niche product, suitable for off-grid rural low power applications where there is a good local wind resource and perhaps a poor solar resource or where the combined use of good wind and solar resources provides greater diversity, thereby reducing the use of backup generation and/or battery storage requirements.

In 1993, Australia's first commercial wind farm comprising nine 225 kW Vestas V27 induction generator Type 1 wind turbines was connected to a 33 kV line 16 km west of Esperance, on the south coast of WA, and integrated into the Esperance diesel power system. Named the Ten Mile Lagoon Wind Farm, its power output needed to be limited at times to maintain the required minimum diesel generator loading. The Vestas V27 units could not be individually power limited, so power limiting was achieved by switching one or more V27 units to pause mode. Ten years later, the Nine Mile Beach Wind Farm, comprising six 600 kW Enercon E-40 Type 4 wind turbines was installed west of the Ten Mile Lagoon Wind Farm [6].

The use of Enercon direct-drive inverter connected Type 4 wind turbines commenced in Australia with the integration of a single 230 kW Enercon E-30 wind turbine into the diesel power system at Denham, Shark Bay, WA in mid-1998. This type of wind turbine, which uses excited variable speed synchronous generators coupled to the grid via rectifier/inverters, can be dynamically power output limited to maintain the required minimum loading of the diesel generators. The integration of wind turbines and diesel power plants was an innovative and novel approach to reduce diesel fuel requirements, while maintaining a reliable electricity to regional communities by covering for the variability or sudden loss of wind power. The success of this project, that achieved 20% average wind energy penetration, lead Western Power and its contractor PowerCorp, with Federal Government grant support, to integrate more Enercon machines at Denham, and 600 kW Enercon E-44 Type 4 wind turbines into the regional diesel grids at Hopetoun and Bremer Bay on WA's south coast and on Rottnest Island. The use of Detroit Series 60 diesel generators, modified by PowerCorp to operate down to 6% loading at these sites, enabled peak wind penetration levels above 90% and annual average wind penetration levels up to 40%. These modified 320 kW diesel generators were referred to as Low Load Diesels (LLDs). Now, the microgrid system in Denham includes 2.61 MW of diesel generation and 1.02 MW of wind generation, after the addition of a fourth larger Enercon machine in 2007, a 330 kW Enercon E-33 [7]. Another example of hybrid wind and diesel generation is at Mawson Station in Antarctica, Australia, which consists of two 300 kW Enercon E-30 wind turbines installed in 2003 [8].

The MW capacity of grid connected utility scale wind farms has progressively increased since the 1980s. Both the MW ratings of wind turbines and their number per wind farm have grown to achieve economies of scale to drive down costs. The 420 MW Macarthur Wind Farm in VIC, operational in 2013, and Snowtown Wind Farm with the capacity of 369 MW in SA, operational in two stages in 2008 and 2014, are examples of very large grid-connected wind farms in Australia. As wind turbines have increased in MW capacity from around 0.2 MW in the 1980s to 2–8 MW now, there are benefits beyond the cost reduction achieved from the economies of scale. Larger units have higher and larger rotors that turn relatively more slowly and harness the better wind energy resource that can exist at higher altitudes. A fewer number of slower rotating larger rotors is visually preferred in the natural environment. Slow rotating rotors do not have the agitating effect of seeing high speed rotors in one's peripheral vision. Although the spacing between machines needs to increase with machine rotor size, the energy density per hectare is increased, thereby improving the utilization of sometimes constrained available sites.

There was an increasing interest in small wind turbines, which are roof-mounted, from 2005 to 2010 but they have not been able to compete with photovoltaic (PV) systems in terms of price. The number of small wind turbines has increased in NSW, since 2009, driven by policies.

6.4 Wind Farms in Australia and Their Differing Technologies

While the earliest wind farms were built in Western Australia, New South Wales and Queensland, South Australia, Victoria and Tasmania have had numerous large wind farms installed to capture the wind energy in the roaring forties, the prevailing westerly winds that flow along the southern edges of the continent.

NSW has the oldest grid connected wind farms, being the 4.8 MW Crookwell Wind Farm (July 1998), which employed 600 kW Type 1 induction generator type wind turbines (Vestas V44 from Denmark), a typical size in the late 1990s. The following year, the 9.9 MW Blayney Wind Farm (October 2000) in NSW commenced operation with 660 kW Vestas V47 wind turbines of the same type. In Northern Queensland, the 12 MW Windy Hill Wind Farm commenced operation in

2000. This was the first grid connected wind farm employing Type 4 variable speed direct-drive inverter type 600 kW wind turbines from Enercon in Germany (Enercon E-40) [9]. Wind farm sites in NSW and Queensland are inland on high ground rather than near the coast. The east coast of Australia does not have the high prevailing winds found on the south and west coast of the southern states.

Victoria's first commercial wind farm, connected to the state grid, is located near Port Fairy on the south coast. The 18.2 MW Codrington Wind Farm commenced operation in June 2001 [9, 10]. It employs Bonus AN 1.3 MW induction generator Type 1 wind turbines [11].

In October 2001, the 21.6 MW Albany Wind Farm commenced operation on the south coast of WA. Stringent technical requirements were imposed by Western Power (WA network utility). To comply with these, Enercon was required to develop and install undervoltage ride-through and voltage control capabilities, within an agreed time after commencing operation. These innovative requirements have become prerequisites for connection of all large wind farms in Australia. (Unlike Type 4 wind turbines, voltage control cannot be provided by induction generator Type 1 wind turbines, so static VAr compensators (STATCOMs) have had to be incorporated into Type 1 wind farms to provide this function.) When it was confirmed that cranes were available in WA to erect 100 m high wind turbines, it was decided to use twelve 1.8 MW Enercon E-66 wind turbines rather than thirty four of the cheaper 600 kW E-40 model, mainly for visual aesthetic reasons. By contrast, a 20 MW wind farm had been planned for the same site 10 years earlier. Had it proceeded, around one hundred 200 kW wind turbines (typical size for the time) would have been needed instead of just twelve 1.8 MW machines. Western Power had a thorough public consultation campaign, however, opponents of the Albany Wind Farm showed the Albany community photos of the Californian wind farms of the 1980s, where hundreds of small (100 kW or less) fast rotating wind turbines "littered" the hilltops of the undulating rural landscape. By selecting the larger 1.8 MW units, the reality was very different, and the wind farm has been highly accepted by the local community. There was little opposition when the local public was again consulted ahead of adding six more 2.3 MW Enercon E-70 wind turbines 10 years later. The Albany Wind Farm, now 35.4 MW in capacity, is connected to the Albany 132/22 kV Substation via three dedicated 22 kV underground circuits 12–15 km long. Wind farms greater than 10 MW are now typically connected directly to a transmission network (66 kV, 132 kV and above), however at the time Western Power owned both the network and the wind farm and so could decide on the least cost connection method that had no visual impact. An underground 132 kV cable to the wind farm site would have been prohibitive in cost.

The largest wind farm in Australia at 420 MW is the Macarthur Wind Farm 260 km west of Melbourne near Hamilton. Commissioned in 2013, it comprises 140 × 3 MW Vestas V112 wind turbines. These are Type 4 variable speed inverter connected wind turbines. They employ a permanent magnet synchronous generator driven by the rotor via a spur/planetary gearbox [12].

The largest West Australian wind farm is the 206 MW Collgar Wind Farm, 25 km south east of Merredin in the Central Wheatbelt of WA. Its 111 × 2 MW Vestas V90 double-fed induction generator Type 3 wind turbines have been derated to 1.856 MW, which helps to increase the wind farm's capacity factor to above 40%. This is the first wind farm on the south-west grid in WA, well away from the coast. The much higher rotors of newer larger wind turbines can access inland overnight winds in the region that significantly increase at altitude—high wind shear. The downside is that traditionally the system load has been relatively low overnight. However, with the increasing use of solar PV, this will not always apply in the future. The double-fed induction generators can adjust their reactive power, however, to comply with Western Power's fast voltage control requirements, 52 MVAr of capacitor banks and 24 × 4 MVAr DVAR IGBT statcoms supplied by US company American Superconductors were installed at the Collgar Wind Farm's substation [9]. The largest wind farm in Tasmania, the 168 MW Musselroe Wind Farm (2013) near Cape Portland also uses 56 × Vestas V90 (3 MW version) Type 3 units, 40 MVAr of capacitor banks and 4 × 4 MVAr DVAR statcoms [9].

The Collgar and Musselroe wind farm designs indicate that the Vestas Type 3 wind turbines still need external reactive power support for voltage control with a portion having dynamic capability to meet technical requirements, albeit less than for Type 1 wind turbines. Type 3 wind turbines do offer variable speed operation at a reduced cost, as the inverter required to excite the generator at varying frequencies is only around 30% of the rating of a Type 4 wind turbine inverter.

Type 1 induction generator fixed two-speed wind turbines from Vestas and Suzlon were commonly used in the older wind farms such as the 89 MW Walkaway Wind Farm (2005) and the 80 MW Emu Downs Wind farm (2006) north of Perth, which both use 1.65 MW NEG Micon NM82 wind turbines (Vestas owned) [5]. These wind farms needed STATCOMs to provide the required voltage control capability as Type 1 wind turbines absorb reactive power from the grid and do not provide any voltage control or support capability. Apart from their suboptimal fixed two-speed operation, which requires a pause in power output while the speed change occurs, this is another reason why Type 1 wind turbines are no longer being used in large Australian wind farms.

The other notable wind farms that use Type 1 units include Snowtown and Hallett wind farms in South Australia.

The 101 MW Snowtown Wind Farm Stage 1 (September 2008) uses 47 × 2.1 MW Suzlon S88 Type 1 wind turbines with one similar Suzlon S95 unit added in 2011. Snowtown Wind Farm Stage 2, built only 6 years later (June 2014), utilizes 90 much more advanced and larger Type 4 Siemens wind turbines rated at 3.0 MW each [5]. These use direct drive permanent magnet synchronous generators, which are inverter connected to the grid. This technology is considered the most advanced in the industry. However, the high cost and supply reliability of the rare earth super magnets has resulted in some manufacturers, like General Electric, to no longer offer permanent magnet synchronous generators. The Hallett Wind Farms all together have 167 × Suzlon S88 Type 1 wind turbines installed between 2008 and 2011 giving a total capacity of 351 MW [5].

Lake Bonney Wind Farm Stage 1 in South Australia, opened in March 2005, uses 46 × 1.75 MW Vestas V66 Type 1 wind turbines [5, 13], however Stages 2 and

3 completed in 2008 and 2009 switched to using 66 × 3 MW Vestas V90 double-fed induction generator Type 3 units, giving a total capacity of 279 MW [5].

The use of Type 4 wind turbines in South Australia began with the start of operation in February 2006, of the 70 MW Mt Millar Wind Farm, which employs 35 × 2 MW Enercon E-70 wind turbines. (Enercon uses the more encompassing term "wind energy conversion systems" or WECs rather than wind turbines.) Construction began in late 2004; however, delays in proving compliance with the network owner's technical requirements caused a prolonged delay in obtaining network access. In the early period of inverter connected generation, utility electrical engineers quite often expressed their poor understanding of the inherently different characteristics of inverter connected generation compared to classical synchronous generators. This led to an overly cautious approach by utility planning engineers in accepting the new technology, which is now understood to not cause frequency instability/pole slipping—especially important on weak networks.

A low level of wind farm construction activity in Australia occurred between 2013 and 2015 due to policy uncertainty from the Federal Government's delay in extending its Renewable Energy Target (RET). Following this delay, the first major wind farm in South Australia using Siemens 3.2 MW Type 4 wind turbines with permanent magnet synchronous generators [4] was opened in three stages between late 2016 and late 2017. This is the 315 MW Hornsdale Wind Farm, north of Jamestown in SA. What makes this wind farm groundbreaking, however, is the on-site addition of the Hornsdale Power Reserve on 1 December 2017, a lithium-ion battery storage facility rated at 100 MW with an energy storage capacity of 129 MWh. Teslas installed and commissioned the unit within 100 days of receiving the order from Neoen, the French company that owns the wind farm. This is the first of a likely common occurrence where energy storage is co-located with variable renewable energy generation (wind and solar) to provide short-term dynamic real power support to the grid. To date, it has been a technical and commercial success.

The largest wind farm in Victoria, the 240 MW Ararat Wind Farm is located approximately 180 km north west of Melbourne [14] and was completed in February 2017. General Electric 3.2 MW wind turbine model GE 3.2-103 was selected for this project. These are double-fed induction generator Type 3 wind turbines, which GE went back to after using permanent magnet synchronous generators [15]. The 55 MW Mumbida Wind Farm, situated around 30 km south-east of Geraldton in WA, utilizes GE's permanent magnet technology. The wind farm has 22 × GE 2.75xl Type 4 wind turbines derated to 2.5 MW [5]. Unlike the Siemens units used at Hornsdale Wind Farm, these wind turbines have gearboxes. The GE units have been programmed to do continuous voltage control of the 132 kV bus voltage at the Mumbida 132/22 kV wind farm substation regardless of their real power output, effectively acting like a STATCOM.

The location of some major wind farms in Australia and the transmission networks for some states are illustrated in Figs. 6.4, 6.5, 6.6, 6.7, 6.8, and 6.9.

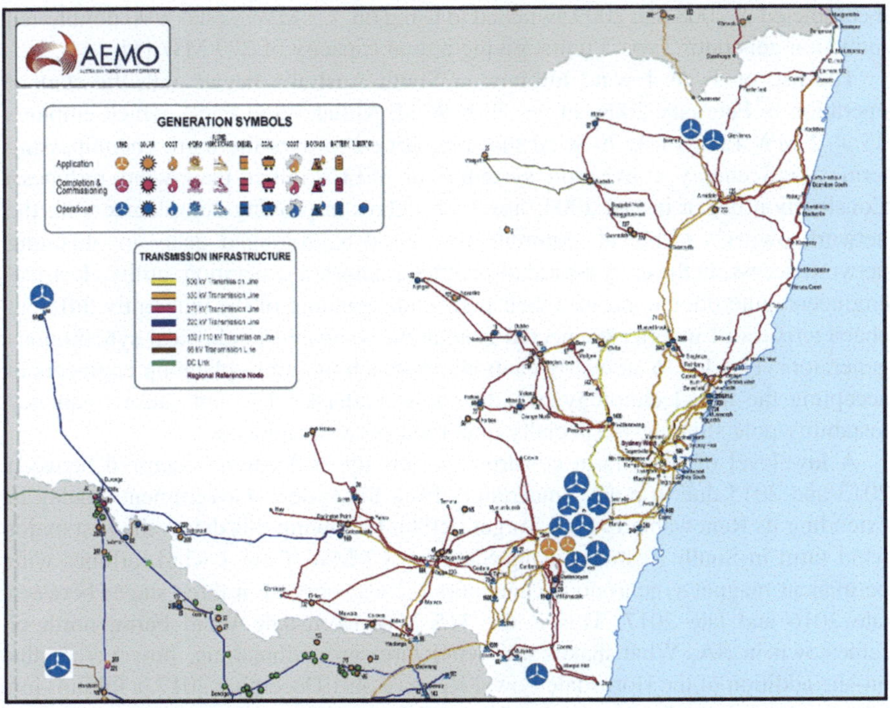

Fig. 6.4 The location of some major Wind Farms in NSW and the transmission networks (Copyright: AEMO, available at: http://wa.aemo.com.au/Electricity/National-Electricity-Market-NEM/Network-connections)

6.5 Environmental and Social Impact of Wind Energy

Australia has a unique environment that, despite its seemingly tough nature, is very sensitive to disturbance. The native birds, animals and plants that make up this are precious national resources that must be both protected and encouraged.

The building, operation and decommissioning of any industrial plant in Australia will affect the local environment and a wind farm is no exception. Despite being a renewable energy resource with environmental benefits, to put 100 m high structures into a site means that some part of that environment will be disturbed. The wind farm designer has an obligation to see that any environmental disturbance is minimized and acceptable to the approving authorities and to the general public.

When developing a wind farm site, the obligation is usually on the wind farm developer to prove whether the disturbance to the area's environment is acceptable. For the wind farm designer this means being aware of the environmental issues that exist and altering the wind farm design to accommodate these. In Australia such work is commonplace and has been very successful and has resulted in the overall environmental impact of wind farm developments in this country being low.

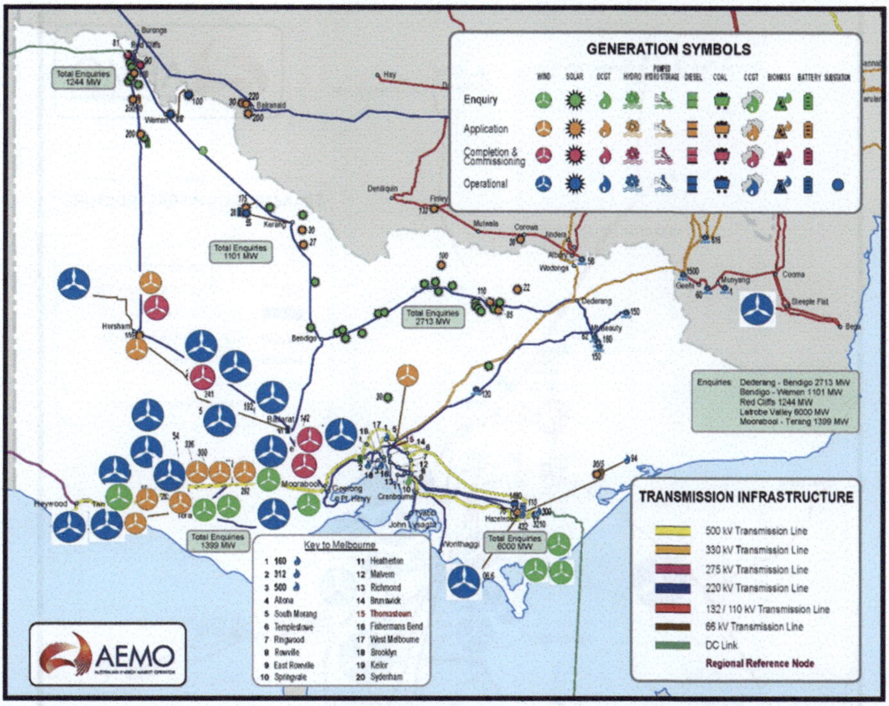

Fig. 6.5 The location of some major Wind Farms in VIC and the transmission networks (Copyright: AEMO, available at: http://wa.aemo.com.au/Electricity/National-Electricity-Market-NEM/Network-connections)

In other parts of the world, this has not always been the case. In other countries wind farms have been implicated in regard to significant effects on birds, soil erosion, land degradation and land contamination through liquid hydrocarbons.

When assessing a prospective site, it is usual practice to undertake a review of existing environmental conditions to ascertain what issues exist and the likely cost of accommodating these. Depending on development scale and the location, this can involve discussions with local authorities, experts or long-term residents in relation to general environmental issues including:

- Environmental status—Protected, locally/nationally/internationally recognized, local planning strategy or use status.
- Flora—endangered or rare species, issues of disease, introduction of weeds and effects on crops.
- Fauna—avian and ground-based fauna, endangered or rare species, migratory patterns, local or adjacent protected areas that could be influenced, rehabilitation, quarantine status for stock protection, and effects of wind farm on animal grazing.
- Land—wind and water erosion, soil borne diseases and issues of access restrictions, seasonal variations in access, and groundwater contamination.

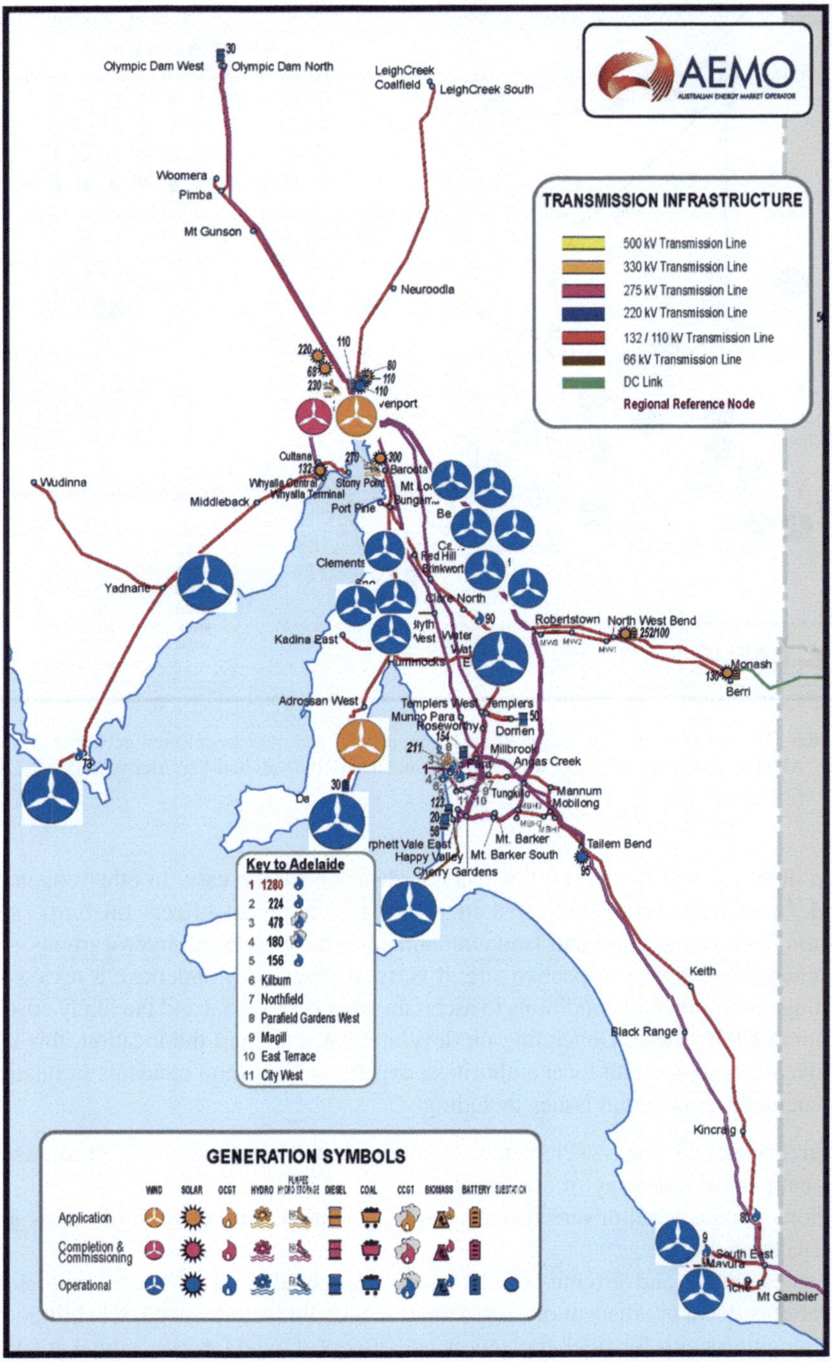

Fig. 6.6 The location of some major Wind Farms in SA and the transmission networks (Copyright: AEMO, available at: http://wa.aemo.com.au/Electricity/National-Electricity-Market-NEM/Network-connections)

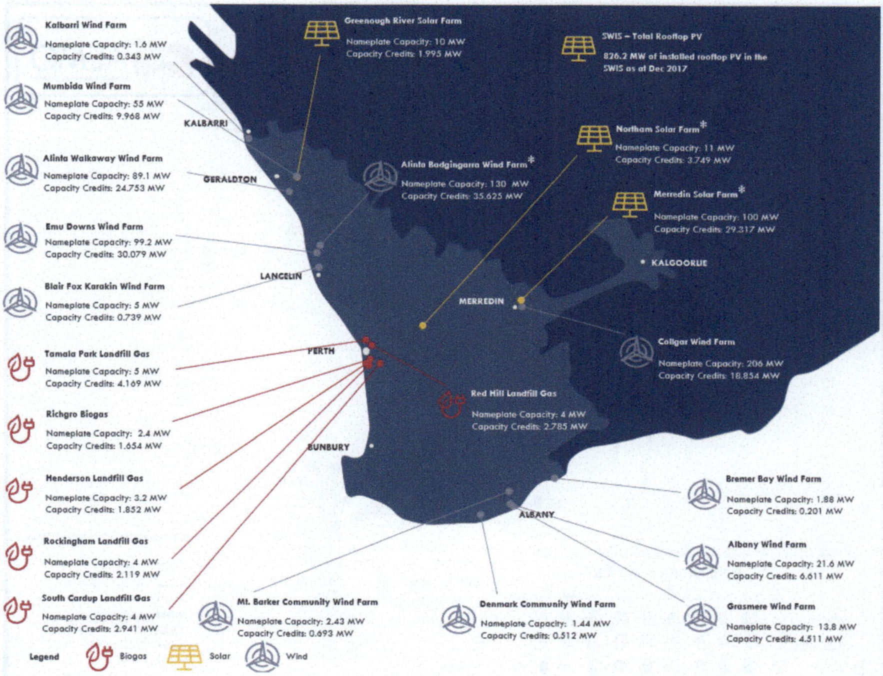

Fig. 6.7 The location of Wind Farms in WA [36]

Such issues can appear overwhelming and can lead to a site being identified as inappropriate to develop.

One high profile case of impact of wind farms on wildlife occurred in 2006 when the Environment Minister of the Liberal-National Coalition led Federal Government at the time, Liberal Senator Ian Campbell, prevented the Bald Hills Wind Farm in South Gippsland, Victoria from being built because of concern of the effect of wind farms on the endangered Orange-bellied Parrot. Research at the time showed that the combined effect of all the wind turbines in SA, VIC and TAS was less than one Orange-bellied Parrot killed per year, compared to the hundred that die each year from other causes for example collision with vehicles, structures, telecommunication towers or killed by household cats. The Minister halted the development of the Bald Hills Wind Farm and declared a State Moratorium on wind farm construction while a new National Code for wind farms was developed. The decision was political, aimed at winning votes in conservative regional areas. Greenpeace spokesman Mark Wakeham stated at the time:

> "*… Senator Campbell appears to be on an anti-wind farm vendetta. What would be the impact on Australia's greenhouse emissions be if the minister went after the coal industry with the same vigour that he is trying to disable the wind industry. Renewable energy like wind power is crucial and must not be used and abused in a game of political football by the Howard government*".

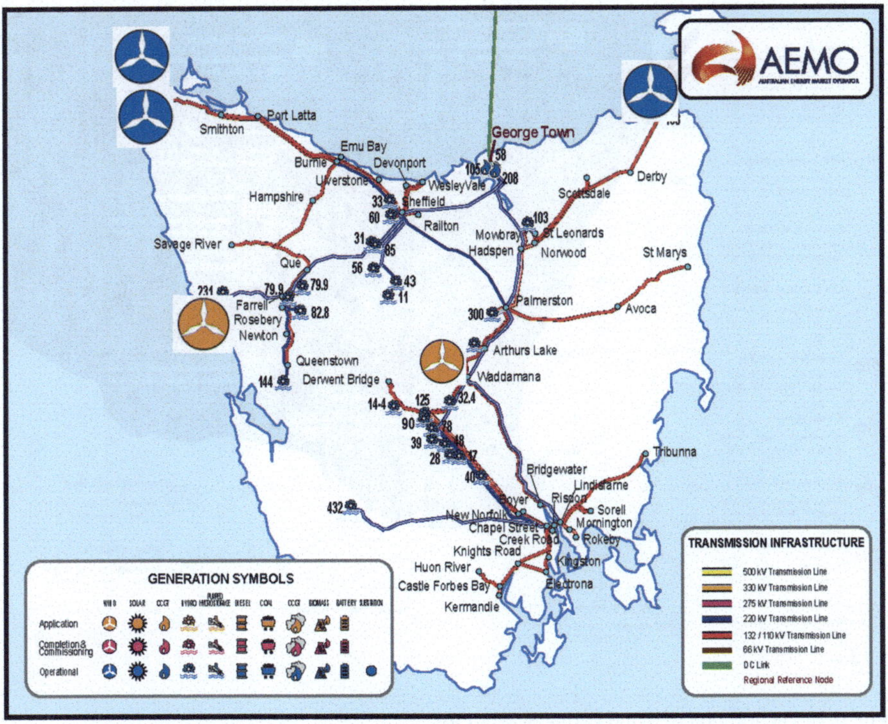

Fig. 6.8 The location of some major Wind Farms in TAS and the transmission networks (Copyright: AEMO, available at: http://wa.aemo.com.au/Electricity/National-Electricity-Market-NEM/Network-connections)

The National Code for wind farms was rejected by States and Territories and is further discussed in Sect. 6.6. Overall, wind farm designers have worked successfully within the sites' environmental constraints and wind farms in Australia have been successfully built in World Heritage Areas, on operational farmland and in areas including sensitive vegetation.

The report by National Health and Medical Research Council (NHMRC) of Australia in 2015 has investigated the evidences on the physical emissions produced by wind turbines, including noise, shadow flicker, and the electromagnetic radiation (EMR) produced by wind turbines [16]. This research states that there is no consistent evidence that the considered physical emissions from wind turbines would be associated with human health. As mentioned in this report, proximity to wind turbines is related to annoyance, or poorer quality of sleep, however, these issues may be due to confounding, bias, or chance. In addition, in the case of Waubra Wind Farm, some local people reported a feeling of nausea along with headaches. The possible cause of this problem was considered by some to be the emission of infrasound by wind turbines, when working. This issue has been called wind turbine syndrome [17, 18].

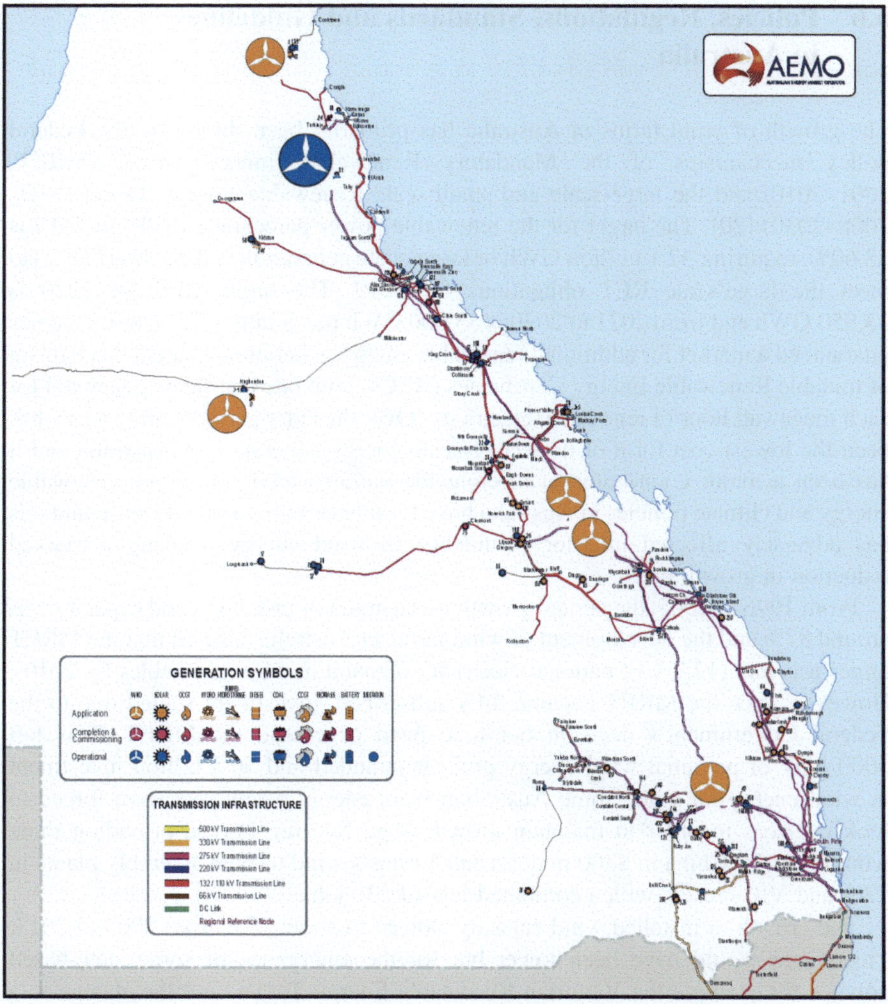

Fig. 6.9 The location of some major Wind Farms in QLD and the transmission networks (Copyright: AEMO, available at: http://wa.aemo.com.au/Electricity/National-Electricity-Market-NEM/Network-connections)

Wind energy has a great influence on reducing greenhouse emissions. Assuming a 0.9 tonne/MWh of carbon dioxide emissions from black coal power generation plants and the generation [19] of 16.3 TWh from wind turbines in 2018, wind generation in Australia reduced emissions by about 14.5 million tonnes in 2018.

6.6 Policies, Regulations, Standards and Guidelines in Australia

The growth of wind farms in Australia has primarily been driven by the Federal policy mechanisms of the Mandatory Renewable Energy Target (MRET, 2001–2010) and the large-scale and small-scale Renewable Energy Target (RET, 2008–2030) [20]. The target for the renewable power percentage (RPP) in 2019 is 18.60%, requiring 32.1 million GWh or large-scale generation certificates (LGCs) to meet the large-scale RET obligations for 2019. The target RPP for 2020 is 33,850 GWh and from 2021 to 2030 is 33,000 GWh per annum [20]. These schemes guaranteed a market for additional renewable energy generation, using a mechanism of tradable Renewable Energy Certificates (RECs) with one certificate generated for each megawatt hour of renewable electricity. Over the latter period, wind energy has been the lowest-cost form of new large-scale energy generation in Australia and it has been a major contributor to meeting the national RET. However, renewable energy and climate policies in Australia have been uncertain, to say the least, and this has adversely affected investor confidence in wind energy, causing a marked reduction in growth rates.

From 1996 to 2005, the annual growth in Australia of installed wind capacity was around 87% and the development of wind farms in Australia ensured that the MRET target was met (12.7% of national electricity demand met by renewables by 2010). However, once the MRET became fully subscribed, investment stalled due to the Federal Government's decision not to expand or extend the MRET. This left 5000 MW of potential wind energy projects stranded and $6.5 billion investment in wind energy on hold. Some Australian wind energy developers were forced to look overseas in order to maintain growth (e.g., Roaring Forties providing three wind farms for China in $300 m deal) and Vestas's wind turbine assembly plants in TAS and VIC closed, with a combined loss of 230 jobs.

The growth in installed wind capacity slowed to about 25% from 2005 to 2013. This figure would have been lower but for the emergence of some state-based policies for example the Victorian Renewable Energy Target, and the change to a Labor Federal Government under Rudd in 2007, which led to the implementation of the RET a year later. With the election of the right-wing Abbott government in 2011, annual growth in Australian installed wind capacity dropped to 10% from 2013 to 2016. The term of the Abbot government, with its statements that "coal was the future of Australia" [21], was a low-point for the renewable energy industry in Australia. The Abbott government's energy policies discouraged investment in renewable energy through the repeal of the carbon tax and created uncertainty in the market through its multiple reviews and possible termination of the RET. In July 2015, the RET was reduced from 41,000 to 33,000 GWh/year of renewable electricity in 2020. In particular, the target will remain the same from 2020 to 2030, with no vision for expansion of the renewable energy market. The policies became personal when Tony Abbott recounted his experience of cycling past the single Enercon 600 kW wind turbine on Rottnest Island. "*...up close, they're ugly, they're* *noisy and they may have all sort of other impacts, which I will leave to the scientists to study.... it's right and proper that we have reduced the Renewable Energy Target because, as things stood, there was going to be an explosion of these things right around our country*" [22]. The situation improved for the renewable energy industry when Malcolm Turnbull became prime minister, after defeating Tony Abbott in a Liberal leadership spill in 2015, and proposed a "National Energy Guarantee" as a scheme to succeed the RET. The internal divisions on energy policy within the Liberal-National Government, with many anti-renewable climate skeptics in its ranks, caused another leadership spill in August 2018 and Scott Morrison is, at the time of writing, the current Prime Minister. The Morrison government's energy plan focuses on reducing electricity prices rather than reducing carbon emissions and this has been criticized by business groups and state governments alike. A federal election must be held by 18 May 2019 and the outcome will be important to the future for the wind industry in Australia. The Opposition has stated that should they come to power, they will adopt the National Energy Guarantee and direct billions of funds towards wind farms through the Clean Energy Finance Corporation [23–25].

In terms of policies related to small wind turbines, the Office of the Renewable Energy Regulator (ORER) in Australia defined a small wind turbine (SWT) as one with a rated capacity of $\leq$10 kWp and a total annual electricity output of less than 25 MWh [26]. In 2009, the Australian Federal Government significantly changed small wind incentives. Prior to June 2009, small wind incentives were only available for off-grid installations (at least 1 km from the main electricity grid) or installations at schools. The mechanism for off-grid installations was provided by the Remote Renewable Power Grants Program (RRPGP), which provided up to 50% of the capital cost of the renewable generating equipment. For a wind installation, this included the wind turbine, tower, controller, dump load, cabling, and circuit protection. By June 2009, the RRPGP program ended in most Australian states except in Western Australia, where the program continued until December 2009. The mechanism for school installations was the National Solar Schools Program, which provided grants of up to $50,000 (up to $100,000 for eligible multi-campus schools) to eligible primary and secondary schools to install solar and other renewable power systems, including small wind turbines. Over $217 million was provided to 5310 schools (or almost 60% of all Australian schools) to install renewable energy systems, rainwater tanks and a range of energy efficiency measures. The National Solar Schools Program closed on 30 June 2013 [27].

On 9 July 2009, the Australian Federal Government introduced the Solar Credits Program, which provided support to households, businesses, and community groups that planned to install small-scale solar PV, wind, and hydro systems. This support took the form of a multiplication of the number of RECs able to be created for eligible installations; thus, additional RECs or "solar credits" were created for an installation. For instance, for the period beginning 9 June 2009 and ending 30 June 2011, the first 1.5 kW of a small wind generation unit was eligible for five times the number of RECs, as would ordinarily be assigned to the unit based on its actual production. Owners of small-scale renewable systems typically would receive a discount on the system or cash payment in return for solar credits. The result was the flooding of the market of a number of these additional RECs, sometimes referred to as "phantom" RECs, resulting in a reduction in the price of RECs. Since both large wind farms and small wind turbines were subsidized by the same REC market, the reduced REC price adversely affected the economics of developing large wind farms. To address this, in January 2011, the Renewable Energy Target was split into two parts; a Large-scale Renewable Energy Target to establish and expand renewable power stations such as wind farms, and a Small-scale Renewable Energy Scheme, which replaced the Solar Credits Program. In its brief period of operation, the Solar Credits Scheme had a significant impact on the small wind industry in Australia, particularly in the state of New South Wales, which offered, from January 2010, a gross feed-in-tariff (FiT) of AUD$0.60/kWh for grid-connected PV and SWTs no greater than 10 kWp. A staggering 471 kWp of new SWT capacity was installed in NSW in 2010 under the combination of the SCS and the FiT, compared to 15 kWp installed the previous year. The short life of such policies (the NSW FiT was reduced to AUD$0.20/kWh in August 2010 and is now a net FiT around AUD $0.06/kWh) did little to develop a sustainable SWT market in Australia. The installation of small-scale renewable energy systems in Australia since 2011 has been dominated by solar PV due to the very low cost of panels and good advocacy, training and regulation on PV offered through bodies such as the Clean Energy Council (CEC) and the Smart Energy Council [20, 28].

Standards Australia has adopted the IEC standard for wind turbines in Australia. Some standards, as listed below, are available on the website of SAI Global.com:

- AS 4959-2010: Acoustics—Measurement, prediction and assessment of noise from wind turbine generators [29];
- AS 61400.21-2006: Wind turbines—Measurement and assessment of power quality characteristics of grid connected wind turbines.

The committee associated with developing this standard is called the "EL-048 committee," comprising several institutes including Australian Industry Group, Clean Energy Council, Clean Energy Regulator, Electrical Regulatory Authorities Council, Energy Networks Australia, and Engineers Australia.

The Call for a National Code for Wind Farms resulted in a Discussion Paper in 2006, which States and Territory governments rejected on the basis of added bureaucracy, since the CEC already had Best Practice Guidelines for Implementation of Wind Energy Projects in Australia since 2002 [30]. In the second half of 2006, perhaps in reaction to the call for a national code, the CEC undertook a review of the Guidelines to ensure that they continued to reflect best practice, by benchmarking them against international and national standards.

In terms of guidelines specifically related to environmental issues relating to wind farm development, the Federal Environment Protection and Heritage Council released draft National Wind Farm Development Guidelines for consultation in July 2010 [31]. This consultation showed that jurisdictions either had existing planning frameworks for managing community concerns or were in the process of developing such frameworks and thus the National Wind Farm Development Guidelines have remained in a draft form as a reference document, while specific guidelines apply to certain jurisdictions for example the New South Wales Wind Farm Guidelines on 2011 [32].

Australia introduced the world's first accreditation scheme for wind farms. The scheme, known as the Wind Farm Australia Auditor Program, provides participating organizations with resources on environmental and social management of wind farms (based on ISO14001). Participants are then audited—this is an independent assessment of their wind farm planning processes or operations against best practice, all Federal and State environmental laws, policies and regulations and the CEC Best Practice Guidelines for Implementation of Wind Energy Projects in Australia.

In more recent times, the Clean Energy Council has focused on the development of community engagement guidelines. In 2013, the CEC released its community engagement guidelines for the Australian wind industry. Developed by independent consultants in the Australian Centre of Corporate Social Responsibility, adhering to the guidelines is a way that the wind industry can show its commitment to involving local communities in the development and management of wind farms. The CEC also released "Wind Farms—A Guide for Communities." This is "an educational tool for communities, councils and landholders about what happens in the development of a wind farm and what they can expect from the developers throughout that process" [33].

6.7 Wind Energy Research in Australia

It is worth noting that Australia does not have a dedicated wind energy association or any conference for academic research in the area of wind energy. The former Australian Wind Energy Association was dissolved and merged with the Business Council of Sustainable Energy to form the Clean Energy Council. The Clean Energy Council promotes a range of renewable energies in Australia and thus the focus on wind energy has been diluted. The Clean Energy Council has an Australian Wind Forum, an exhibition and conference for the Australian wind industry. With the focus on industry for this forum, wind energy academics in Australia often need to travel to present at the European Academy of Wind Energy (EAWE) or American Wind Energy Association (AWEA) conferences.

At Murdoch University, there is a small wind turbine centre (NSWTC) for testing wind turbines and providing consumer labels for each machine. Also, there is a "wind energy research group" in Newcastle University, working on instrumentation of small wind turbines and quantifying the highly turbulent wind resource in the built environment. In addition, Centre for Energy Technology at the University of Adelaide covers the design of wind farms. Furthermore, Australian National University has an "Energy change institute," which focuses research on remote sensing and turbulence. Moreover, Centre for Renewable Energy and Power Systems at the University of Tasmania provides research on wind turbine integration. RMIT University offers a course titled 'Wind and Hydro Power'.

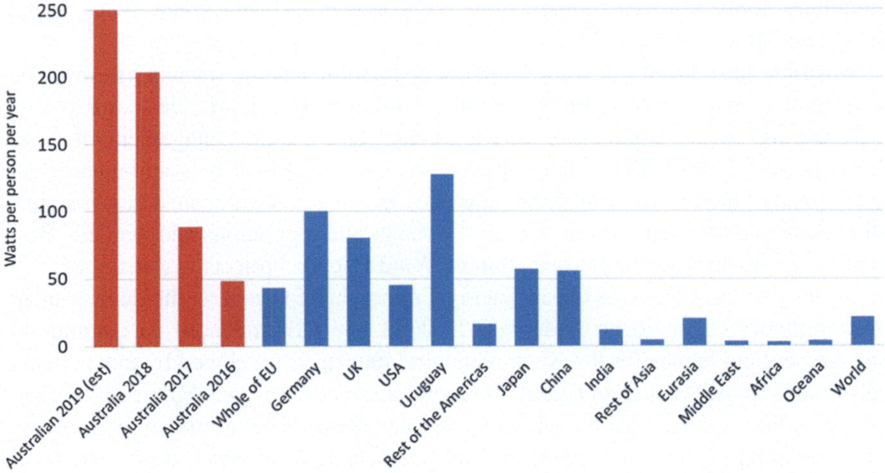

Fig. 6.10 Renewable installation per year [19]

6.8 Wind Energy Outlook

Based on the RET scheme, from 2021 to 2030, the annual generation of renewable energy in Australia should be at least 33,000 GWh [20], so a consistent installation and operation of renewable energy including wind energy is essential. At the moment, the installation of renewable energy harnessing technologies in Australia is four to five times faster per capita than European Union (EU), Japan, China, and the USA, as seen in Fig. 6.4 [19].

In January 2019, one 130 MW wind farm in the Badgingarra Renewable Facility, WA has received the accreditation, which is 46.1% of total accredited renewable power station in Australia. Also, the total capacity of the committed large-scale wind energy projects is 3296.5 MW, as of the end of January 2019. The committed projects refer to those projects that have been granted development approval and where a final investment decision has reached according to the commercial understanding of the term. There are some wind energy projects with a high degree of confidence to commit with a total capacity of 664 MW. The list of committed wind turbine projects is provided in [34], in which the major wind power installations will happen in VIC (Fig. 6.10 and Table 6.1).

6.9 Conclusion

This chapter addresses the situation of wind energy history, harnessing, and associated technology and standards in Australia. The target of renewable energy in Australia is to add 33,000 GWh to the generation of renewable energy to which

Table 6.1 Committed wind energy projects across Australia as of 31 January 2019 [3]

Project name	State	MW capacity
Bulgana Green Power Hub	VIC	194
Cattle Hill Wind Farm	TAS	144
Cherry Tree Wind Farm	VIC	57.6
Coopers Gap Wind Farm	QLD	453
Crowlands Wind Farm	VIC	80
Crudine Ridge Wind Farm	NSW	135
Dundonnell Wind Farm	VIC	336
Elaine Wind Farm	VIC	83.6
Granville Harbour Wind Farm	TAS	112
Kennedy Energy Park (Wind)	QLD	43.2
Lincoln Gap Wind Farm Stage 1	SA	126
Lincoln Gap Wind Farm Stage 2	SA	86
Moorabool North Wind Farm	VIC	150
Murra Warra Wind Farm Stage 1	VIC	225.7
Stockyard Hill Wind Farm	VIC	536
Warradarge Wind Farm	WA	180
Yandin Wind Farm	WA	210
Yendon Wind Farm	VIC	144.4
Total		3296.5

wind energy has made the greatest contribution. The total capacity of committed wind farms in Australia, as of January 2019, is 3296.5 MW.

References

1. P.A. Coppin, K.A. Ayotte, N. Steggel, Wind resource assessment in Australia—a planners guide, CSIRO (2003), http://www.cmar.csiro.au/e-print/open/coppin_2003a.pdf
2. 100 per cent renewables study—modelling outcomes, Australian Energy Market Operator (AEMO), Australia (2013), https://www.environment.gov.au/system/files/resources/d67797b7-d563-427f-84eb-c3bb69e34073/files/100-percent-renewables-study-modelling-outcomes-report.pdf
3. Australian energy statistics, Department of the Environment and Energy (2019), https://www.energy.gov.au/government-priorities/energy-data/australian-energy-statistics
4. Windpower on Rottnest island, Soft Technology: Alternative Technology in Australia, no. 13, pp. 11–13 (1983).
5. D. Clarke, Wind power and wind farms in Australia. https://www.ramblingsdc.net/Australia/WindPower.html
6. Wind farm near nine mile beach, esperance, vol. Bulletin 647 Environmental Protection Authority Perth, Western Australia, Australia (1992), http://www.epa.wa.gov.au/sites/default/files/EPA_Report/711_B647.pdf
7. Synergy, Synergy's renewable energy integration 30 years on (2017), http://www.ipsconnect.org/wp-content/uploads/2017/11/Chris-Dowe-Synergy%E2%80%99s-Renewable-Energy-Integration.pdf

8. A. Heather, The Mawson station wind turbines, Department of the Environment and Energy (2014), http://www.antarctica.gov.au/living-and-working/stations/mawson/this-week-at-mawson/2014/this-week-at-mawson-4-july-2014/1
9. AMSC, Australian case studies applications of d-var statcom (2015), http://wapp.com.au/wp-content/uploads/2018/04/Australian-Case-Study-Highlights-2016.pdf
10. Pacific Hydro, Codrington wind farm, http://www.pacifichydro.com.au/english/projects/operations/codrington-wind-farm/
11. Wind Turbine Models, An bonus 1300/62, https://en.wind-turbine-models.com/turbines/59-an-bonus-1300-62
12. AGL, Macarthur wind farm, https://www.agl.com.au/about-agl/how-we-source-energy/macarthur-wind-farm
13. Wind Turbine Models, Vestas v66-1.65, https://en.wind-turbine-models.com/turbines/15-vestas-v66-1.65
14. Renewable Energy Systems, Ararat wind farm, http://www.ararat-windfarm.com/about/
15. J. Quilter, After 10 years GE goes back to DFIGs (2012), https://www.windpowermonthly.com/article/1153928/10-years-ge-goes-back-dfigs
16. T. Merlin, S. Newton, B. Ellery, J. Milverton, C. Farah, Expert review of the evidence on wind farms and human health (2015), https://nhmrc.gov.au/about-us/publications/expert-review-evidence-wind-farms-and-human-health#block-views-block-file-attachments-content-block-1
17. N. Pierpont, *Wind Turbine Syndrome* (K-Selected Books, Santa Fe, 2009), p. 294
18. L. Harrison, Wind farm and health issues, Waubra foundation (2016), https://waubrafoundation.org.au/resources/harrison-l-wind-farm-health-issues/
19. A. Blakers, M. Stocks, B. Lu, *Australia: The Renewable Energy Superstar* (Australian National University, Canberra). http://re100.eng.anu.edu.au/publications/assets/100renewables.pdf
20. The renewable energy target (ret) scheme (2019), https://www.environment.gov.au/climate-change/government/renewable-energy-target-scheme
21. K. Von Strokirch, N. Economou, M.O.J. Ryan, P. Barratt, V. Blokker, A. Bronowski, H. Ware, *Social Alternatives*, vol 35 (University of the Sunshine Coast, Australia, 2016). https://socialalternatives.com/sites/default/files/pdf/issue/free/socialalternatives35_2_small_0.pdf
22. J. Glenday, Tony Abbott launches another attack on ‘ugly’, ‘noisy’ wind turbines, ABC News (2015), https://www.abc.net.au/news/2015-06-12/tony-abbott-launches-another-attack-against-wind-farms/6541952
23. D. Crowe, ‘Ad hoc and extreme’: peak business group slams Morrison government on energy policy, The Sydney Morning Herald (2018), https://www.smh.com.au/politics/federal/ad-hoc-and-extreme-peak-business-group-slams-morrison-government-on-energy-policy-20181123-p50ht8.html
24. K. Murphy, Malcolm Turnbull convinces party to unite on energy policy, The Gaurdian (2017), https://www.theguardian.com/australia-news/2017/oct/17/malcolm-turnbull-convinces-party-to-unite-on-energy-policy
25. A. Probyn, M. Clarke, Bill Shorten to adopt Malcolm Turnbull's energy policy, pledges to throw billions at renewables, ABC News (2018), https://www.abc.net.au/news/2018-11-21/labor-energy-policy-ahead-of-election/10519018
26. Small-scale renewable energy scheme, Australian Government, Clean Energy Regulator (2018), http://www.cleanenergyregulator.gov.au/RET/About-the-Renewable-Energy-Target/How-the-scheme-works/Small-scale-Renewable-Energy-Scheme
27. National solar schools program, International Energy Agency (2013), https://www.iea.org/policiesandmeasures/pams/australia/name-24327-en.php?s=dHlwZT1lZSZzdGF0dXM9T2s,&return=PG5hdiBpZD0iYnJlYWRjcnVtYiI-PGEgaHJlZj0iLyI-SG9tZTwvYT4gJnJhcXVvOyA8YSBocmVmPSIvcG9saWNpZXNhbmRtZWFzdXJlcy8iPlBvbGljaWVzIGFuZCBNZWFzdXJlczwvYT4gJnJhcXVvOyA8YSBocmVmPSIvcG9saWNpZXNhbmRtZWFzdXJlcy9lbmVyZ3llZmZpY2llbmN5LyI-RW5lcmd5IEVmZmljaWVuY3k8L2E-PC9uYXY-

28. S.J. Ross, M.P. McHenry, J. Whale, The impact of state feed-in tariffs and federal tradable quota support policies on grid-connected small wind turbine installed capacity in australia. Renew. Energy **46**, 141–147 (2012)
29. Standards Australia, As iec 61400.2-2013: wind turbines design requirements for small wind turbines (2013), https://infostore.saiglobal.com/en-au/standards/as-iec-61400-2-2013-120057_SAIG_AS_AS_251614/
30. Clean Energy Council, Best practice charter, Clean Energy Council (2018), https://www.cleanenergycouncil.org.au/advocacy-initiatives/community-engagement/best-practice-charter
31. Environment Protection and Heritage Council, Future of the national wind farm development guidelines, Environment Protection and Heritage Council (2010), http://www.nepc.gov.au/resource/ephc-archive-future-national-wind-farm-development-guidelines
32. NSW Government, The NSW government has a wind energy framework, Planning and Environment (2019), https://www.planning.nsw.gov.au/Policy-and-Legislation/Renewable-Energy/Wind-Energy-Framework
33. Clean Energy Council, Community engagement, https://www.cleanenergycouncil.org.au/advocacy-initiatives/community-engagement
34. Clean Energy Regulator, Large-scale renewable energy target supply data, Australian Government, Clean Energy Regulator (2019), http://www.cleanenergyregulator.gov.au/RET/About-the-Renewable-Energy-Target/Large-scale-Renewable-Energy-Target-market-data/large-scale-renewable-energy-target-supply-data
35. Geoscience Australia, Energy resources and market, Australian Government (2018). https://aera.ga.gov.au/#!/energy-resources-and-market
36. AEMO, Electricity statement of opportunities, Australia (2018), https://www.aemo.com.au/-/media/Files/Electricity/WEM/Planning_and_Forecasting/ESOO/2018/2018-WEM-ESOO-Report.pdf

Chapter 7
Hybrid Wind Energy Solutions Including Energy Storage

J. K. Kaldellis

7.1 Introduction: Remote Energy Consumers and High RES Potential

Electrical energy is assumed nowadays as a vital element of our life and it is a prerequisite for upgrading the living standards all around the world, even in remote regions and small islands far away from central electrical networks. Actually, according to official statistics [1], almost two billion people worldwide have no direct access to reliable electrical networks with 500,000 of them living in the European Union and other financially developed countries. Unfortunately, isolated consumers are facing a dramatically insufficient infrastructure status [2, 3] as far as their electricity consumption needs is concerned, based mainly on outdated, high cost and heavy polluting internal combustion engines. In this context, hybrid wind-based power systems have proven to be one of the most interesting and environmentally friendly technological solutions for the electrification of remote consumers, especially in the presence of high wind potential [4, 5] in collaboration with the appropriate energy storage technologies.

One of the most representative areas where several decades of remote islands are located is the Aegean Archipelagos. In this European area of the SE Mediterranean there exist more than thirty autonomous thermal power stations (APS) of various sizes [6], starting from 50 kW up to decades of MW (Fig. 7.1). To this end, all these APS are operating using remarkable quantities of diesel or heavy oil (Fig. 7.2), while the corresponding marginal production cost is extremely high, exceeding 1000€/MWh in certain small islands (Fig. 7.3). As a general picture the average electricity production cost for the entire Aegean Sea area varies between 250 and 300€/MWh, being five times higher than the corresponding cost of the Greek mainland [7].

J. K. Kaldellis (✉)
Lab of Soft Energy Applications and Environmental Protection, Mechanical Engineering Department, University of West Attica, Athens, Greece
e-mail: jkald@uniwa.gr

A. Sayigh, D. Milborrow (eds.), *The Age of Wind Energy*, Innovative Renewable Energy, https://doi.org/10.1007/978-3-030-26446-8_7

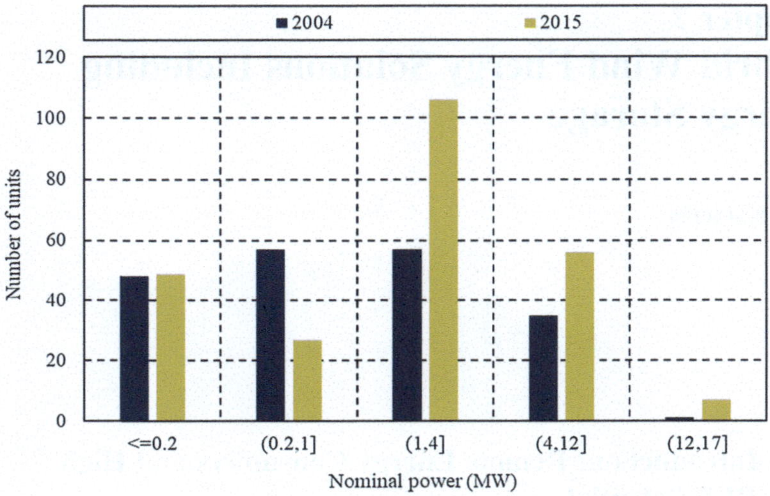

Fig. 7.1 Nominal power of thermal units employed in Aegean Archipelagos islands

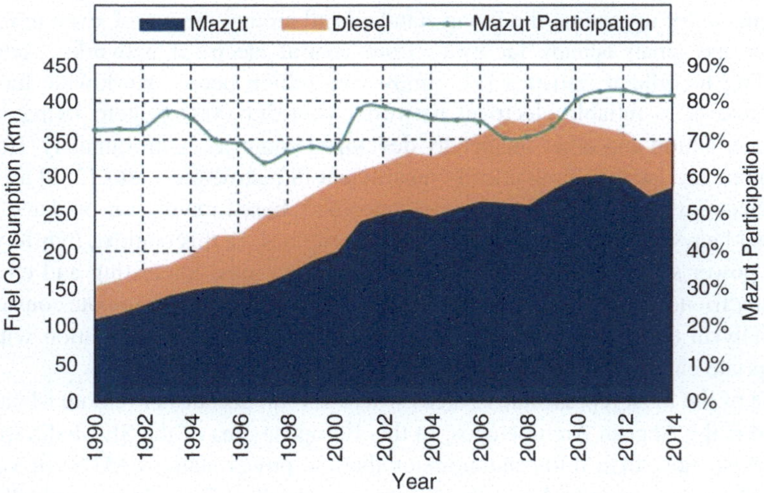

Fig. 7.2 Long-term evolution of fuel consumption for APSs of Aegean Sea

On the other hand, most of remote islands possess excellent wind and solar potential, while one may find acceptable biomass potential and very high geothermal fields/reservoirs in specific island locations. For example, according to the extended long-term measurements by PPC [8], the Hellenic Meteorological Agency [9], CRES [10] and private companies, one may easily conclude that the average wind speed in the Aegean Archipelago varies between 8 and 9.5 m/s, while the corresponding annual solar potential ranges between 1500 and 1850 kWh/m^2 at horizontal plane, see also Fig. 7.4.

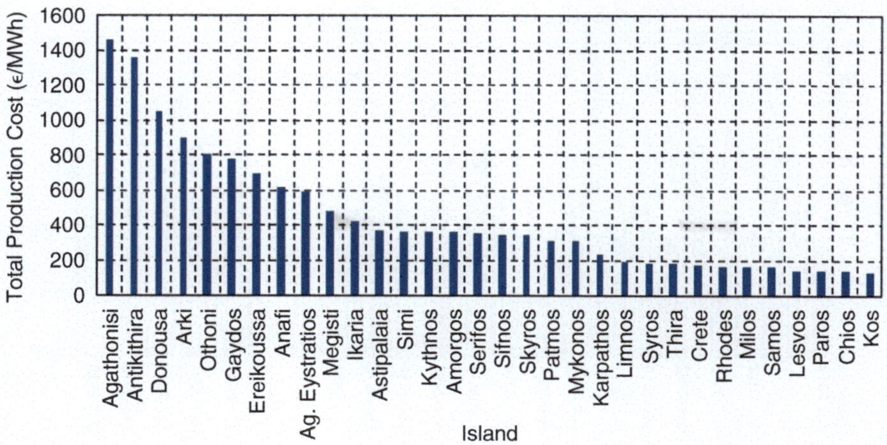

Fig. 7.3 Electricity production cost of Aegean Sea APSs, 2017

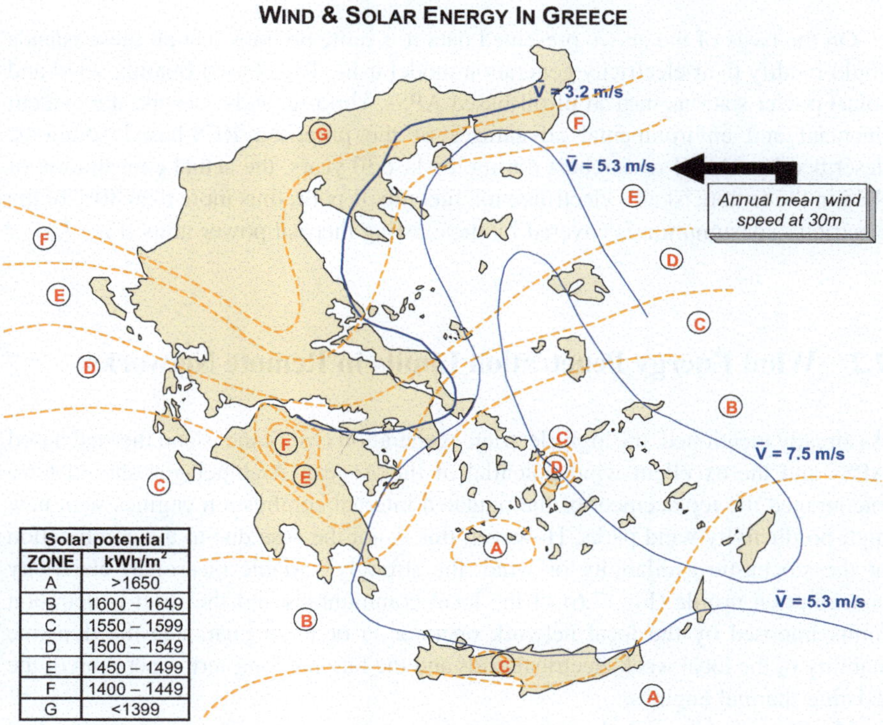

Fig. 7.4 Wind and solar potential in Greece

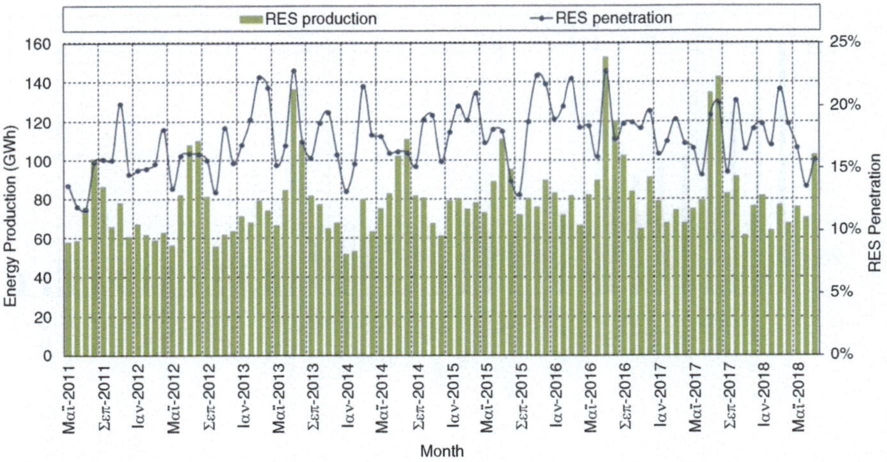

Fig. 7.5 Long-term evolution of RES-based electricity generation and RES contribution in the Aegean Archipelagos Electrical Networks

On the basis of the above presented data it is quite obvious that all these islands could modify their electricity generation model using RES-based (mainly wind and solar) power stations instead of oil-based APSs. Unfortunately, despite the evident financial and environmental advantages of the proposed RES-based solutions, described by several researchers during the last 30 years, the actual contribution of RES in the remote islands electrification fuel mix is poor, thus more than 80% of the electricity consumption is covered by the existing thermal power units (Fig. 7.5).

7.2 Wind Energy Penetration Limits in Remote Networks

As already mentioned, the high electricity generation cost of the island thermal based APSs and the excellent wind potential of the Aegean Archipelagos should have encouraged the replacement of the outdated internal combustion engines with new high-productivity wind parks. However, this is not the case due to the combination of the stochastic availability of wind, the strongly variable (seasonal) electricity consumption profile (Fig. 7.6) of the local communities and the strict penetration limits imposed by the local network operator in order to guarantee the dynamic stability of the local weak electrical grids and the rational long-term operation of the existing thermal engines.

More specifically, in order to estimate the maximum wind power penetration, that is the wind power absorbed by the local grid, without remarkable operational problems of the entire system, one should know the local grid load demand "$P_L(t)$" vs. time along with operational characteristics of the existing thermal units, while for bigger island regions, the topology of the local grid is also required.

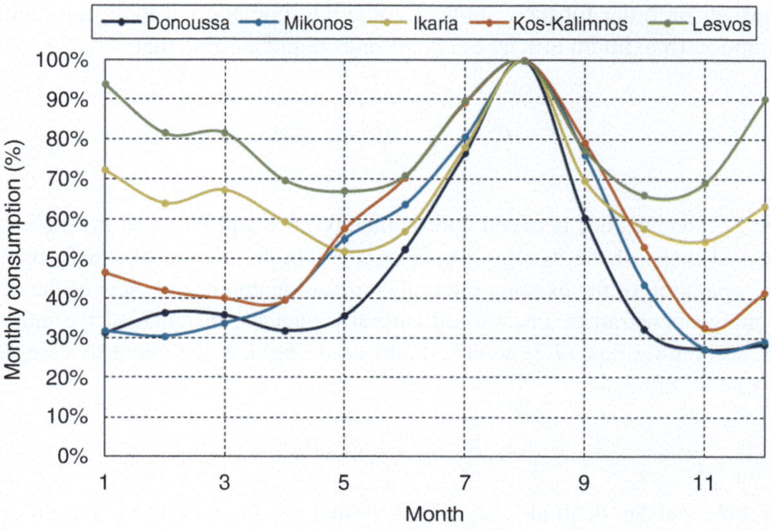

Fig. 7.6 Annual variation of electricity consumption of selected representative islands

Furthermore, it should be noted that thermal units are not permitted to operate below a certain limit, to avoid increased wear and maintenance requirements [11, 12]. This limit is mentioned as the "technical minimum" of each engine; hence the minimum output power of the "in operation" thermal units "$P_{d_{\min}}$" is calculated as:

$$P_{d_{\min}} = \sum_{i=1}^{i=i_{\max}} P_{\mathrm{d}_i}^{\min} = \sum_{i=1}^{i=i_{\max}} k_i \cdot P_{\mathrm{d}_i}^{*} \tag{7.1}$$

where the technical minimum of each engine is expressed via an appropriate factor "k_{i}" and the rated (or maximum) output power "$P_{\mathrm{d}_i}^{*}$" of the unit. Typical values of "k_{i}" are 30–50% for heavy-oil units and 20–35% for diesel-fired engines (including gas turbines), depending a lot on the age and the overall condition of the engine. On top of this, the annual maintenance plan of the system, affecting the number ($i_{\max}$) of engines "in operation" during the year, should also be considered. In addition, due to the stochastic nature of wind one cannot disregard the probability of an unexpected loss of a significant part of the "in operation" wind parks. To avoid (or to minimize) loss of load events [12, 13] in similar situations, the system operator should maintain full spinning reserve, which suffices to cover the total load demand, that is, the "in operation" thermal units should be able to cover "P_{L}", thus:

$$\sum_{i=1}^{i=i_{\max}} P_{\mathrm{d}_i}^{*} \geq P_{\mathrm{L}} \tag{7.2}$$

However, in order for these units to come into operation at their minimum fuel consumption (maximum efficiency) point one should assume that:

$$P_{\mathrm{w}}^{*} + \sum_{i=1}^{i=i_{\max}} \xi_i \cdot P_{\mathrm{d}_i}^{*} \geq P_{\mathrm{L}} \tag{7.3}$$

where "ξ_i" takes values between approximately 65% and 80%, on the basis of the operational maps of the existing diesel engines. In an attempt to satisfy the "economic" operation of the existing internal combustion engines in view of the desired wind energy penetration, one should carefully plan the dispatch of thermal units. Hence, combining Eqs. (7.2) and (7.3) and assuming that one may use a single "ξ_i" value, one finally gets:

$$P_{\mathrm{w}}^{*} \leq (1 - \xi) \cdot P_{\mathrm{L}} \tag{7.4}$$

For practical applications, Eq. 7.4 is written in the following simplified and widely used form, that is,

$$P_{\mathrm{w}}^{*} \leq \lambda_1 \cdot P_{\mathrm{L}} \tag{7.5}$$

where "$P_{\mathrm{w}}{}^{*}$" is the maximum acceptable wind power by the local network and "λ_1" is the corresponding maximum instantaneous participation limit, based on the operational characteristics of the existing thermal power units. Finally, in order to avoid annoying system frequency excursions and increasing wear of the existing thermal power units, an additional penetration limit is also imposed, dictated by the instantaneous rate that the "in operation" units can compensate any power deficit of the system. This dynamic penetration limit [11, 13] is characteristic of the network as well as of the spatial distribution and the type of the system's wind turbines. In general, this limit "λ_2" is determined by the system operator (incorporating also a certain level of subjective assessment) and is up to now empirically set in the range of 20–40%. In case of emergency, this value may drop down to 15% or even to zero. In this context, the dynamic penetration constraint is expressed as:

$$P_{\mathrm{w}}^{*} \leq \lambda_2 \cdot P_{\mathrm{L}} \tag{7.6}$$

On the basis of the above analysis, the maximum absorbed wind energy "$P_{\mathrm{w}}^{*}(\mathrm{t})$" by the local electrical system can be estimated according to the following equations, that is,

$$\text{if } P_{\mathrm{L}}(t) \leq P_{d_{\min}}(t) = \sum_{i=1}^{i=i_{\max}} k_i \cdot P_{\mathrm{d}_i}^{*} \quad \text{then } P_{\mathrm{w}}^{*}(t) = 0 \tag{7.7}$$

In this case, there is no wind energy absorption by the local network; hence all the wind energy production is rejected.

$$\text{if } P_{d_{\min}}(t) \leq P_{\mathrm{L}}(t) \leq (1+\lambda)\cdot P_{d_{\min}}(t) \quad \text{then } P_{\mathrm{w}}^{*}(t) = P_{\mathrm{L}}(t) - P_{d_{\min}}(t) \tag{7.8}$$

where "λ" is the wind power upper participation limit depending on the optimum operation of the system thermal power units "λ_1" and the dynamic stability of the local network "λ_2," that is,

$$\lambda = \min\{\lambda_1, \lambda_2\} \tag{7.9}$$

Finally,

$$\text{if } P_{\mathrm{L}}(t) \geq (1+\lambda)\cdot P_{d_{\min}}(t) \quad \text{then } P_{\mathrm{w}}^{*} \leq \min\{\lambda \cdot P_{\mathrm{L}}(t), P_{\mathrm{L}}(t) - P_{d_{\min}}(t)\} \tag{7.10}$$

In this last case, the wind energy penetration is bounded by the upper wind power participation limit "λ" and the instantaneous load demand of the system. In most practical application cases the "λ" value is taken (as a thumb rule) $\leq 30\%$. Applying Eqs. 7.1–7.10, one has the ability to estimate the maximum wind energy penetration in the local grid. However, in cases of low wind energy production, the real wind energy contribution is quite lower (i.e., the maximum penetration limit is not approached), further reducing the contribution of clean energy to the balance of remote islands.

The results of all these constraints are clearly demonstrated in Fig. 7.5, where for almost 8 years the monthly average RES penetration limit (including photovoltaic generators) hardly exceeds 20%, while at least one third of this percentage corresponds to solar power stations providing maximum power during summer, that is, the high touristic period.

In view of the above, the Hellenic Regulatory Authority of Energy (RAE) characterizes all autonomous island networks as saturated, something that the investors are already aware of, thus no new wind parks have been installed in the Aegean Sea islands during the last 10 years. Moreover, due to the economic recession, the local electricity demand has not increased significantly; thus, there is almost no possibility for additional wind power installation during the next few years under the given wind energy exploitation status.

7.3 Wind-Based Hybrid Energy Solution Using Energy Storage

Stand-alone systems based on wind potential exploitation have been proved both interesting and environmentally friendly technological solutions for the electrification of remote consumers. However, the first installation cost is quite high [14], while in some occasions the life cycle cost is also high [15] mainly due to increased energy storage requirements in case of long calm spell periods. In this context, in order to limit the relatively high operational cost and increase the system reliability, several authors suggest the reinforcement of the stand-alone solution with the parallel exploitation of more than one RES, that is, installation of stand-alone hybrid energy systems based on the available renewable potential of each candidate region.

Actually, a hybrid energy system incorporates two or more electricity generation options based either on pure RES or utilizing also a small thermal power unit (e.g., diesel–electric generator or a small gas turbine) along with an appropriate energy storage bank and the corresponding electronic devices. In this context, a hybrid energy system combines the potential of more than one RES, that is, wind/solar/hydropower or even biomass, while the utilization of geothermal and wave energy is also expected in the near future.

The main advantages of wind-based hybrid energy systems include:

- Increased reliability of the hybrid energy installation, since it is based on more than one electricity generation sources.
- Reduction of the energy storage capacity, especially in cases that the different RES utilized present complementary behavior.
- Limited operation and maintenance (O&M) cost, especially in cases that the installation of photovoltaic panels replaces classic energy storage devices, like lead-acid batteries [16].
- Optimum environmental behaviour, especially in cases that the hybrid energy system does not use any fossil fuel (exclusively RES-based hybrid energy systems) [17].
- Minimum levelized life cycle electricity generation cost, not dependent on the fossil fuels price volatility, especially in cases that the hybrid energy system is based on optimum design techniques.

On the other hand, the hybrid energy systems present also some disadvantages, like:

- In most cases the hybrid energy system is over-sized, since the system designers try to make each system component able to cover the load demand without the contribution of the other participating energy sources. This aspect is however faced by the new existing sizing algorithms.
- The first installation cost is rather high, although the long-term cost is normally low. This high installation cost discourages some of the potential investors.
- The application of different technologies introduces a complication degree (especially in the electronic control devices and in the O&M procedures) in the stand-alone installation, a serious problem especially for remote consumers.
- The introduction of thermal units (e.g., diesel–electric generators) and the utilization of batteries are both related with environmental impacts, thus decreasing the environmentally friendly attributes of these RES-based systems.

On the basis of the existing information in the international literature one may find several wind-based hybrid energy configurations, such as the following:

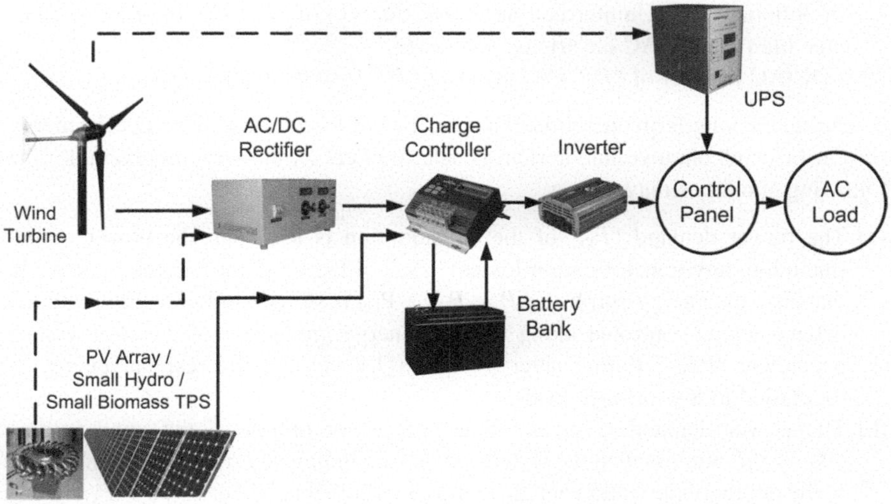

Fig. 7.7 Typical hybrid wind-based system

- Wind–diesel systems
- Wind–photovoltaic based systems
- Wind–hydro installations
- Wind–biomass-based installation
- Wind–photovoltaic and diesel-based systems
- Wind–hydro and diesel-based installations
- Wind–hydrogen–fuel cell hybrid energy systems

Similar power stations are able to cover the electricity needs starting from single remote consumers up to isolated villages and remote islands, with minimum fossil fuel consumption.

Accordingly, a typical wind-based stand-alone hybrid energy system, see Fig. 7.7, is based on:

1. One or more (usually small) wind converters of "P_o" kW.
2. A photovoltaic array of z panels ("P^+" maximum/peak power of every panel) properly connected to feed the charge controller with the voltage and the power required, or a small hydro turbine able to meet the remote consumer load demand, or even a small thermal power station based either on biomass (biogas or biofuel) or consuming fossil fuels.
3. An appropriate energy storage device (e.g., a lead-acid battery storage array) able to guarantee "h_o" hours of autonomy, or equivalently with energy storage capacity "Q_{max}" and maximum discharge capacity "Q_{min}",
4. An AC/DC rectifier of "P_r" kW in case that the energy storage installation operates on DC current,
5. A charge controller of "P_c" kW,

6. An optional UPS (Uninterruptible Power Supply) of "P_p" kW in order to guarantee high quality AC electricity generation,
7. A DC/AC inverter of "P_p" kW, in case of AC load demand.

During the long-term operation of a typical wind-based stand-alone hybrid energy system (a wind–photovoltaic system is selected here as the working example) the following situations may appear:

(a) The power demand "P_D" of the consumption is less than the power output (including any transformation losses) "P_W" of the wind turbine ($P_W > P_D$). In that case the energy surplus ($\Delta P = P_W - P_D$) is stored via the rectifier and the battery charge controller along with the energy production of the photovoltaic generator, "P_{PV}." If the battery is full ($Q = Q_{max}$), the residual energy is forwarded to low-priority loads.
(b) The power demand is greater than the power output of the wind turbine ($P_W < P_D$) but less than the sum of power (including any transformation losses) of the photovoltaic station and the wind converter, that is, $P_W + P_{PV} > P_D$. In this case, the remaining load demand is covered by the photovoltaic station via the DC/AC inverter. Any energy surplus from the photovoltaic station is stored in the battery via the charge controller. If the battery is full ($Q = Q_{max}$), the residual energy is forwarded again to low-priority loads.
(c) The power demand is greater than the power output of the two renewable stations, that is, $P_W + P_{PV} < P_D$, where $P_W + P_{PV} \neq 0$. In similar situations the energy deficit [$\Delta P = P_D - (P_W + P_{PV})$] is covered (along with the corresponding losses) by the batteries via the DC/DC controller and the DC/AC inverter. During this operational condition, special emphasis is laid on the management plan of the three electricity production subsystems.
(d) There is no renewable energy production (e.g., low wind speed, machine not available and zero solar irradiance), that is, $P_W + P_{PV} = 0$. In that case, all the energy demand is covered by the battery—DC/DC controller—DC/AC inverter subsystem under the condition that $Q > Q_{min}$. In cases (c) and (d), when the battery capacity is near the bottom limit, an electricity demand management plan should be applied; otherwise the load would be rejected.

Recapitulating, the scope of a stand-alone system is to meet the electricity demand of a remote consumer at a rational cost and under a given loss of load (or reliability level) constraint. Depending on the importance and on the installation to be served one may demand no-load rejection (i.e., the load should be fulfilled at any case) operation or may permit a maximum (predefined) number of hours without electricity load coverage [18].

As already mentioned, interest in the use of wind energy has significantly grown during the last years, mainly as a reaction to concerns about the environmental impact from the use of fossil and nuclear fuels, along with the oil and natural gas price instability in the international market. On the contrary, RES and especially wind energy have demonstrated their independence from economic fluctuations,

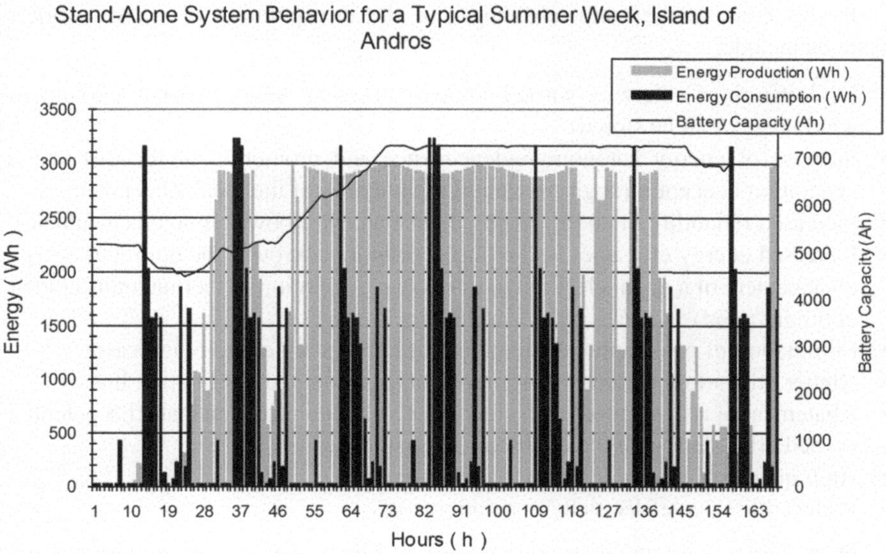

Fig. 7.8 Energy storage contribution in the energy balance of a hybrid power system

while in most cases, an initial cost reduction in the course of time is encountered; see for example the considerable price reduction of photovoltaic panels.

However, due to the stochastic behaviour of wind, wind generation cannot provide firm capacity to an electrical power system [11, 13]. Additionally, these fluctuations can—in some cases—cause problems on a distribution network related to stability, harmonics or flickering. Such issues pose serious obstacles, as also described in Sect. 7.2, to the extensive establishment of wind-only power systems for the electrification of remote consumers and small (weak) power grids [12, 19]. On the other hand, an energy storage system, when sized appropriately, can match (see Fig. 7.8) the stochastic wind power production to a generally variable and hardly predictable system demand, remarkably limiting the energy production cost (e.g., generating capacity savings).

To this end, the vast majority of stand-alone and hybrid energy systems use several energy storage devices in order to store wind energy during high wind and low consumption periods and provide electrical energy during low wind and high load demand periods. In the following one may find the main pros and cons of the most commonly applied energy storage solutions for wind-based stand-alone and hybrid energy systems [20, 21], including among others:

- Lead-acid batteries
- Pumped hydro
- CAES (Compressed Energy Storage)
- Flywheels

In this context, the main advantages of the incorporation of energy storage systems include:

- Exploitation of otherwise wasted amounts of energy (e.g., rejected amounts of wind energy can be stored)
- Increase of energy autonomy/independence and promotion of the distributed generation concept through maximum exploitation of the local RES potential
- Increased reliability of energy supply (since an extra power source is available)
- Increased energy efficiency and reduced emissions through the optimum energy management of a given electricity system (e.g., operation of thermal units at their optimum point)
- Elimination of peak demand and deferral of electricity capacity increase
- Higher utilization and decongestion of transmission and distribution lines
- Abatement of risk entailed by the fuel price volatility provided that RES potential is used efficiently with the contribution of storage
- High quality of power delivered to end-users
- Reduced life cycle electricity generation costs

On the other hand, the main disadvantages of these systems are the following:

- High initial cost required in most cases
- Inherent transformation and other types of losses
- Considerably lower energy densities than fossil fuels
- Introduction of environmental (especially for bulk energy storage like pumped-hydro) and safety concerns (e.g., toxic wastes in the case of certain battery types)
- Additional energy use in the first place in order to build/construct a new energy storage system/device (may however ameliorate an entire stand-alone system energy payback through maximum exploitation of RES energy production)
- Anticipation of advances in other scientific and technological fields required for some energy storage technologies to develop.

7.4 Optimum Sizing-Financial Evaluation of Wind-Based Hybrid Energy Systems

Wind-based hybrid energy systems constitute an interesting option for covering the electrification needs of a wide range of applications, starting from remote telecommunication stations up to medium sized (island) communities. In previous studies the authors have extensively investigated a large variety of wind-based hybrid energy systems, including wind–hydro (using water pumping as energy storage technique) [22, 23], wind–photovoltaic (using batteries) [4], wind–diesel [15], wind–CAES [24], wind–hydrogen–fuel cell [25], and even wind–photovoltaic–diesel configurations [26].

In all these cases analyzed for the optimum sizing of the hybrid energy solutions, one needs the following input data:

1. The power demand (energy consumption) of the consumer, with special emphasis on the average hourly load, the peak load, and the diurnal and seasonal energy demand variation.
2. The available wind potential, including if possible long-term detailed wind speed values (and direction), wind speed probability density profiles on annual and monthly basis, maximum calm-spell, and maximum wind speed values.
3. The available solar and hydro potential, with the same details as described above for the wind potential case. The existence of biomass or even wave energy and geothermal potential is also welcome.
4. The existence and the operational characteristics of the nearby electrical networks.
5. The possibility to use as a backup option an appropriate diesel–electric generator and the maximum diesel oil consumption desired (zero oil consumption is also an option).
6. The energy storage alternatives, depending on various parameters (e.g., storage duration (minutes, hours, weeks, etc.), energy storage capacity, input–output power, depth of discharge, round-trip efficiency, service period (operational cycles), initial and maintenance and operation (M&O) cost, land usage as well as special requirements (i.e., topography, water reservoirs, air caverns, environmental impacts, hazards risk, etc.)).

On the basis of the above described information one can present selected results of published work dealing with representative wind-based hybrid power stations. The first example presented concerns the wind–hydro solution (Fig. 7.9) investigated for the medium-sized island of Ikaria [22, 27]. It is important to note that a similar

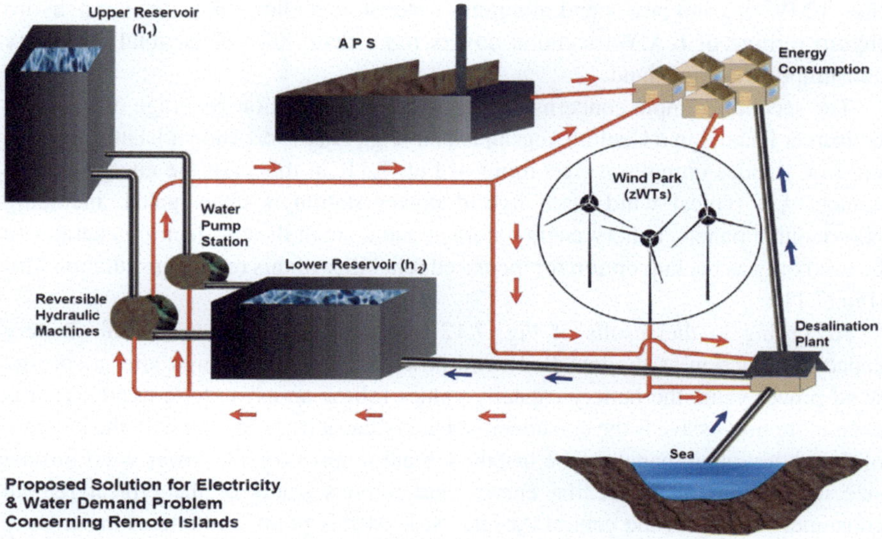

Fig. 7.9 Integrated wind–hydro energy/water solution for remote islands

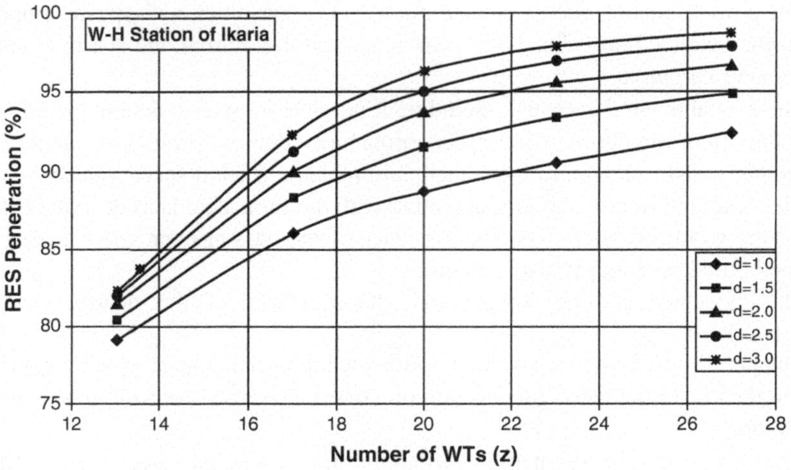

Fig. 7.10 Wind-based hybrid power system contribution in the local electrical consumption of Ikaria Island

project is under construction in Ikaria island during the last 15 years and its operation is expected to start during 2019.

As we can see from Fig. 7.10 by using [22] a rational number (e.g., 20 × 300 kW) of small wind turbines, that is, installed wind power of 6 MW, one may cover 88% up to 96% of the Ikaria island electricity consumption using two water reservoirs of 400,000 m^3 ($d_o = 1$, one typical day energy autonomy) each, up to 1,200,000 m^3 ($d_o = 3$). Keep in mind that a water reservoir of almost 1,000,000 m^3 at elevation of more than 700 m above sea level already exists on the island. Moreover, with less than 4 MW of wind power and minimum water storage dimensions the wind–hydro (hydro turbine of 6 MW) solution covers more than 80% of the total electricity consumption of the island.

The second example concerns the electricity consumption coverage of a remote consumer located in a Central Aegean island (e.g., Naxos or Andros islands belonging to Cyclades complex). According to Fig. 7.11, in this case the energy performance of a typical wind-based hybrid power station is investigated, including photovoltaic panels, battery energy storage and a small diesel–electric generator to be used only as backup option for increased reliability of this remote installation [26] (Fig. 7.11).

According to the results of Fig. 7.12 one may note the considerable battery capacity reduction as the installed wind power increases, although after a specific wind power value the battery capacity values follow an asymptotic trend. What is even more impressive is the considerable battery capacity reduction with the increase of the photovoltaic panels to be installed, leading also to quite lower wind turbine size in order to guarantee the energy autonomous status of the typical remote consumer leaving in the central Aegean Sea. At this point it is interesting to note that the results presented require zero oil contribution.

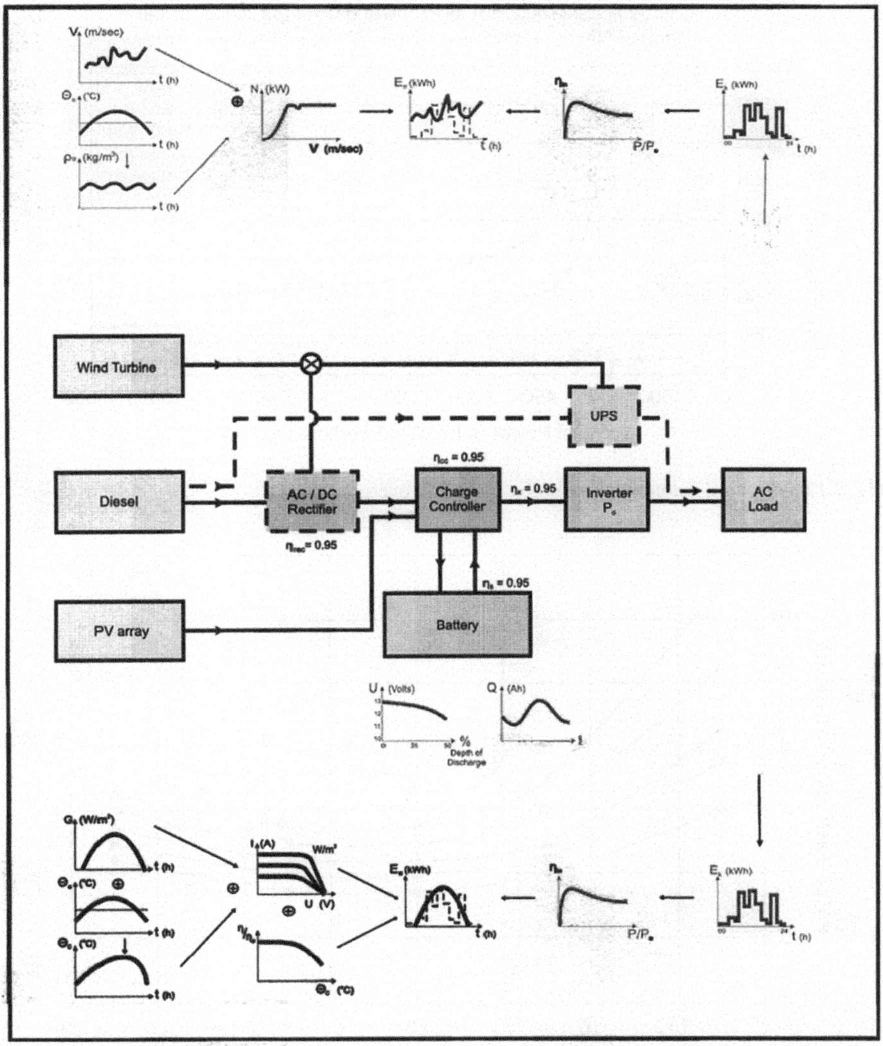

Fig. 7.11 Proposed autonomous wind-based hybrid energy system

One of the more important issues to be analyzed in order to support such wind-based hybrid solutions is their financial performance in comparison with the corresponding cost of the existing oil-based ones. For this purpose we shall be based on the previous described cases, while for an extended financial analysis one may refer to several cited journal papers [4, 5, 14, 15, 23]. To this end, in Figs. 7.13 and 7.14 one may find the minimum payback and the maximum financial efficiency of the wind–hydro hybrid energy solution for the island of Ikaria. Actually, although the results have been presented almost 20 years ago, they are still valid, at least on a comparative basis. Thus in Fig. 7.13 we can find that the minimum

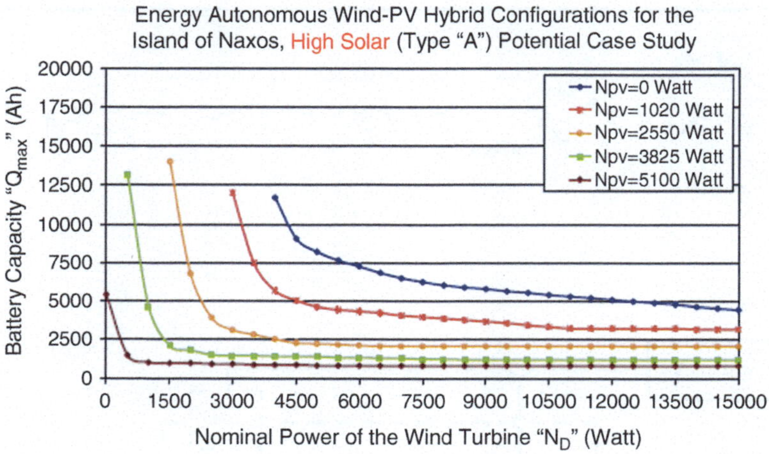

Fig. 7.12 Proposed autonomous wind-based hybrid energy system

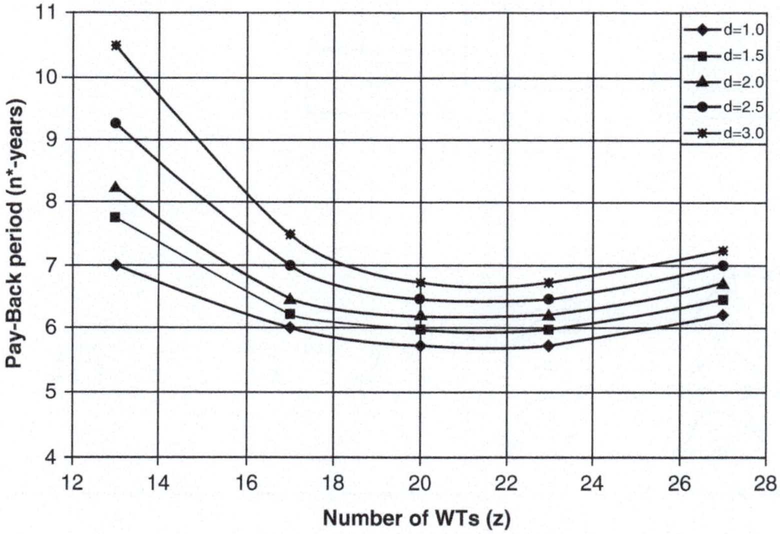

Fig. 7.13 Wind-based hybrid power system payback period for the Ikaria Island configuration

payback period is expected by installing 5–7 MW of wind power, while the energy storage dimensions affect fairly the corresponding payback value.

On the other hand, the optimum financial efficiency (i.e., the ratio of net present value with the initial capital invested) is noted for 6–7 MW of wind power and for medium sized energy storage water reservoirs. Of course there are several parameters affecting the financial performance of such installations, including the technology evolution-cost reduction, the oil and electrical energy price, the local market capital cost, etc.

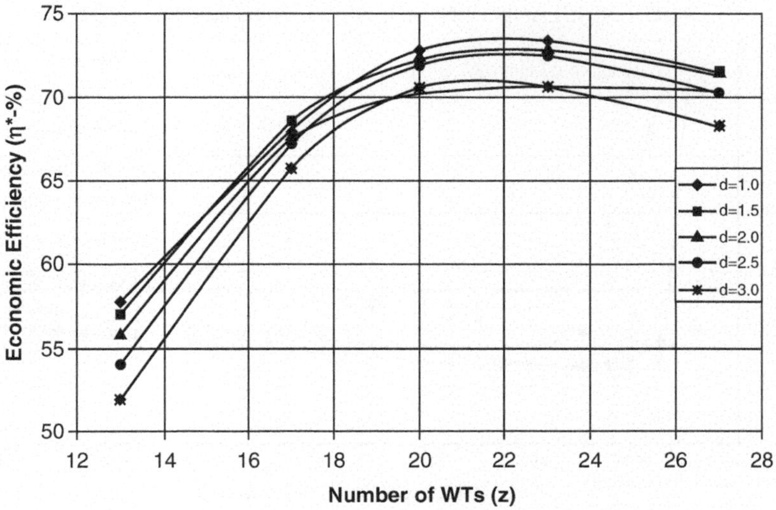

Fig. 7.14 The financial performance (PR or $\eta*$) for the wind-based hybrid power system of the Ikaria Island

The next case analyzed concerns a wind–diesel hybrid energy system used to cover the energy demand of a remote consumer located at a high wind potential area of Cyclades (e.g., the island of Andros). Applying the proposed solution for the Andros case, one may obtain the necessary wind power-battery size configurations that guarantee 1 year energy autonomy for various typical annual oil quantities (i.e., $M_f = 0$ kg/year up to $M_f = 1000$ kg/year). Bear in mind that approximately 2000 kg of oil are necessary in order for the diesel–electric generator to meet the electricity requirements of the consumer under investigation without any other additional energy source. Rationally, the dimensions of the hybrid system are remarkably reduced as the contribution of diesel oil is increased. In fact, this reduction is relatively greater when small quantities of diesel oil are used, while for larger oil quantities the battery bank size is only slightly decreased for a given wind turbine rated power.

In order to obtain a clear cut idea concerning the feasibility of a similar wind-based hybrid power system able to face the annual electricity consumption of a typical family, in Fig. 7.15 one has the opportunity to investigate the 20-year cost variation for selected representative cases of the proposed solution. More specifically, Fig. 7.15 presents:

1. The autonomous wind (only)-battery solution ($M_f = 0$ kg/year)
2. The diesel only solution ($M_f = M_{fmax} = 2000$ kg/year)
3. The 5% annual diesel-oil penetration ($M_f = 100$ kg/year)
4. The 12.5% annual diesel-oil penetration ($M_f = 250$ kg/year)
5. The 25% annual diesel-oil penetration ($M_f = 500$ kg/year)
6. The 50% annual diesel oil penetration ($M_f = 1000$ kg/year)

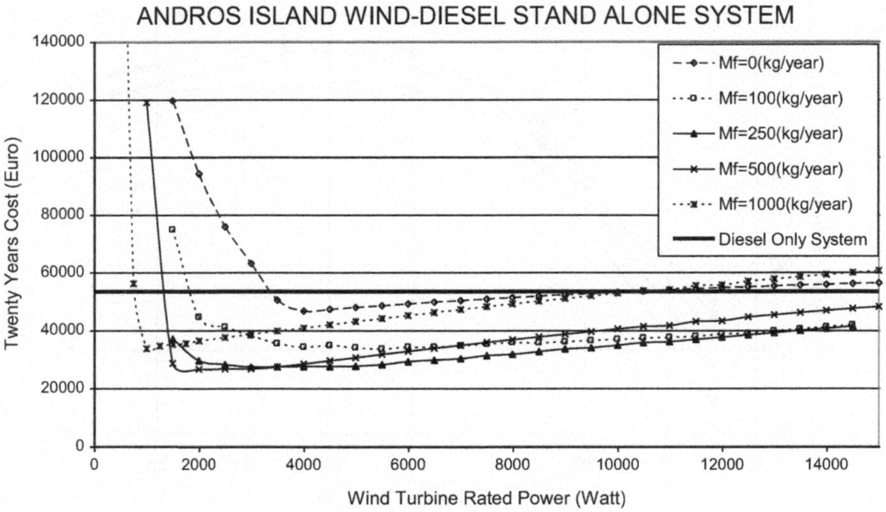

Fig. 7.15 Twenty-year cost analysis of a wind–diesel hybrid system

After a closer inspection of the calculation results and considering the numerical values used by the authors of this financial evaluation, we may state the following comments for the 20-year cost solution:

1. The optimum zero-oil solution should be based on a 2.5 kW wind turbine and 1700 Ah battery capacity, while the corresponding 20-year cost in present values is 26,300 €, less than 50% of the one corresponding to the diesel only solution.
2. In any case, even the autonomous wind power solution seems to be less expensive than the diesel-only system operation, although the external cost is excluded from the data presented.
3. By increasing the diesel oil contribution the 20-year cost is remarkably reduced, being quite lower than the diesel-only solution. This situation is inversed after a minimum cost point is achieved.
4. For each $M_f = ct$ configuration there is a minimum cost area, which leads to lower battery capacity and wind turbine rated power as the diesel-oil penetration increases.

Interesting conclusions may also derive by analyzing (breakdown) the 10-year minimum cost distribution; see Fig. 7.16. As it originates from Fig. 7.16, for low diesel-oil penetration, the main cost contribution is due to the high battery cost (including the variable M&O cost-battery replacement) and the fixed M&O cost. One cannot also disregard the wind turbine purchase cost contribution, which represents approximately 15% of the total system cost. On the other hand, for high diesel-oil penetration, diesel-oil purchase cost represents over 50% of the entire system cost. On top of this, for the optimum system configurations, the diesel-oil and the battery bank correspond to 40% and 35% of the total system cost in Andros island.

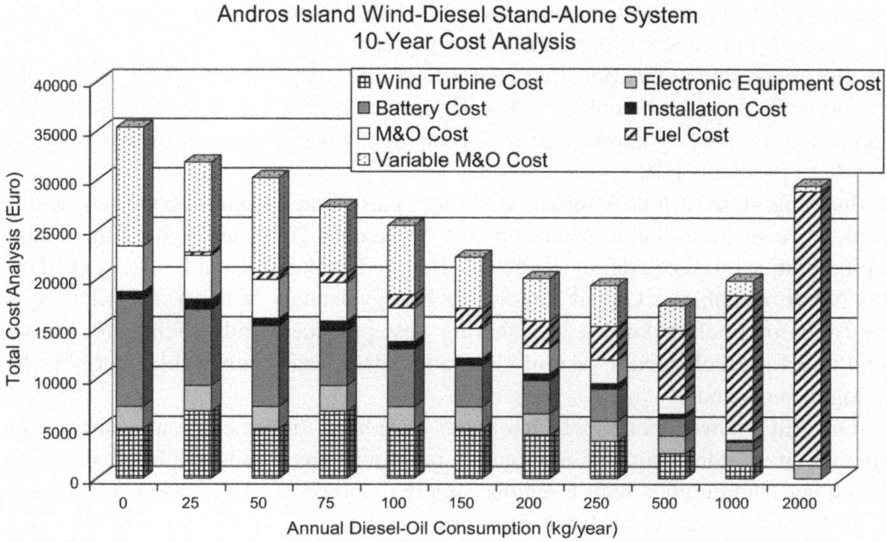

Fig. 7.16 Total 10-year cost analysis of a typical wind–diesel hybrid energy system, high wind potential case

Recapitulating, one may state that in most cases, with even fairly well wind potential, a wind-based hybrid energy system presents a competitive advantage in comparison with every diesel only electrification solution. Furthermore, the proposed solutions save over 75% of the fuel required by a diesel-only system annually in order to obtain full energy autonomy of the installation, while the corresponding 10- or 20-year total operational cost ranges between 60% and 50% of the diesel-only solution respectively. Finally, one should not disregard the significant environmental benefits of adopting clean-green solutions instead of imported oil-based ones.

7.5 Selected Applications of Wind-Based Hybrid Power Systems

Locally generated and distributed energy projects are now increasingly seen as a viable and promising alternative to the traditional model and seem capable of delivering benefits that range from increased security of supply for stakeholders to local economic benefits as well as reduced environmental impacts. The provision of electricity to remote, rural communities is a challenging issue. The marginal cost of grid extension is greatly dependent on the distance from the existing electrical network, the size of the community and the electricity consumption profile. Due to their isolated nature, energy supply in islands is more difficult and expensive [28, 29]. Finally, the overseas fuel transportation demand raises significantly the cost of energy supply. Costly fossil fuel imports from distant locations have serious impacts on every island's economic activity and prevent investments towards the direction of the local socioeconomic development. The exploitation of local RES (mainly wind and solar) potential can reduce these expensive imports and create important business and employment opportunities. Islands facing similar challenges can benefit by pooling knowledge and optimum practices for successfully overcoming these problems [30].

In recognition of this, Ministers and other participants from 48 countries held a conference in St. Julian's, Malta on 6–7 September 2012, and issued the Malta Communique on Accelerating Renewable Energy Uptake for Islands. They called on IRENA to establish a Global Renewable Energy Islands Network (GREIN) as a platform for pooling knowledge, sharing best practices, and seeking innovative solutions for accelerated uptake of clean and cost-effective renewable energy technologies on islands.

Off-grid renewable energy technologies have been implemented in islands in an attempt of readdressing the economic, social and environmental balance, while providing communities with a regular supply of power as well. Many developing countries have been electrifying rural areas and islands in this way for decades, acknowledging the advantages that such practices can have over grid extension (e.g., reduced transmission losses, lower capital requirements, lower operating and environmental costs, cheaper peak-time generation, and more employment opportunities for the local workforce). With respect to the integration of RES on islands, various technical aspects must be taken into consideration, such as the type of storage system that should be provided. The system may be the continental power network itself for cases where such an interconnection is feasible or the implementation of storage means as well.

7.5.1 El Hierro (Canary Islands, Spain)

El Hierro, is the western, southernmost and smallest of the Canary Islands (an autonomous community of Spain), in the Atlantic Ocean, off the coast of Africa, with a population of almost 11,000 inhabitants. Like the entire island complex, due to their size and remoteness, El Hierro had to deal with energy problems, in terms of total dependency on fossil fuels, high energy costs and often unreliable energy supply.

The annual energy demand of the island is about 42 GWh and it was met by a 13.3 MW diesel power plant located at Llanos Blancos, on the western side of the Island connected to a small distribution network of medium voltage. It was estimated that 6000 tonnes of diesel were consumed annually, thus over of 1.8 million Euros were spent per year to meet the energy needs and consequently, environmentally wise, greenhouse gas emissions of 18,700 tonnes of CO_2, 100 tonnes of SO_2, and 400 tonnes of NO_x were produced every year [31].

In 1997, El Hierro adopted a sustainable development plan to protect its environmental and cultural abundance making the island a self-sustained location. The

Fig. 7.17 The main components of the El-Hierro wind-based hybrid power station

"El Hierro Hydro-Wind Plant," approved in 2002 and Gorona del Viento El Hierro S.A., a public–private partnership that was established for the implementation of the hybrid wind-pumped hydro energy system. The hybrid system comprises a wind farm of 11.5 MW total installed power, a 6 MW pumping station, un upper reservoir of 380,000 m^3 capacity, hydro turbines of 11.3 MW, a lower reservoir of 150,000 m^3 and penstocks (Fig. 7.17).

By the summer of 2014, El Hierro's wind-pumped hydro power plant operated and achieved covering 25% of the electricity demand during peak hours in August. The goal set by Gorona for the first year was to make sure that 80% of the energy provided annually to the grid would be renewable (green energy).

7.5.2 *Ikaria Case (Aegean Archipelagos, Greece)*

A similar project is the one being developed in Ikaria island (North Aegean Prefecture) by PPC Renewables since the early 2000s [27]. Ikaria is a small-medium-sized island (population approximately 6500 habitants, area of 255 km^2) in the East Aegean Sea, located approximately 240 km from Athens. Its major town is Agios Kirikos with 2500 habitants, and the main economic activities of the local society are agriculture, fishing, merchant marine and tourism. The annual energy production of the local APS was 26 GWh for 2015. The peak-load demand, approximately 6.7 MW, occurs also during summer, while the corresponding minimum value is approximately 1 MW. The island has an excellent wind potential, since for several locations the annual mean wind speed approaches 10 m/s, at 50 m height. Besides,

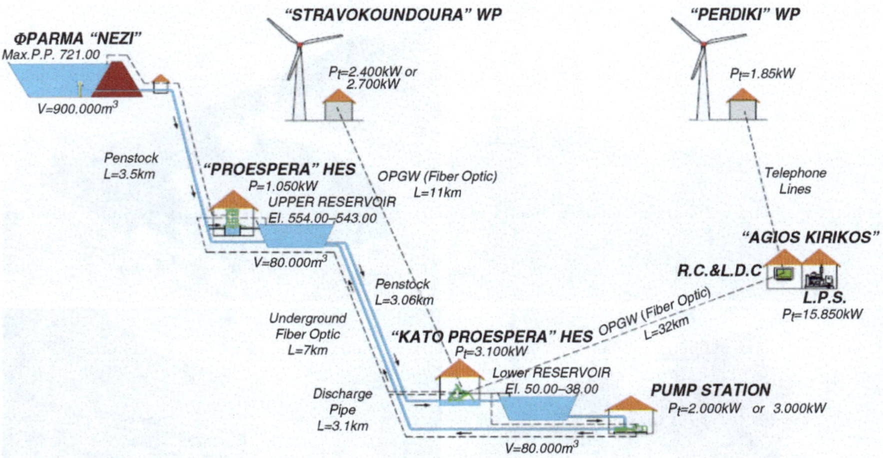

Fig. 7.18 Ikaria wind–hydro hybrid power general plan

there is a remarkable natural water reservoir at almost 700 m elevation, which can be used as the basis for the application of the proposed wind–hydro solution.

On the other side, the evolution of the local APS production cost presents an annual increase of 4%, while an important part of it (~50%) is due to the fuel cost. The APS of Ikaria consists of seven internal-combustion engines along with their electrical generators, and their specific fuel consumption is 270 g/kWh (+0.5 g/kWh lubricants).

On the basis of the information given, the corresponding wind-based hybrid power station is almost ready to operate. In this case, one combines the high wind potential (average wind speed of the order of 9 m/s) and the hydro potential of the island. Note that in Ikaria there is a water reservoir of almost 1,000,000 m^3 at an elevation of 715 m [32]. Thus PPC Renewables has installed (Fig. 7.18) a wind park of 2.7 MW and two Pelton hydro turbines (1.05 and 3.1 MW respectively) in order to exploit the available renewable energy potential. The new element of this installation is the installation of a 2 MW water pumping station at the low reservoir of the hybrid power station in order to use the wind energy surplus to store energy via water pumping. During the commercial operation of this hybrid power station, the contribution of the oil-based electricity in the island will be minimized in the range of 20–30%.

7.5.3 Isle of Eigg (Scotland)

Isle of Eigg is the second largest island of the Small Isles Archipelago in the Scottish Inner Hebrides. It is located 20 km off the Scottish west coast, south of the Isle of Skye, which makes it uneconomical to connect to the national grid. The island is 9 km long from north to south and 5 km long from east to west. It has almost

Fig. 7.19 The main components of the *Isle of Eigg (Scotland)* wind-based hybrid Power station

100 inhabitants with 38 households and five commercial properties. It is reached by ferry line from Mallaig.

Since the 1 February 2008, the island started covering its power demand through a unique system comprising of RES (wind + photovoltaic), backed up by diesel generators, (Fig. 7.19). The electricity is distributed around the island through an underground micro-grid system that supplies energy for 24 h a day. Before that the island did not have electricity supply and most residents used individual diesel generators, while few of them relied on a small hydroelectricity plant. Batteries/ inverters have been commonly used to ensure electricity access. The Isle of Eigg off-grid electrification system shows that an off-grid system can support the electrical energy needs of modern life [33]. The residents of the island are enjoying a reliable supply of electricity that meets their requirements effectively, but more importantly, their carbon footprint has reduced considerably since around 90% of their electricity comes from RES. In this context, it is reported that the CO_2 emission per household on the island is 20% lower than the rest of the UK.

7.5.4 Tilos Island (Greece)

One of the most well known hybrid power stations in Europe is the wind-based hybrid power station of Tilos island (Dodecanese complex, NW of Rhodes island) [34]. Actually, in Tilos island a medium-size wind turbine of 800 kW along with a small photovoltaic park of 160 kW_p have been installed in the frame of the TILOS Horizon 2020 research program (Fig. 7.20). In the same island, innovative $NaNiCl_2$ batteries have been installed with energy storage capacity of almost 3 MWh. According to the data provided by the corresponding research team [35] one may cover 75% of the total annual energy consumption of the island on the basis of renewable energy resources, while the vast majority of the electricity provided (almost 65%) comes from the wind turbine of the hybrid power station.

Recapitulating, it is clear that a carefully designed hybrid wind-based system can be an effective electrification option for any developed or developing country. The

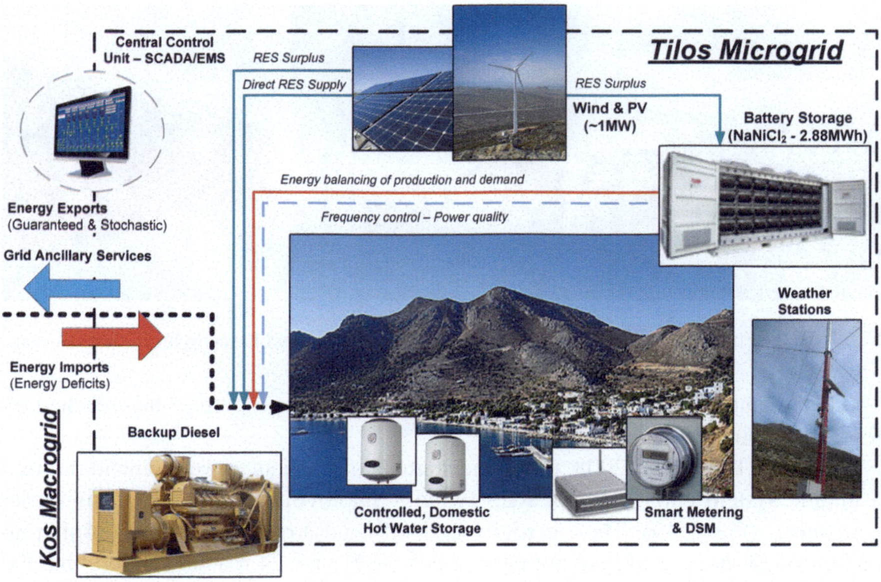

Fig. 7.20 Schematic presentation of TILOS Island wind-based hybrid power station

islanders have the opportunity to receive constant, 24 h electricity supply with acceptable reliability. This demonstrates that a RES-based supply does not need to be a temporary or a pre-electrification option. This is an important conclusion to be used by all countries in their decision making process. From the above described and from several other cases around the world, one may notice that wind-based electrification is not just a temporary solution for the developing world. Furthermore a properly designed hybrid energy system can be a viable alternative to grid extension in remote areas. With the main part of electricity generated from RES and back-up systems being used on an occasional basis, local grids can successfully reduce carbon emissions of electricity generation. Demand management plays also an important role and requires active participation of the users.

7.6 Conclusions and Proposals

The current chapter extensively investigates the prospects and the challenges of using wind-based hybrid energy solutions in order to increase the wind energy contribution in the electrical energy mix of remote areas. For this purpose, the electricity production characteristics of autonomous/remote energy consumers are examined with a special focus on remote island communities. Accordingly, the wind and the renewable (in general) potentials of all these areas are analyzed in order to underline the capability of the wind energy to cover the energy needs of local consumers.

Subsequently, the operational issues of high wind energy penetration/contribution in existing electrical grids (stochastic production, grid stability, backup power systems, etc.) are presented, taking into consideration various, long-term data from several international studies. In this context, the solution of developing integrated hybrid energy systems is discussed in detail, including also energy storage applications.

Next, the most representative hybrid power configurations are investigated, providing also some interesting results from the application of numerical models used for the operational simulation and optimum sizing of the existing wind-based hybrid power systems. Hybrid energy solutions are accordingly evaluated on a financial basis, mentioning also the major environmental benefits of their operation. The estimated electricity production cost is compared with historical data, as well as with the current electricity generation cost on the basis of existing thermal power units.

The current chapter concludes with selected examples of wind-based hybrid energy applications from all around the world, supporting the technical, environmental and economic advantages of clean energy solutions. According to the results obtained, the proposed wind-based solutions are able to satisfy the electricity needs of remote communities of various sizes with rational cost, high reliability, and minimum environmental impacts. Actually, it may be the only integrated solution that can significantly increase the renewable energy contribution in the fuel mix of our planet.

References

1. European Commission (EC), A Plan for Action in Europe. Wind Energy—the Facts, Brussels, Belgium (1999)
2. J.K. Kaldellis, A. Gkikaki, E. Kaldelli, M. Kapsali, Investigating the energy autonomy of very small non-interconnected islands. A case study: Agathonisi, Greece. Energy Sustain. Dev. **16** (4), 476–485 (2012)
3. T.L. Jensen, *Renewable Energy on Small Islands*, 2nd edn. (Forum for Energy & Development, Copenhagen, 2000)
4. J.K. Kaldellis, D. Zafirakis, K. Kavadias, Minimum cost solution of wind-photovoltaic based stand-alone power systems for remote consumers. Energy Policy **42**, 105–117 (2012)
5. M. Kapsali, J.S. Anagnostopoulos, J.K. Kaldellis, Wind powered pumped-hydro storage systems for remote islands: a complete sensitivity analysis based on economic perspectives. Appl. Energy **99**, 430–444 (2012)
6. J.K. Kaldellis, D. Zafirakis, Present situation and future prospects of electricity generation in Aegean Archipelago Islands. Energy Policy **35**(9), 4623–4639 (2007)
7. G. Tzanes, Ef. Zafeiraki, Chr. Papapostolou, D. Zafirakis,K. Moustris, K. Kavadias, K. Chalvatzis, J.K. Kaldellis, Assessing the status of electricity generation in the non-interconnected Islands of the Aegean Sea region. *Renewable Energy Integration with Mini/Microgrids, REM 2018*, Rhodes, Greece, 29–30 Sept 2018
8. Public Power Corporation, *Wind Speed Measurements for Greece, 1980–8* (PPC, Athens, 1985)
9. Hellenic Meteorological Agency, *Meteorological data of Greek meteorological stations for the period 1955–1998* (Department of Climatology, Athens, 2001)

10. Centre for Renewable Energy Sources (CRES-Greece), Wind potential map for Greece (2019), http://www.cres.gr/kape/
11. A. Papathanassiou St, N.G. Boulaxis, Power limitations and energy yield evaluation for wind farms operating in island systems. Renew. Energy **31**, 457–479 (2005)
12. J.K. Kaldellis, Evaluating the maximum wind energy penetration limit for weak electrical grids. *European Wind Energy Conference, Conference Proceedings*, Bella Centre, Copenhagen, 2001, pp. 1215–1218
13. J.K. Kaldellis, Maximum wind potential exploitation in autonomous electrical networks on the basis of stochastic analysis. J. Wind. Eng. Ind. Aerod. **96**(8–9), 1412–1424 (2008)
14. J.K. Kaldellis, K.A. Kavadias, Cost–benefit analysis of remote consumers' electrification on the basis of hybrid wind-diesel power stations. Energy Policy **35**, 1525–1538 (2007)
15. J.K. Kaldellis, E. Kondili, A. Filios, Sizing a hybrid wind-diesel stand-alone system on the basis of minimum long-term electricity production cost. Appl. Energy **83**, 1384–1403 (2006)
16. J.K. Kaldellis, P. Kostas, A. Filios, Minimization of the energy storage requirements of a stand-alone wind power installation by means of photovoltaic panels. Wind Energy **9**(4), 383–397 (2006)
17. J.K. Kaldellis, E.M. Kondili, A.G. Paliatsos, The contribution of renewable energy sources on reducing the air pollution of Greek electricity generation sector. Fresen. Environ. Bull. **17**(10a), 1584–1593 (2008)
18. J.K. Kaldellis, An integrated feasibility analysis of a stand-alone wind power system, including no-energy fulfillment cost. Wind Energy **6**(4), 355–364 (2003)
19. J.K. Kaldellis, The wind potential impact on the maximum wind energy penetration in autonomous electrical grids. Renew. Energy **33**(7), 1665–1677 (2008)
20. J.K. Kaldellis, D. Zafirakis, Optimum energy storage techniques for the improvement of renewable energy sources-based electricity generation economic efficiency. Energy **32**(12), 2295–2305 (2007)
21. J.K. Kaldellis, D. Zafirakis, K. Kavadias, Techno-economic comparison of energy storage systems for island autonomous electrical networks. Renew. Sustain. Energy Rev. **13**(2), 378–392 (2009)
22. J.K. Kaldellis, K. Kavadias, E. Christinakis, Evaluation of the wind-hydro energy solution for remote islands. Energy Convers. Manage. **42**(9), 1105–1120 (2001)
23. M. Kapsali, J.K. Kaldellis, J.S. Anagnostopoulos, Investigating the techno-economic perspectives of high wind energy production in remote vs interconnected island networks. Appl. Energy **173**, 238–254 (2016)
24. D. Zafirakis, J.K. Kaldellis, Autonomous dual-mode CAES systems for maximum wind energy contribution in remote island networks. Energy Convers. Manage. **51**(11), 2150–2161 (2010)
25. J.K. Kaldellis, K. Kavadias, D. Zafirakis, The role of hydrogen-based energy storage in the support of large-scale wind energy integration in island grids. Int. J. Sustain. Energy **34**(3–4), 188–201 (2013)
26. J.K. Kaldellis, G. Th. Vlachos, K.A. Kavadias, Optimum sizing basic principles of a combined photovoltaic-wind-diesel hybrid system for isolated consumers, *EuroSun 2002 International Conference, Paper W141*, Bologna, Italy, June 2002
27. P. Mastorakis, G. Betzios, J. Kaldellis, A proposal of installation of a combined wind–hydro station for the islands of Aegean Sea. *NTUA-RENES National Conference on the Application of Soft Energy Sources*, Athens, 1998
28. J.K. Kaldellis, D. Zafirakis, E. Kaldelli, E. Kondili, Combined photovoltaic and energy storage systems. An integrated electrification solution for small islands. Int. J. Environ. Technol. Manage. **10**(2), 123–149 (2009)
29. C. Rae, F. Bradley, Energy autonomy in sustainable communities - a review of key issues. Renew. Sustain. Energy Rev. **16**, 6497–6506 (2012)
30. N. Duic, G. Krajacic, M. Carvalho, RenewIslands methodology for sustainable energy and resource planning for islands. Renew. Sustain. Energy Rev. **12**, 1032–1062 (2008)

31. R. Godina, E.M.G. Rodrigues, J.C.O. Matias, J.P.S. Catalão, Sustainable energy system of El Hierro Island. *International Conference on Renewable Energies and Power Quality (ICREPQ'15) La Coruña*, Spain, 25–27 March, 2015, Renewable Energy and Power Quality Journal (RE&PQJ), ISSN 2172-038 X, No.13 (2015)
32. PPC Renewables, Hybrid systems (2019), https://www.ppcr.gr/en/hybrid-systems
33. Z. Chmiel, S.C. Bhattacharyya, Analysis of off-grid electricity system at Isle of Eigg (Scotland): lessons for developing countries. *Renew. Energy* **81**, 578–588 (2015)
34. Technology Innovation for the Local Scale Optimum Integration of Battery Energy Storage (TILOS), Horizon 2020 Program, No 646529 (2018), http://tiloshorizon.eu
35. J.K. Kaldellis, D. Zafirakis, P. Ktenidis, TILOS project—technology innovation for the local scale optimum integration of battery energy storage, *14th International Conference on Nanosciences & Nanotechnologies—NN17*, Thessaloniki, Greece, 4–7 July 2017

Chapter 8
Risk Analysis in Wind Energy: An Alternative Approach for Decision-Making

Isabel Ferraris and Mario D. de la Canal

8.1 Introduction

8.1.1 Risk Analysis

Modern formal approaches for risk analysis are useful and proven tools to ensure the design, construction and operation of engineering enterprises to which wind energy farms belong.

Nevertheless, a conceptual and practical review and improvement is claimed by the engineering community itself as the associated science and technology become more developed, databases grow in complexity and variables related to human behavior are recognized to influence and condition any technical situation.

Engineering processes are open universe problems; the whole can never be completely captured.

Three basic characteristics to determine risk engineering are:

- Uncertainties associated with all available information
- Limited knowledge of the whole and the details of all problematic approach
- Human factor necessary for all stages

Risk concept is always associated with the future, with possibilities, with events which have not happened yet. Risk is a natural consequence of uncertainty and inherent to all human activities. This is the reason why zero risk does not conceptually exist.

Although uncertainties may be classified using different criteria and a variety of types can be found in the bibliography, they are in essence random or epistemic.

In risk analysis it is widely spread the idea that risk and probability are synonymous perhaps due to the enormous success of formal and sophisticated techniques based on Probability Theory used for quantification. These approaches of the

I. Ferraris (✉) · M. D. de la Canal
Facultad de Ingeniería, Universidad Nacional del Comahue, Neuquén, Argentina

A. Sayigh, D. Milborrow (eds.), *The Age of Wind Energy*, Innovative Renewable Energy, https://doi.org/10.1007/978-3-030-26446-8_8

bivalent mathematics are grounded on the assumption that variables are random leaving aside the epistemic aspects.

On the other hand, risk analysis implies three aspects on which this chapter focuses:

- Risk analysis approach: it is widely accepted that a useful methodology to address risk analysis is based on Probability Theory grounded on bivalent mathematics. Methods have grown over time in accuracy and sophistication. Probabilistic and semi-probabilistic tools have been developed in a degree that risk and probability have become almost the same. However variables treated this way must be random. At the same time there is consensus among technical and scientific communities that not all the involved variables are random. Trying to represent random and not random variables an alternative approach to handle risk is presented.
- The modeling process: the assessment of the relevant variables and those which may be left aside in order to handle the problem. Modelization incorporates an uncertainty that can be thought of as a "conscious ignorance." The "unconscious ignorance" is due to the limitation of human knowledge which always is out of the analysis. Model border conditions, sometimes forgotten or misled, must be always assessed over time to keep the process on track. In this way algorithms might be validated.
- The human factor: all stages of the processes of risk analysis are embedded in and are conditioned by individual and internal and external organizational behavior. Political, social, economic contexts influence decisions and actions. Assumptions related to training, skill, performance and other human attributes levels are implicitly or explicitly made. Control of HF must be carried on in order to check if the initial suppositions are fulfilled. Reaching established accepted levels of human characteristics would permit validate outputs as well.

All this process is shown in Figure 8.1.

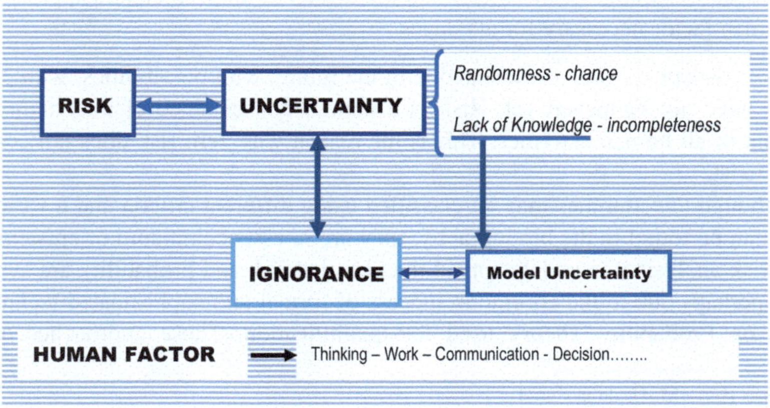

Fig. 8.1 Risk, HF, model uncertainty

ICOLD, in its Bulletin 130 [1] points out as a recommendation: "*human cognitive processes must be brought to bear on both the collection and verification the collection and verification of the data and in the construction of the behavioral models. Therefore, despite their numerical characteristics, the results of the analysis are not absolute but reflective of a mix of physical data, cognitive analysis and the judgment of the engineers involved. Being human judgments, it is impossible for the engineers to separate their analytical judgments from the personal values that govern their practice. Thus, in reality, it is impossible to obtain the "complete data set", to have accurate and precise comprehensive behavioral models, and to eliminate the human element from the analysis process.*"

As an example extensively used in system reliability studies, fault tree analysis offers the ability to focus on an event of importance, such as a highly critical safety issue, and intervene to minimize its occurrence or consequence. Fault tree analysis is performed using a top-down approach. It begins by determining a top-level event, and then work down to evaluate all the contributing events that may ultimately lead to the occurrence of the top-level event.

The probability of the top-level event can then be determined by using mathematical techniques. However the algorithm used to link different values of P_f (probability of failure) does not reflect the important and inevitable interactions between variables.

However the resulting fault tree diagram is a graphical representation of the chain of events in the system or process, built using those events and logical gate configurations.

The probabilistic point of view is normative. It does not allow handling different types of uncertainties as that of modeling and due to human behavior presented above.

In another aspect, although risk is a collective conception at the same time has different interpretations. The approach adopted here takes as its starting point the definition given by ICOLD [1].

Risk assessment is the essential anticipatory element that underpins the safety management process. The result of risk analysis is a transparent mathematical construct of the uncertainty in the future performance of a dam, the most common form of this statement of uncertainty been in terms of probability. In risk assessment the results of the risk analysis and risk evaluation processes are integrated and recommendations are made concerning the need to reduce risk. Figure 8.2 shows this.

This conception can be fully extrapolated to wind energy enterprises.

Anyway risk analysis can be addressed using an alternative approach using formal soft tools capable to represent different kind of uncertainties [2].

On the other hand, risk assessment is the basis for management control. It is the process in which rests decision-making.

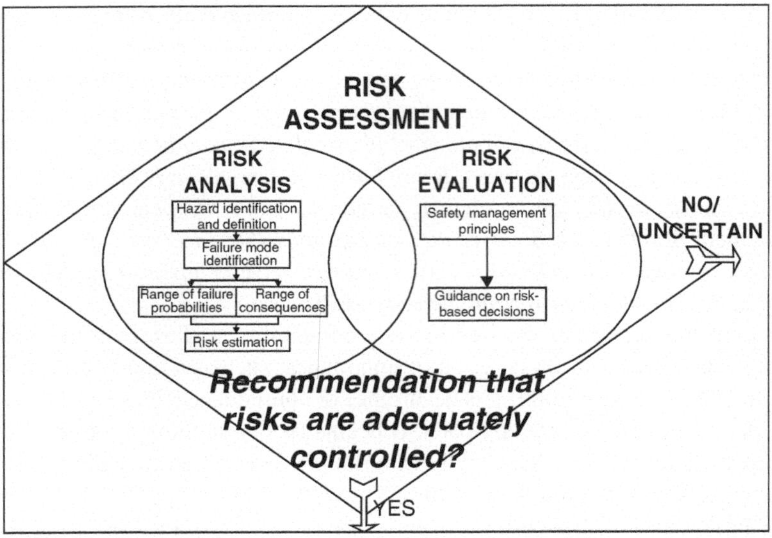

Fig. 8.2 Risk assessment process [2]

8.2 Complexity, Ignorance, Uncertainty

8.2.1 Complexity

Size, multidisciplinarity, and novelty are common characteristics of engineering problems in general and of renewable energy projects in particular. These three categories describe different aspects of the complexity this activity addresses. Engineering artifacts are prototypes, it is not always appropriate to extrapolate similar designs, especially in situations where the context influences significantly.

The novelty of a work poses a challenge to engineering.

Complexity is one of the crucial issues to address when determine viability. In the following scheme three sources has been chosen to summarize and illustrate the magnitude of the problem (Table 8.1).

The study and design of a wind facility need information from different sources and with varying degrees of precision (multidiciplinary). The starting point will be to assume that the information is limited (ignorance), incomplete and uncertain. Uncertainty, ignorance, and complexity, the main characteristics of this problem, are briefly described in the following section.

If it is the case, the complexity may be assessed by the definition, design and calculation of indicators that evaluate them for a particular wind generation project. The global complexity can be represented using a hierarchical holistic structure. Each holon in the hierarchy can be presented in more detail by an expert, seeking to explain the logic of operation. An aggregation of this information with its associated uncertainty, allows quantifying complexity [3].

Table 8.1 Complexity

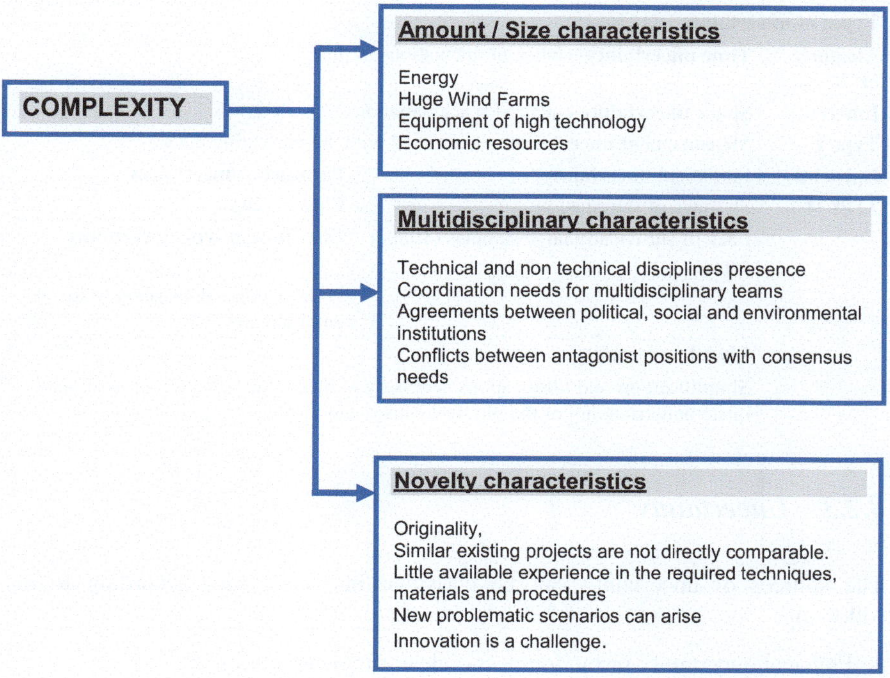

8.2.2 Ignorance

Innovative projects incorporate the concept of ignorance [4]. Wind Energy projects are examples of that. The universe of discourse in science and engineering is restricted and what is known or have been tested is the plausibility or the degree of truth of phenomena. The so-called model uncertainty derives from the assumption of a closed universe containing all the relevant factors that are known or thought that influence on the problem under study. Not only can the so-called simplifying hypothesis remain outside the model but also the unknown facts. The latter may be significant in large-scale projects. The more we are aware of ignorance, the more we have to be cautious about its presence when a problem is addressed.

Incorporating all available information as well as its associated uncertainty leads to more rational solutions in accordance to reality. The mathematical tool that allows addressing this information and can also include the unknown aspect should go beyond the traditional bivalent approaches. Fuzzy perspectives, less accurate but with richer capability of description, better fit this requirement.

Table 8.2 Different types of uncertainties

Types of uncertainty		
Aleatoric or	**Time uncertainty**—inherent fluctuations in time	
Inherent	**Space uncertainty**—natural space variations	
Type I	**Measurement uncertainty**—associated with the measuring device	
Epistemic Type II	**Statistical uncertainty**	**Parameter uncertainty**
	Shortage of information	Limited data
	Lack of sufficiently large samples of input data	**Distribution type uncertainty**
		How theoretical distribution fits to empirical data
	Model uncertainty	
	Simplifications and idealizations necessary to model the behavior, or to an inadequate understanding of the physical causes and effects.	

8.2.3 Uncertainty

The sources of uncertainty can be classified into categories according to the following:

1. Physical uncertainty or randomness—aleatoric uncertainties
2. Understanding or knowledge—epistemic uncertainties

Aleatoric uncertainty concerns to an event (regarding its occurrence), that has random results in an experiment repeated almost in an unlimited number of occasions under constant boundary conditions. This may be described and addressed using the methods of probability analysis. On the other hand, if boundary conditions are apparently subject to arbitrary fluctuations or a comprehensive system overview is lacking or the number of observations is only available to a limited extent an information deficit exits. This type of uncertainty is referred to as informal uncertainty. In contrast to the latter, epistemic uncertainty is characterized by linguistic variables representing quantified verbal postulations. Table 8.2 resumes their main characteristics.

The quality of the information must be statistically secured with a large enough sample values. Inaccuracies, unreliable data, or uncertainty that cannot be described or insufficiently described statistically can thus only be accounted for approximately. Bayes theorem serves as a suitable probabilistic method for processing subjective information. This approach makes use of prior distributions based on subjective probabilities and generally contains objective and subjective information. There is no way of separating objectivity and subjectivity in these results. So probabilistic methods may only be applied to a limited extent.

The theory of fuzzy random variables permits modelling uncertain parameters (which partly exhibit randomness) but which cannot be described using random variables without an element of doubt. The randomness is “disturbed” by a fuzziness component. The reasons for the existence of fuzzy randomness might be [5].

- Although samples are available for a structural parameter, these are only limited in number. No further information exists concerning the statistical properties of the universe.
- The statistical data material possesses informal uncertainty, that is, the sample elements are of doubtful accuracy or were obtained under unknown or non-constant reproduction conditions.
- The available sample elements were generated under reproduction conditions which were non-constant but nevertheless known in detail.

Other interesting uncertainty is introduced by Elms [6]. He identifies a surprisal uncertainty, which covers matters that are unexpected, those things that neither random variability nor limitations of model quality will cover. There are many sources of surprise. Virtually all arise from human factors (HF) [7] from errors, slips and lapses, but this is not the only cause of failure. More details can be seen in Sect. 8.3.

8.2.3.1 Model Uncertainty

The first step for thinking about reality it is to define a model. That means to take a simplified view of reality, leaving outside an amount of factors that for the designer does not take an important role in the performance study.

Theoretical models used in the design process, may represent the reality in a higher or lower degree, but it will never be the same.

In general, code takes into account this uncertainty using a partial safety factor in the algorithm of calculation. This factor comes from probabilistic analysis.

The difference between models and reality is an epistemic uncertainty, which means essentially lack of knowledge.

Model uncertainty should assess the appropriate and inappropriate model aspects, so a bivalent approach is not correct.

Sometimes, a measurement shows that a model cannot be used to represent a reality (we are outside the domain of validity of the model or that it can hardly be built) (Table 8.3).

Table 8.3 Model uncertainty

Model	**Physical**	Material and structure
	Human	Human factor and organization that embodies what the model represents
	Surroundings	Environmental characteristics where the structure is located

8.3 Human Factor

Human Factor, HF, is an organized and high complex system that intervenes in the process of design, execution and maintenance of an engineering device, where context factors impact strongly.

HF can be thought of as the system where engineering is embedded and processes are performed and developed. When HF is addressed, problems become more complex and therefore analysts have to incorporate high complex operations to account for them and at the same time new uncertainties are introduced.

There is consensus within engineering community that HF plays a major role in failure mechanisms; all these are based on statistics and research extracted from [7].

- Human rather than technical failures now represent the greatest threat to complex and potentially hazardous systems.
- Managing the human risks will never be 100% effective. Human fallibility can be moderated, but it cannot be eliminated.
- Different error types have different underlying mechanisms, occur in different parts of the organization, and require different methods of risk management. The basic distinctions are between:
  - Slips, lapses, trips, and fumbles (execution failures) and mistakes (planning or problem solving failures). Mistakes are divided into rule based mistakes and knowledge based mistakes
  - Errors (information-handling problems) and violations (motivational problems)
  - Active versus latent failures. Active failures are committed by those in direct contact with the patient, latent failures arise in organizational and managerial spheres and their adverse effects may take a long time to become evident.
- HF problems are a product of a chain of causes in which the individual psychological factors (i.e., inattention, forgetting, etc.) are the last and least manageable links.
- People do not act in isolation. Their behavior is shaped by circumstances. The same is true for errors and violations. The likelihood of an unsafe act being committed is heavily influenced by the nature of the task and by the local workplace conditions.
- Safety significant errors occur at all levels of the system, not just at the sharp end. Decisions made in the upper echelons of the organization create the conditions in the workplace that subsequently promote individual errors and violations. Latent failures are present long before an accident and are hence prime candidates for principled risk management.

These statements are set up as starting points in an attempt to address HF in risk analysis.

As was commented, "zero human errors" is conceptually not possible. HF is complex and unpredictable with not defined behavioral patterns. People usually work in isolation but belong to a team and the team belongs to an organization. A context is around all these, like a kind of nested array. Human actions are difficult to estimate. So, what sort of formal approach can be chosen to model HF, as those based on bivalent, crisp mathematics are definitely not suitable?

Hard engineering problems have solutions. However in problems involving HF the concept of solution must be replaced by control.

The process of HF evaluation may be capable of the following:

- Detect the natural variability of human behavior. Information is required during the process
- Represent the interactions between the principal variables. Those, based on evidence that better describe HF should be chosen.
- Account for the lower limit of acceptability of human behavior in a specific activity.
- Link the obtained values with an action plan (decision-making process) which in case could modify unaccepted levels of HF

Given HF characteristics, measuring in this case means to look for proper parameters which represent and describe relevant aspects and performance against failure scenarios. They have to be measured with some precision as well. The measurement has to be partial in order to be able to modify the current state if necessary. This leads to intervene on the organization. In this way the reaction capacity of the system to the actions implemented would keep it on track and can be monitored and control by a subsequent measurement. The intention is to maintain the chosen parameters around certain values previously agreed as acceptable. Figure 8.3 sketches this process which was outlined in [8, 9].

Monitor and control is a continuous task as can be seen. However and going further running this course of action, limit values must not be reached up as they guarantee a good performance. These maximum values may be established through a

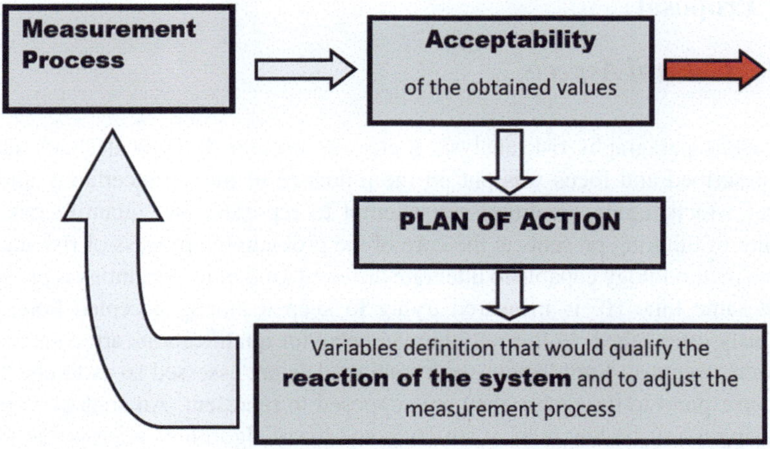

Fig. 8.3 Process of HF evaluation and control

process of acceptability. In case those numbers were reached, unacceptable functioning is assumed and the process is interrupted. If that is the case, initial or border conditions of the problem have to be revised and eventually changed. The red arrow in Fig. 8.3 represents this situation. It can be interpreted that the system must be reconfigured. And when that is already done, a practical dictatum to follow might be "what looks acceptable today may not look so tomorrow" so monitoring should go on.

Human errors occur during the progress of a system or project due to blunders or mistakes by an individual or individuals. In general, human errors are difficult to estimate. However, measures such as education, a good work environment, a reduction in task complexity, and improved personnel selection as well as control measures such as self-checking, external checking, inspections, and legal sanctions have proven to be successful in reducing human errors.

If the human errors are inevitable and ever-present, defenses will have to be generated to manage and acquire competence to face them. This is a part of the target of this paper.

It is desirable to develop abilities, within the working groups, to deal with adverse scenarios as well as interpersonal problems. This not only works in decreasing human errors but also, facing the occurrence of them, diminishing the consequences and increasing the capacity of the human group recovery. This information may be used in future and similar situations. A data base in this way could be created, useful to develop the capacity of reaction of individuals and groups in case they have to face confusing, adverse or unexpected situations. This concept is what in neuroscience is called resilience.

Permanent control HF measures would warrant that risk analysis process is performed as was conceived and planned.

8.4 Proposal

8.4.1 General Aspects

In previous paragraphs risk analysis topic was presented. Typical characteristics were described and focus was put on the influence of model uncertainty and HF, features which highly condition any attempt to represent and quantify risk. The structure of thinking presents at the core of the problem the process of risk analysis where a methodology capable to integrate different kind of uncertainties is proposed. At the same time HF is measured trying to keep it among accepted boundaries previously established. In this way human behavior qualifications are controlled as they were assumed. Simultaneously formal models are assessed so as to check that they correspond to the reality they are supposed to represent. An analogous procedure using a holistic hierarchical structure and fuzzy algorithms is proposed for the three tasks as Fig. 8.5 sketches.

8.4.2 Hierarchical Structure: Holistic Approach

Based on an observational point of view Koestler [10] argues that living organisms as well as social structures are organized in hierarchical schemes. These arrangements can be thought of as ascendant series of levels of increasing complexity. The components show as "parts" or "wholes" within the hierarchy in a way that depends on how they are thought of. A "part" as commonly the word is used, means something fragmentary, which cannot exist for itself. Conversely a "whole" is something complete. Koestler points out that they do not exist in such absolute way. Based on this he proposes the name of "holon" to that component, from the Greek *holos* (whole) plus *on* at the end suggesting part or particle as in electr*on* or prot*on*. In that way, the inherent characteristic of a holistic structure is that it is much more than an aggregation of parts. Holons of upper levels have high conceptual content together with low precision while going down to lower levels less concept elements with and more precision can be found which conform a net of strong inter and intra level relationships.

Under this perspective the whole is much more than an aggregation of parts. The idea of union or convolution is present here. On the other hand, holons can be hard or soft.

The holistic approach proposed provides a structured framework to think about the whole problem and the same time to pay attention to the details. It can be thought of as "the law of the process" that shows the network of components with their relationships.

Based on it two factors influence on risk assessment: the possibility of the occurrence of an unwanted event, the threat, and the characteristics and proneness to be affected of the exposed community and environment. It is usual to associate the risk only to the threat. However when the context is analyzed, it is clear that risk may be different for diverse surroundings although the threat is the same. A simple example is the presence or not of people nearby.

Risk implies the study and determination of the threat (holon **Threat**) and the Vulnerability of the Community and Environment (holon **Consequences**) in an integrated form. Risk (holon **Risk**) results from the convolution of both. Figure 8.4 shows a hierarchical structure extracted from [2].

This hierarchical like-tree array permits to establish a common formal language among the different disciplines as they have diverse ways of thinking, priorities and points of view. The multidisciplinary aspect of the problem can in this way be addressed through an analogous procedure. A qualified group of experts carry out this task. Each expert has a common working protocol to express and communicate lines of reasoning, from lower and precise levels of definition with objective data to higher levels with more conceptual and subjective contents.

One of the challenges in the formal modeling is the mixture of information. Fuzzy numbers are used to characterize numerical and linguistic variables with their uncertainties. The first are converted through a fuzzification operation and the second using labels to translate words into numbers. The hierarchical structures

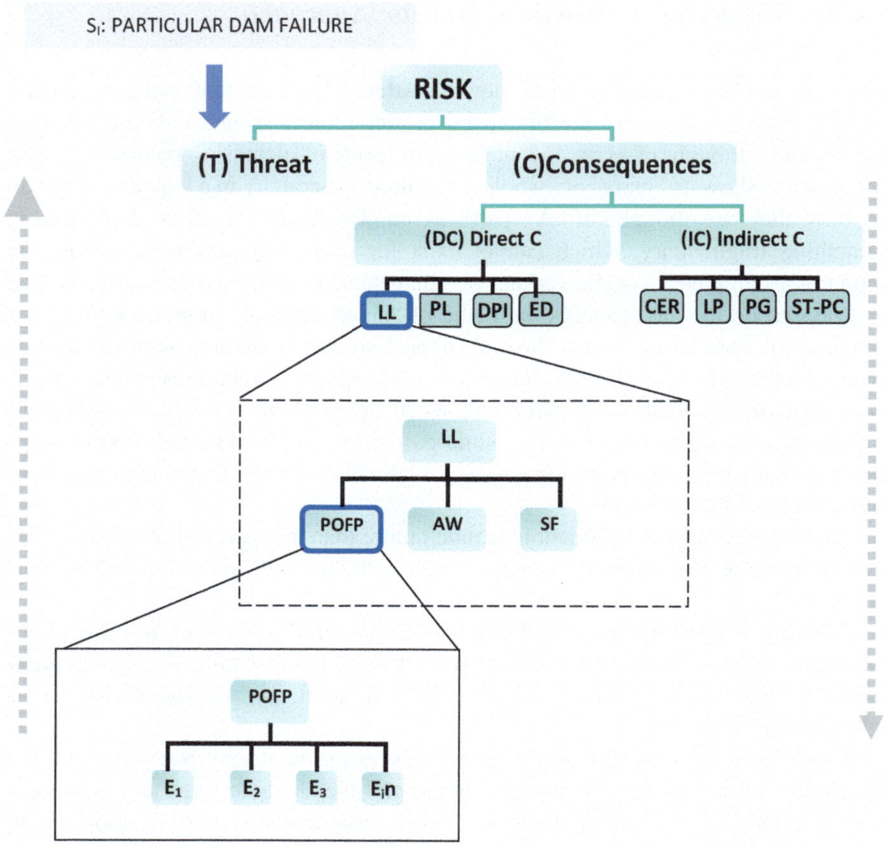

Fig. 8.4 Hierarchical arrangement: holistic approach

are loaded al the lowest levels and using basic fuzzy arithmetic operations three indexes for Risk Analysis, Human Factor and Model Uncertainty, performed in parallel, are obtained. The stringency and constrains are introduced through a filter operation using fuzzy arithmetic as well. Then the degree of acceptability (ga) is calculated [11]. The output is in fact the structure in itself, a map that shows synthesis an analysis at the same time complexity and uncertainty can be captured. The whole picture and the details can be recognized and vulnerable or weak points identified to control, intervene, remedy or reformulate if necessary.

It is important to comment that outputs are obtained at a certain time *t*. Circumstances, conditionings, overall or detailed factors may change over time. The process can then be managed, and if it is the case, once corrective actions have been taken, the procedure can be repeated in a kind of a dynamic feedback to improve performance [3].

Using this structure of thinking and representation, Risk analysis, HF and Model Uncertainty can be managed. Synthesis of the steps of the risk process evaluation represented in Fig. 8.5:

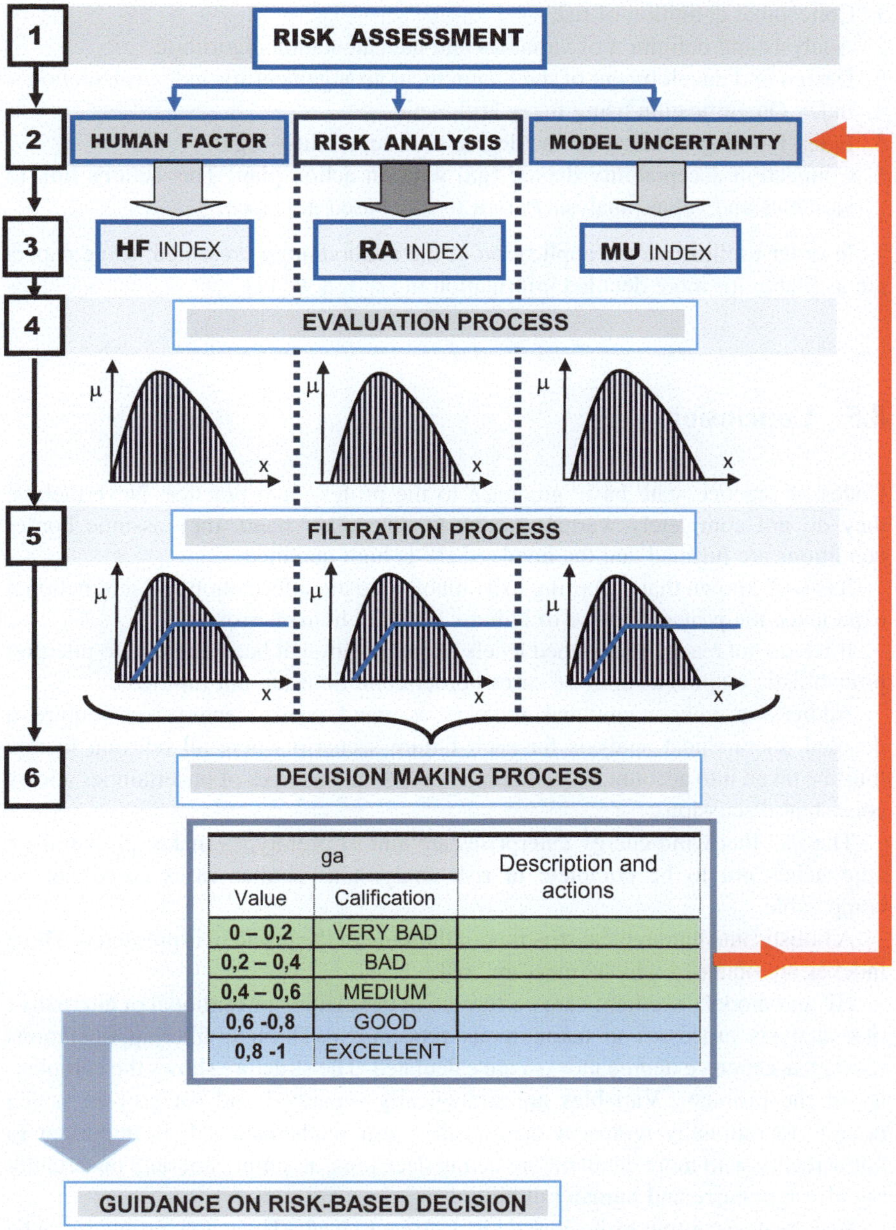

Fig. 8.5 Holistic hierarchical procedure

1. Conceptual definition of risk
2. Analysis and definition of variables and the calculation algorithm.
3. Design and development of the hierarchical arrangement for index calculation
4. Index Quantification using fuzzy arithmetic.
5. Index Filtering to calculate the degree of acceptability—(**ga**).
6. Connection acceptability degree (ga) with an action plan. This actions link to variables and values analysis **two** or to risk-based guidance.

In order to illustrate the application of the methodology presented, some papers are available for more detailed information in [2, 3, 8, 9, 11].

8.5 Conclusions

Codes of practice state basic guidance to the professional practice. Nevertheless, they do not completely warrant safety. On the other hand, they assume border conditions are fulfilled and the involved HF is high qualified.

It is well known that according to available statistic information and international experience the principal cause of failures is due to human errors.

If HF do not reach the assumed levels of qualification, it becomes hard to interpret probabilistic outputs as in these cases border conditions are not fulfilled

Addressing nonconventional projects as wind energy enterprises require a detailed, hierarchical analysis for each failure scenario where all relevant factors must be taken into account. Considering all types and sources of uncertainties allows consistent discussions.

The fact that wind energy enterprises are almost prototypes makes probabilistic approaches not to be complete in risk analysis as similar cases could not be comparable.

A holistic and hierarchical structure of thinking of the problem is presented. Three indexes are obtained which "measure" risk.

HF and model uncertainty are performed in parallel to conventional or alternative risk analysis processes in autonomous procedures. Through filtering operations (fuzzy), acceptance degree indexes are calculated. The structure shows the complexity of the problem. Variables are strategically separated and categorized which permits to rationally represent them with a soft mathematics. It is important to know reality with more detail before taking decisions, assuming not only uncertainty but also ignorance and human fallibility.

Risk analysis output is the hierarchical structure itself. The whole problem can be seen and the details as well. Good and bad performance points can globally be detected and actions can be taken based on objective matters. The final target is to contribute to the decision-making process.

Acknowledgements The authors of this chapter, directed by engineer Arturo Bignoli, started a research group at the Engineering Faculty of National University of Comahue. The working topic under study was at the beginning structural safety in general. Then it was applied to dam safety and at present renewable energy projects are addressed. In all the cases the core of the activities was the analysis and treatment of the uncertainties in general and those associated to soft variables in particular. Alternative logics (fuzzy logic, interval probability) have been used to represent and incorporate them in the proposed algorithms and procedures. The authors appreciate engineer Arturo Bignoli's intellectual and human generosity during the years.

References

1. ICOLD Bulletin 130, Risk assessment in Dam Safety Management—a reconnaissance of benefits, methods and current applications, 2005
2. I. Ferraris, M.D. de la Canal, Risk analysis in engineering facilities—a holistic approach. 1 ISUQSM, Maresias, Brazil (2012)
3. I. Ferraris, M.D. de la Canal, C. Labriola, L. Bertani, Uncertainty, ignorance and complexity: viability of a large-scale engineering project. ICRVAM ISUMA uncertainties 2018, Brasil (2018)
4. B. Ayyub, *Elicitation of Expert Opinions for Uncertainty and Risk* (CRC Press, Boca Raton, 2001)
5. B. Möller, M. Beer, *Fuzzy Randomness—Uncertainty in Civil Engineering and Computational Mechanics* (Springer, Berlin, 2004)
6. D. Elms, Balancing uncertainty in structural decisions, in *International Forum on Engineering Decision Making Second IFED Forum*, Canada, 2006
7. J. Reason, *Human Error* (Cambridge University Press, New York, 1990)
8. I. Ferraris, M.D. de la Canal, F. Pellón, *El Factor Humano en elProceso de Evaluación de la Seguridad Estructural* (AIE, Buenos Aires, 2014)
9. M.D. de la Canal, I. Ferraris, Human factor evaluation of a validation of risk assessment, in *ICVRAM 2014*, Liverpool UK, 2014
10. A. Koestler, *The Ghost in the Machine* (Arkana, London, 1967)
11. M.D. de la Canal, I. Ferraris, C. Labriola, L. Bertani, Bajo del Gualicho Argentina-Evaluación de la Aceptabilidad del Proyecto. CLADI, Argentina, 2017

Chapter 9
Wind Energy in Argentina: Actuality and Prospects

Carlos Labriola

9.1 Introduction: Argentina, A Country with a Variety of Climates and Types of Orography

9.1.1 Geographic Situation

Argentina is in southern hemisphere with Chile, Bolivia, Paraguay, Brazil, and Uruguay as bordering countries. He owns part of the Antarctic continent, being the closest country to it. Figure 9.1 [1] shows the most relevant regions:

1. *Andes*: Where the Andes Mountain Range is from North to South; it has several mineral holdings.
2. *Mesopotamia*: surrounded by the rivers Parana (Big Hidro: Yacyretá, 3200 MW), Uruguay (Big Hidro: Salto Grande, 1890 MW) and Iguazú.
3. *Pampas*: plains and wetlands for grazing and cereal crops (wind farms of Cooperatives).
4. *Comahue*: Northwest zone of Patagonia, (West of Neuquén and Río Negro) same orography with renewable sources (wind, sun, 4000 MW of Big Hydro) and non-renewable (Vaca Muerta Non Conventional hydrocarbons)
5. *Patagonia*: include Comahue Region; steppe and plateau zone rich in hydrocarbons and wind energy (Neuquén, Rio Negro, Chubut and Santa Cruz: 5000 MW wind power by 2025)

C. Labriola (✉)
Faculty of Engineering, National University of Comahue, Neuquén, Argentina

A. Sayigh, D. Milborrow (eds.), *The Age of Wind Energy*, Innovative Renewable Energy, https://doi.org/10.1007/978-3-030-26446-8_9

Fig. 9.1 Regions of Argentina

9.1.2 Demographic Situation

The highest demographic concentration [2] is in the Pampas (60% of the population, 50% of electricity demand), while the rest of the inhabitants are in the other regions: 30% in the Northwest (Jujuy, Salta, Rioja, and Catamarca), Cuyo (San Juan San Luis and Mendoza), North Corrientes and Misiones), and only 10% in Patagonia where there are usually 5–10 inhabitants per km^2.

9.1.3 First Applications of Renewable Energy Sources (RES) in Argentina [3]

During 1977–1978, the development of Renewable Energy technology begun in Argentina with PV cells. San Miguel Institute, dependent of Salvador University. This institution grouped experts to develop photovoltaic cells and training in Solar Energy. Then part of this Human Resources (HHRR) was absorbed by the Space Research Institute dependent on the Argentine Air Force.

In 1984, during the democratic government of Dr. Raúl Alfonsin, RES Regional Centers were created to give impetus to RE development in all the country. These centers were located in the most representative provinces of the Renewable Resources that have (See Sect. 9.2).

Individually, the provinces began to study RES applications in isolated rural areas:

- Neuquén started to make geothermal prospecting in 1985 in Caviahue-Copahue area, near "Las Mellizas" lagoons (1990: first South American Geothermal Power Plant of 670 kW). Also, since 1988, it implemented a Rural electrification plan for dispersed inhabitants with PV sets.
- Since the 1990s, Chubut has developed its dispersed rural electrification plan by installing small wind turbines and photovoltaic panels.
- During 1990s European companies installed with approval of National Centre of Wind Energy of Rawson, wind farms of tens of kW in Chubut (Rio Mayo) and Santa Cruz (Pico Truncado).
- Since 1995, a national electrification plan for dispersed rural population (4,000,000 inhabitants) has been started. With this plan, private companies were created, particularly in the North-West of Argentina, which installed and maintained PV panels and some wind systems, with a subsidized rate of 50%, but after the 2001 crisis the companies began to be non profitable.
- Rural Electrification [4] throughout the country was continued in 2004 through the PERMER (Renewable Energy Project for Rural Markets), financed by the World Bank, joining the provinces including those with their own plans.

Table 9.1 Regional centers of RES development

Regional center	Place	Characteristics
Wind Energy Regional Center (WERC)	Rawson, Chubut http://organismos.chubut.gov.ar/cree	This institution developed the second Wind Map of Argentina. It is also an actual international consultant for study of wind resources and location of wind farms around the world
Geothermal Energy Regional Center (GERC)	Neuquén, 1985–1991. Today is represented by the Investment Development Agency	It implemented first geothermal venture of South America, "Las Mellizas" power plant in Caviahue-Copahue of 670 kW, managed by provincial energy Agency of Neuquén (EPEN). Since1997, it is out of service, due to lack of maintenance
Regional Center of Micro-Hydro (MHRC)	Oberá, Misiones	Today National University of Obera is working like this center, where they continued developing Banki and Pelton mini and micro-turbines
Solar Energy Regional Centre (CRES)	Salta-INENCO	This center became the Institute for Non-Conventional Energy of the National University of Salta. Fundamentally, it develops solar–thermal equipment

The local industry of biofuels is created since 1984–1991. It begun by means of alco-naphtha for cars using bio-ethanol obtained from sugarcane. During 1990s the production of alco-naphtha was suspended because it was not profitable.

In summary: Argentina has wind farms (since 1990), PV (since 1985), mini and micro-Hydro (since 1960), macro Hydro (since 1930), Biomass applications (alco-naphtha, since 1984), Geothermal Prototype power station (1990–1997 "Las Mellizas" 670 kW).

There are other RES which are in study particularly in Academic Institutions, such as:

- In the case of Tidal energy, during 1924–1957, three projects were developed in Valdez Peninsula (Romero-Erramuspe-Loscazoff). Two of these projects were discarded during 1990s because of UNESCO declared this peninsula protected area for marine fauna.
- Marine Currents and Wave energy: Since 2008 the potential and technology to be applied in the Atlantic Coast of Argentina, from Rio Negro to Santa Cruz, is studied and analyzed in Engineering Faculties of several National Universities (see Sect. 9.9.2). INVAP industrial company with the National University of Mendoza and Balseiro Institute develop prototypes of hydrokinetic turbines.

Table 9.2 First wind installations in Argentina

Year	Wind farm	Turbines	Nr.	Power	Situation
5/1995	Pico Truncado, Santa Cruz	VENTIS 20–100, 100 kW	10	1 MW	Dismantled
2/1990	Río Mayo, Chubut	Aeroman, 30 kW	4	120 kW	Dismantled

9.2 Regional RES Centers and First Wind Installations

9.2.1 Regional RES Centers [4]

The Argentinean Government created in 1984 the first institutions at national level regarding Renewable Energy Sources. These institutions were the Regional Centers for the Development of Renewable Energy Sources (RES), located in the most representative Provinces of Mini-Hidro, Wind, Solar and Geothermal resources. These Centers are describes in Table 9.1.

9.2.2 State Wind Power Installations

In the case of wind farms, since the early 1990s, Argentina has been installing wind equipment in Patagonia. The first state electrical facilities had the problem that the corresponding maintenance was not carried out, resulting in a short useful life (3–5 years) (Table 9.2).

During 1990s decade Dr. Bastianon design and constructed a wind turbine of 10 kW with Navy economical support. This was the first prototype of wind turbine made in Argentina.

9.3 Facilities in Cooperatives and Municipalities

9.3.1 First Wind Farms Connected to Argentinean Interconnected Electrical System (AIES)

During mid-1990s, as a result of greater openness to the international market, a new impulse was given to Wind Energy projects, where several Public Services Cooperatives and Municipalities began to install new European wind turbines (Neg-Micon, Enercon, Bonus, etc.). Wind farms installed in this period are connected to AIES. Most of these facilities are still in service today and cooperatives have been the pioneers of private wind farms in the country. Comodoro Rivadavia wind farm of 16.56 MW is currently out of service (O/S) since 2010 due to lack of maintenance. Also, Rada Tili turbine of 400 kW is O/S. These wind farms are showed in Table 9.3.

Table 9.3 Wind installations in the 1990s of cooperatives and municipalities [5]

Location	Province	In service since	No. Turb.	Power (kW) manufacturer	Total (kW)	Accumulated (kW)
C. Rivadavia	Chubut	1-July-94	2	250—Micon	500	500
Cutral Co	Neuquen	1-Oct.-94	1	400—Micon	400	900
Punta Alta	B. Aires	1-Feb.-95	1	400—Micon	400	1300
Tandil	B. Aires	1-May-95	2	400—Micon	800	2100
Rada Tily	Chubut	1-Mar-96	1	400—Micon	400	2500
C. Rivadavia	Chubut	1-Sep.-97	8	750—Vestas	6000	8500
Mr. Buratovich	B. Aires	1-Oct.-97	2	600—BONUS	1200	9700
Darregueira	B. Aires	1-Oct.-97	1	750—Vestas	750	10,450
Punta Alta	B. Aires	1-Dec.-98	3	600—bonus	1800	12,250
Claromeco	B. Aires	1-Ene-99	1	750—Vestas	750	13,000
P. Truncado	Santa Cruz	1-Nov.-00	2	600—Enercon	1200	14,200
C. Rivadavia	Chubut	1-Jul.-01	16	660—Gamesa	10,560	24,760
Gral. Acha	La Pampa	1-Mar.-02	2	900—Vestas	1800	26,560

9.3.2 Synchronous Generation/Pitch Control Versus Asynchronous Generation/Ställ Control [6]

Most of the turbines installed during 1990s had asynchronous generators, with control of loss of load in fixed blades (Ställ) (AS-SC) [6]. Few turbines had synchronous generator and pitch control (SG-PC) (Enercon-Pico Truncated) that were installed. Practically the result has been that SG-PCs have few O/S situations. However, AS-SGs have had problems especially premature microcracks in blades. Also the Argentinean companies that own them did not have enough money for their maintenance due to the economic crises which begun in 2001. Nowadays wind turbines in the world which have a power more than 2 MW, are made with Pitch Control, regardless if the generation is synchronous or asynchronous. As a conclusion, it could be seen that blades with Stall Control, have a very premature fatigue with the Patagonian winds because of its media wind speed and index of turbulence particularly near of Comodoro Rivadavia (V_w = 8–9 m/s at 10 m of height—T_{index}: 0.17/10 min). This regional characteristic has caused a change of blades up to three times during the first 10 years of operation. Other inconvenient is premature wear in a quadrant of the crown where the pinion of orientation system is displaced of turbines near Comodoro Ricadavia. This is a very expensive piece and it is redesigned in four assemblable parts to reduce costs. This wear occurs because of wind direction during a gust in this place, usually changes 60% of times up to 45° from the original direction of the wind.

9.4 Training in Wind Energy Ventures

9.4.1 Work Force Training

During 1990s there were no trained local technicians and work force for installations of large RE installations except for Big Hydro ones because of all the installations of Big Hydro Plant since 1970s (El Chocón-Cerros Colorados 1450 MW, Alicura 1000 MW, Piedra del Águila 1400 MW, Salto Grande 1890 MW, Yacyreta 3200 MW, etc.). The author of this chapter returns to his country during 2000 after his postgraduate degree in Renewable Energy and The Environment at the University of Reading, and he realizes that there are no skilled workers in installation, operation, and maintenance of renewable energy equipment. During 2003, he created the Technique Course (Technicature) in Renewable Energy and Environment in Academic Unit of Plaza Huincul of National Technological University [7], which allowed to have the first 55 Technicians in Renewable Energy and Environment of the country in 2005. Provinces by means of their Educational Ministry or other National Universities created other "techicatures" in renewable and non-renewable sources. Today, practically the majority of the provinces with renewable resources have at least tertiary level "technicatures" in Renewable Energy.

9.4.2 Postgraduate Training

During 1990s, the lack of HHRR specialized in installations, especially in the maintenance of Renewable Energy devices, was critical. From 1977 to 1990 there was postgraduate training in Solar Energy (Salvador University) and eventually courses in wind energy were given. Since 1998, a Master's degree in Renewable Energy is offered by INENCO of National University of Salta. It is first postgraduate course in Argentina and it has workshop modality. Then this postgraduate course was developed in different places (Buenos Aires, Comodoro Rivadavia) facilitating the mobility and attendance of the assistants of different regions. Since 2008, postgraduate trainings were developed in National Universities:

- 2008: Master Degree in RES, National University of Rosario (UNR).
- 2013: Master Degree in RES, National Technological University of Buenos Aires (2013: UTN-FRBA).
- 2018: Specialization in Wind Energy, National University of Comahue (UNCo). This postgraduate course is dedicated to wind farms design and applications in domestic homes, buildings of some floors and industrial ones.

Particularly National University of Southern Patagonia (UNPA), Santa Cruz, in Caleta Olivia Academic Unit (UACO) begun in 2008 with Bachelor's degree in Electromechanical Engineering Orientation Renewable Energy. Its curricula includes vector Hydrogen for energy accumulation and practical laboratory work and tests in Hydrogen Plant of Pico Truncado [8], a pioneer in Central and South America with all the technology for Hydrogen production, accumulation and distribution (tanks, pumps, cars, etc.) to be used in transport and gas pipelines.

Since 2010 there is a great demand for training of technicians, engineers and postgraduates in RES, particularly in Wind Energy.

9.5 Legal Framework: Environment and Situation of the RES

Since middle 1990s onwards [4], awareness began to be raised in Argentina on the care of the Environment. It is initiated in education from primary and secondary school, including the advantages in Renewable Energy Sources. During early 1984, only natural gas was being used to generate electricity in thermal power plants to AIES. Also Lead tetra-ethyl was no longer used in naphtha as antiknock. Fiscal Oilfields Company (YPF), produces its own anti-knock benign with environment. In the late 1990s, the government commissioned a task force of International Panel of Climate Change (IPCC) experts to study the influence of overheating on Argentina for the next 100 years. Argentina was the only country in the Americas to face this type of study. Four scenarios are defined and it is observed that the Southern Hemisphere and particularly Argentina will be 2 °C less than the Northern Hemisphere by 2100.

But the advance in the mitigation of GHG in electricity generation using natural gas is left aside in 2005, produced by deficit of extraction of Hydrocarbons in the country making agreements with OPEP countries (Venezuela) and USA (liquid gas) to import hydrocarbons for thermal power plants and gas turbines. During 2016, the exploration and exploitation of gas (Vaca Muerta, Neuquén) in Argentina is promoted increasing the use of natural gas from it in thermal-electric generation reaching during 2019 for full hydrocarbons self-supply. Parallel to 2015/2016, renewable energy sources were promoted to reduce GHG emissions in electricity generation and electrical trains. These actions are complemented by a legal framework that is summarized in Table 9.4.

9.6 National Plans of Facilities for Renewable Energy Sources

9.6.1 Situation of the Electricity Market and the RES Since 2004

Taking into account some previous paragraphs, it is possible to explain in following items about the evolution of National Plans of development for RES, especially Wind Energy, with chronology of political and economical events:

Table 9.4 Laws related to RES

Law/Decree year	Topic	Remarks	Restriction
Law 25019/1998 (impulse electrical cooperatives RES investments) [9]	Promotion regime for wind and solar energy	Proposes to reduce taxs in investments in solar and wind energy. Create promotion fund from those sources	Only generators or electric power distributors could install this type of installations
Law 26123/2006[a] [10]	Promotion of the hydrogen for development of technology, uses and applications (fuel and energy vector) from wind energy	The National Fund for promotion of hydrogen and the promotional tax regime is created. Inclusion of 5% of hydrogen in gas pipelines	It is still not regulated
Decree No. 134/2015	Emergency of the National Electric Sector: It attempts to strengthen the electric system from 2005 to 2015 focused on the increase in thermal generation	Fossil fuels are imported. During 2016, the interest in production of unconventional hydrocarbons (Vaca Muerta) is reinforced	2005–2015 electricity is again obtained from fuel oil, gas and gas oil. Argentina imports hydrocarbons including gas. Dirty GHGs are generated
Law 27190/2015 and 27191/2016 [11]	National Development Regime is created for using RES	Argentina open its internal market to import RE technology. Conditions: RE farms have to have 30% of national purchase	2016–2017 internal and external market is stabilizing

[a]Law 26191/2016 of Promotion of RES installations allowed initiating the national private wind industry that was favored by protections to imported technology

- Since 1994/1995 to 2008: Operation of private wind farms begin in AIES. Wind power installation for 2008: 30 MW (0.1% of demand).
- 2004–2016: Crisis of imbalance between generation and demand, produced by non profitable political tariff in WSEM for private energy utility companies.
- 2010: State Plan of GEN-REN 1000 MW for Wind, Solar, Biofuels and Mini/Micro Hydraulic [12]: 51 projects were offered with 1436.5 MW, which are: Wind 895 MW-84% (17 wind farms in Chubut, Santa Cruz and Buenos Aires), Biofuels 12% (Buenos Aires, Corrientes, Santa Fe), Mini-Hydro 1% (Jujuy, Catamarca, Mendoza), PV 2% (San Juan). Few projects begun installations (130 over 745 MW). Price for profitable wind installation 140 U$S/MW. The wind projects had to be only with new turbines.
- 2010–2016. In spite of the GEN-REN and incorporation of conventional thermoelectric installations, the generation deficit remains and this restriction required special provisions for cogeneration, for example for specific events of large electrical loads, such as shows, football matches, etc. that could overload the AIES, it was necessary to use generator sets. In addition, the promotion of electrical domestic devices (Split air conditioners, electric cookers, etc.) and lack of maintenance on distribution grid due to non profitable political tariff,

Table 9.5 Wind power suppliers of the GEN-REN 2010 (755 MW offered—130 MW installed until 2013) [13–15]

Wind Farm	Company	MW	Situation
Malaspina I	IMPSA	50	GENREN: in Project
Puerto Madryn Oeste	Energías Sustentables S.A.	20	GENREN: in Project
Malaspina II	IMPSA	30	GENREN: in Project
Puerto Madryn II	Emgasud Renovables S.A.	50	GENREN: in Project
Puerto Madryn I	Emgasud Renovables S.A.	50	GENREN: in Project
Rawson I	Emgasud Renovables S.A.	50	In operation since September 2011
Rawson II	Emgasud Renovables S.A.	30	In operation since January 2012
Rawson III	Emgasud Renovables S.A.	30	In operation since June 2015
Puerto Madryn Sur	Patagonia Wind Energy S. A.	50	GENREN: in Project
Puerto Madryn Norte	International New Energies S. A.	50	GENREN: in Project
Koluel KaikeI I	IMPSA	50	GENREN: in Project
Koluel Kaike II	IMPSA	25	GENREN: in Project
Loma Blanca I	ISOLUX S.A.	50	GENREN: in Project
Loma Blanca II	ISOLUX S.A.	50	GENREN: in Project
Loma Blanca III	ISOLUX S.A.	50	GENREN: in Project
Loma Blanca IV	ISOLUX S.A.	50	In operation since July 2013
Loma Blanca I Base	Sogesic S. A.	49.5	GENREN: in Project
Loma Blanca II Base	Sogesic S. A	49.5	GENREN: in Project

produce unforeseen failures in Medium and Low Voltage networks during summer, season of maximum demand (instead of winter).

- 2016: Law 27191 (revision of 26190) allows the implementation of RENOVAR Plans, reaching to 8% of ER in Network for 2018 and 20% in 2025). RENOVAR-1, ENEW 1.2 and RENEW 2 were delayed and some projects were rejected due to economic changes. In all RENOVARs, only new turbines manufactured abroad or in the country are installed Table 9.5.

There are two local manufacturers of wind turbines in Argentina impulse since 2005 and by GENREN, which are:

- IMPSA Wind [16]: Part of Pescarmona Industries S.A., Mendoza. It has built a 2.5 MW turbine with synchronous generator of Nd-Fe-Bo permanent magnets and converting AC voltage with variable frequency because of wind variation to DC, and then to AC with fixed network frequency. It has installed this kind of wind turbines in Arauco Wind Farm, Province of La Rioja and has offered it for the Malaspina Koluel Kaike projects.
- NRG [17]: from Comodoro Rivadavia (CR), Chubut. This company is a group of companies for oil services that have built a prototype of 1.5 MW, tower height of

Table 9.6 Wind farms not included in GENREN but included in Plus Energy (Energía Plus) program

Wind farm	Wind turbine company	Power (MW)	In operation since
Arauco I and II—La Rioja [12]	IMPSA	50.4	2011–2013
Diadema—Chubut [16]	CAPSA oil company	6.3	2011

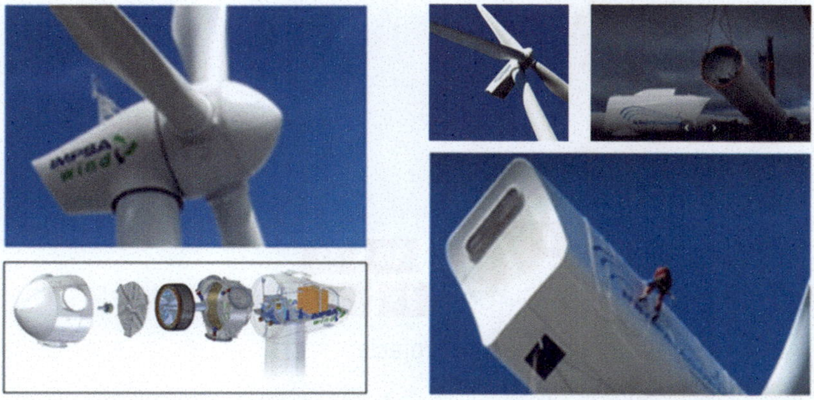

IMPSA Wind, Mendoza

NRG S.A. Industries, Com. Rivadavia, Santa Cruz

Fig. 9.2 Argentinean Wind Turbine Companies [16, 17]

80 m and rotor diameter of 82 m. It is installed and working near CR since 2010. This company produces spare parts for wind turbines for international market. During 2018 NRG signed a contract with the Castelli Cooperative, for provision of several 1.5 MW turbines for a wind farm in Cerro de la Gloria, Province of Buenos Aires (Fig. 9.2).

Table 9.6 shows two wind farms which adhere Energy Plus tariff during 2012, which is better than the Government regulated one in the WSEM. In south of CR are Diadema I wind farm, of 6.3 MW with Enercon turbines 3 × 2.1 MW. This project was originally for Hydrogen production, but as the Law was not regulated till today, the owner company joined the Energy Plus. Nowadays, Diadema II (27.3 MW) is in construction and installation.

Figure 9.3 shows the wind farms that were in production since 2013. It shows that most of wind farms are located in Patagonia and southern of Buenos Aires province. The Veladero/Barrick Gold mine (7) in San Juan has the world's highest wind power installation (3000 m), of 2.5 MW (ρaire: 1.23 kg/m^3) /1.6 MW (ρaire: 0.8 kg/m^3). Facilities 8–18 correspond to cooperatives.

It do not exceed 3 MW per facility, except the Antonio Moran Park in Comodoro Rivadavia, Chubut. Several cooperatives since 2001 started a financial crisis and did not perform the corresponding maintenance and today some machines remain out of service (7).

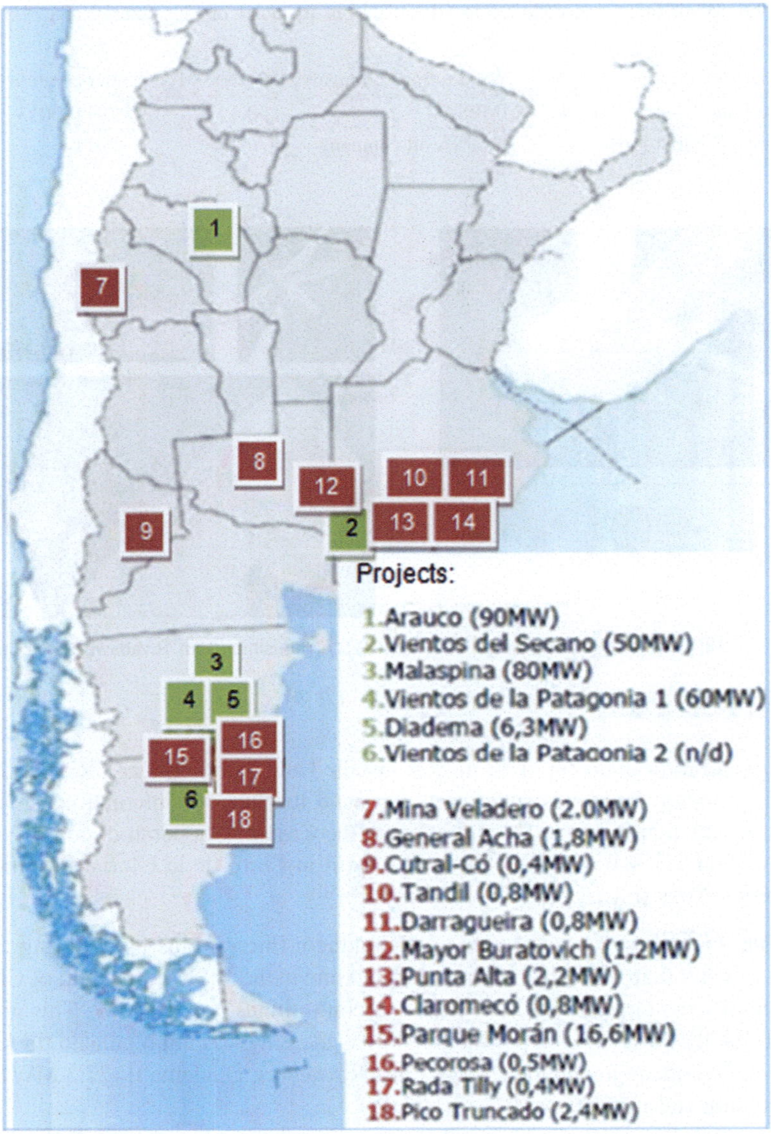

Fig. 9.3 GENREN 2010: No. 1–6—Existing wind farms: No. 7–18 [13]

9.6.2 RENOVAR [18, 19]

During 2016, Argentinean Government made changes in electricity tariff market, by means of Law 27191. The RENOVAR was implemented. It proposes to have 8% renewable energy in the network by 2018 and 20% by 2025. The first stage was

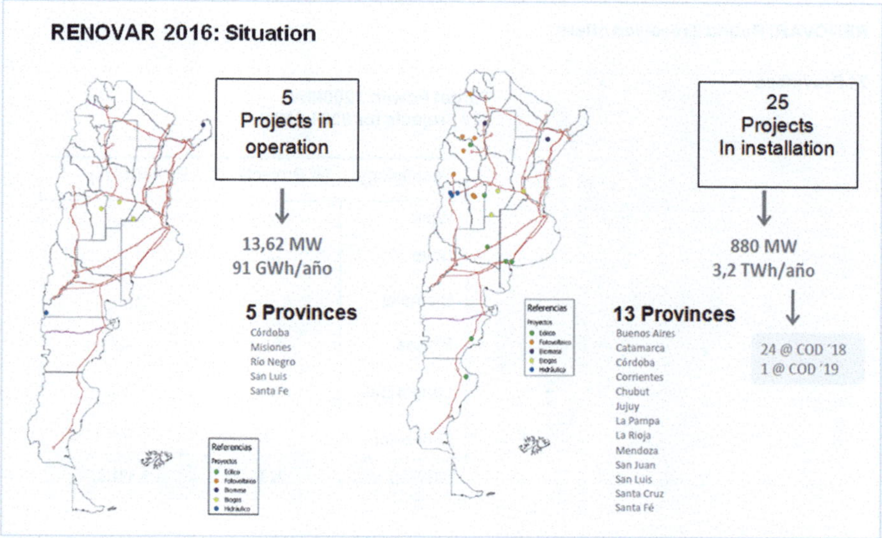

Fig. 9.4 RENOVAR 1 + 1.5 (AIES 500 kV with red color)

RENOVAR-1 of 1000 MW required (Fig. 9.4), which had 123 offers with a total of 6343 MW. They qualified 103 offers (5209 MW), Wind (3468 MW), PV (2811), Biofuels and Biogas (11 MW) and small hydroelectric plants (5 MW). Eighteen offers did not qualified: 598 MW of Wind, 506 MW of Solar and 30 MW of Biomass. Renovar-1 gave a great impulse to the use of renewable energy sources that allowed to continue even with projects of the GENREN that were not implemented during 2010. In the case of Patagonia, because of the transmission limit of the 500 kV network, installed power is limited to 600 MW in Comahue and it is limited to 400 MW in Southern Patagonia (Fig. 9.4).

These rounds provide wind farms that are in operation or finishing their construction by 2018, which are:

Province of Chubut [16]*:* Garayalde of Pan American Energy Company (100 kM of CR, 24.15 MW with Vestas Turbines), Manatiales Behr of YPF (100 MW: near CR), ALUAR Wind Farm (50 MW: 14 × 3.6 MW Vestas, near Puerto Madryn).

Province of Santa Cruz [17]*:* "El Bicentenario Wind Farm," built by Petroquimica Comodoro Rivadavia (100 MW, 17 km southeast of Jaramillo on Route 3); "La Deseada" wind farm, from Eólica Pico Truncado S.A. company (100 MW, 15 km from the Santa Cruz Norte Transformer Station); Viento Austral wind farm (97 MW, near Comandante Luis Piedra Buena); "Vientos Los Hércules," from the company Eolia Sur, (97.2 MW, south of Pico Truncado), and finally "Puerto Deseado Wind Farm," from KAIKOS SA, (39 MW near Puerto Deseado), Santa Cruz would contribute 434 MW of power to the AIES by means of the RENOVAR 2 (12.51% of the offers presented).

RENOVAR: Round 2 received offers

21 Provinces

Buenos Aires
Catamarca
Chaco
Chubut
Córdoba
Corrientes
Formosa
Jujuy
La Pampa
La Rioja
Mendoza
Misiones
Neuquén
Río Negro
Salta
San Juan
San Luis
Santa Cruz
Santa Fe
Santiago del Estero
Tucumán

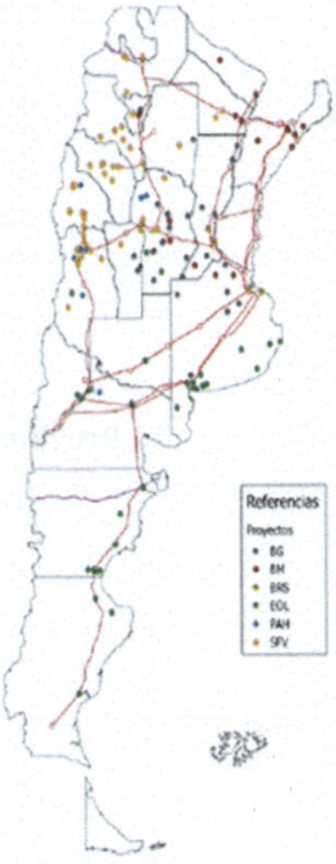

Target Power: 1200MW
228 projects for 9391.3MW

Technology	Nº Projects	Power(MW)
Wind	58	3.810,5
Solar	99	5.290,6
Biomass	20	186,6
Biogas	32	56,8
Landfil Gas	4	15,1
Mini-Hydro	15	31,6
TOTAL	228	9.391,3

Fig. 9.5 RENOVAR Round 2, [20]: Location of 228 projects that include 58 wind farms

It can be seen that in Rounds 1 and 1.5 (text near Fig. 9.4), proposed wind projects surpass proposed PV ones in 1741 MW. On the other hand, in Round 2 (Fig. 9.5) is quite different, offering almost 1300 MW more of solar energy than wind power. Figure 9.6 shows the projects qualified or accepted by the Argentine Government in RENOVAR 2.

9.6.3 Analysis of the Price of Energy in Argentinean Whole Sale Electricity Market [19, 20]

Since 2005–2010 (GENREN) profitable tariff for wind projects was 140–120 U$S/MWh but WSEM had a less regulated valuefor it. During RENOVAR rounds (1–1.5–2), the average prices of energy in MWh offered in wind and solar projects (Fig. 9.7) were lower than the reference required of the government. During 2017, 59 projects awarded in the RENOVAR 1 will be completed. The average energy price in RENOVAR I was U$S109.9/MWh generated by RES. In RENOVAR II the average price is U$S60/MWh generated by RES, which is a slightly higher value than the international value (US $40). It is relevant to see that price of MWh value generated by RES has dropped from U$S140/120 in 2005–2015 to approximately a half of this value in 2017.

Table 9.7 shows an improvement in the price of MWh in the wholesale market in Fig. 9.8

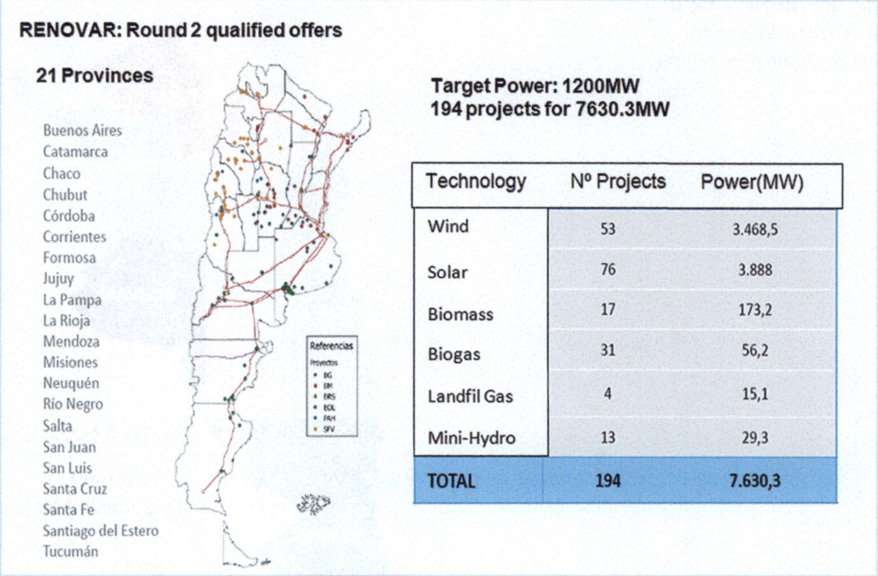

Technology	Nº Projects	Power(MW)
Wind	53	3.468,5
Solar	76	3.888
Biomass	17	173,2
Biogas	31	56,2
Landfil Gas	4	15,1
Mini-Hydro	13	29,3
TOTAL	194	7.630,3

Fig. 9.6 RENOVAR Round 2 [20]: Qualified offers

9.6.4 RENOVAR III-MINIREN (2018) [21]

This proposal of Argentinean National Government promotes 400 MW of generation from RES in in the working days of the week. Also it fixes the wind/solar combination limiting it to 20 MW per province, except Buenos Aires which is 60 MW. This round of bidding permits to interconnect to AIES through 13.2 kV / 33 kV/66 kV lines. There are quotas for technology by regions and provinces (Fig. 9.9) and differentiated quotas without regions: Mini/micro-hydro 10 MW, Biomass 25 MW, Biogas 10 MW, landfill gas 5 MW. For these projects the maximum power will be 10 MW and minimum 0.5 MW.

There were some delays in the construction of RES application projects of RENOVAR 1 and 2 due to the economic changes of 2018 and the lack of 132 kV lines in the AIES to connect them to 500 kV system. Consequently during 2018 several companies involved in RENOVARs ask for extra time to finish the installation of RE farms. Also the electrical demand is not growing because of economical recession which produce during 2018 the fall of the gross domestic product (GDP-2.5%). So it will make possible to have all the RE projects carried out in short, medium and long term works for future demand when GDP is expected to grow (2019–2020).

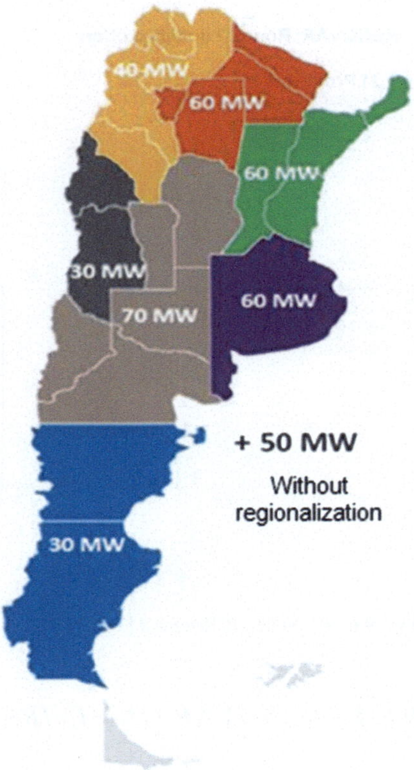

Fig. 9.7 RENOVAR III MINIREN: Maximum power required by region

Table 9.7 Prices (U$S/MW) of the average MW.h offered RENOVAR I (left) and RENOVAR II (right)

	RENOVAR I Price U$S/MWh			RENOVAR II Price U$S/MWh				
	Offered		Maximum	Power MW			Prices U $S/MWh	
Technology	Min.	Med.	awarded	Target	Qualified	Awarded	Min.	Max
Wind	49.1	69.5	82	550	3458.5	665.8	37.3	41.2
Solar	59.0	76.2	90	450	3888.0	556.8	40.4	43.5
Biomass	110.0	114.6	110	100	173.2	117.2	92.0	106.7
Biogas	118.0	177.8	160	35	56.2	35	150.0	156.8
Landfill gas	–			15	15.1	13.1	128.0	129.2
Mini-hydro	111.1	114.5	105	50	29.2	20.8	89.0	98.9

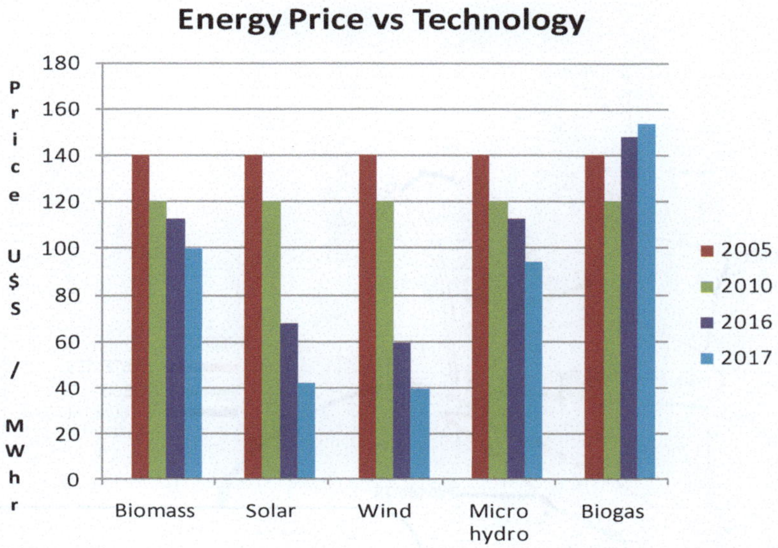

Fig. 9.8 Evolution of the wholesale price of energy in Argentina since 2005–2017 (average values)

9.7 Evolution of the Argentine Energy System for Connection of New Electrical Generation

RENOVAR 1 and 2 could saturate some 500 kV of AIES lines. That is why RENOVAR III includes only generation of small RE systems in a global value up to 400 MW. CAMMESA is the company that fixed the daily commission of power plants in AIES. Also it has a Technical Office that simulates AIES for different conditions of demand and generation. It has established how AIES could be growing with different scenarios and for different periods. The next items show possible scenarios for AIES growing with Renewable and Non-renewable generation.

9.7.1 Actual AIES and AIES in Short Term [22]

9.7.2 Medium and Long-Term AIES Situation [22]

Patagonia will produce 5000 MW from wind–hydroelectric and the existing 1000 MW wind–thermal. It will require a DC link, with a capacity equal to the Base Power of the City of Buenos Aires (CABA) and Greater Buenos Aires (GBA).

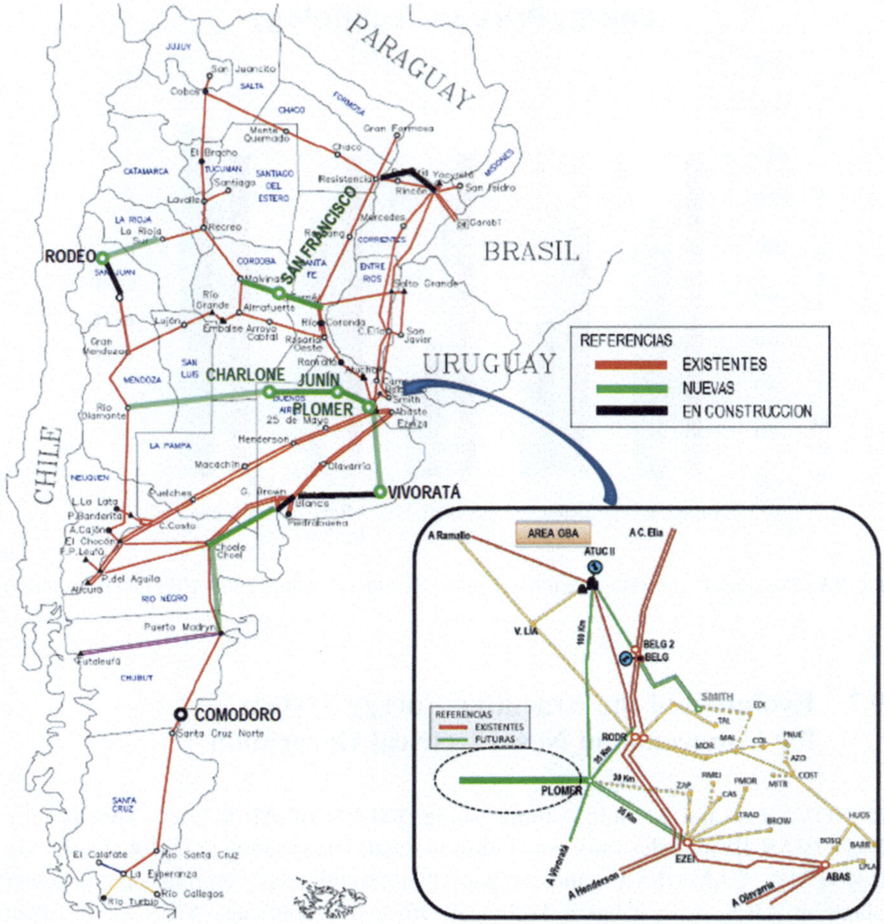

Fig. 9.9 AIES in short Term (next 2 years) [22]. (Existing: Red 500 kV—Violet 330 kV—Green and Black 500 kV to be made in 2018–2019)

From the Andean Northwest area, it is expected to have other 5000 MW. The existing lines and future ones to be installed in the short and the medium term will be able to evacuate this power to Central, Comahue, Northeast, and Pampas Regions. It is expected to have the AIES circuit described in Fig. 9.10 for 2025.

Other scenarios for Patagonia extra Power of 5000 MW are studied with two systems of 500 KV lines, one near Atlantic Sea and other near Los Andes Mountain Range, but Fig. 9.10 shows which will be technically and economically possible.

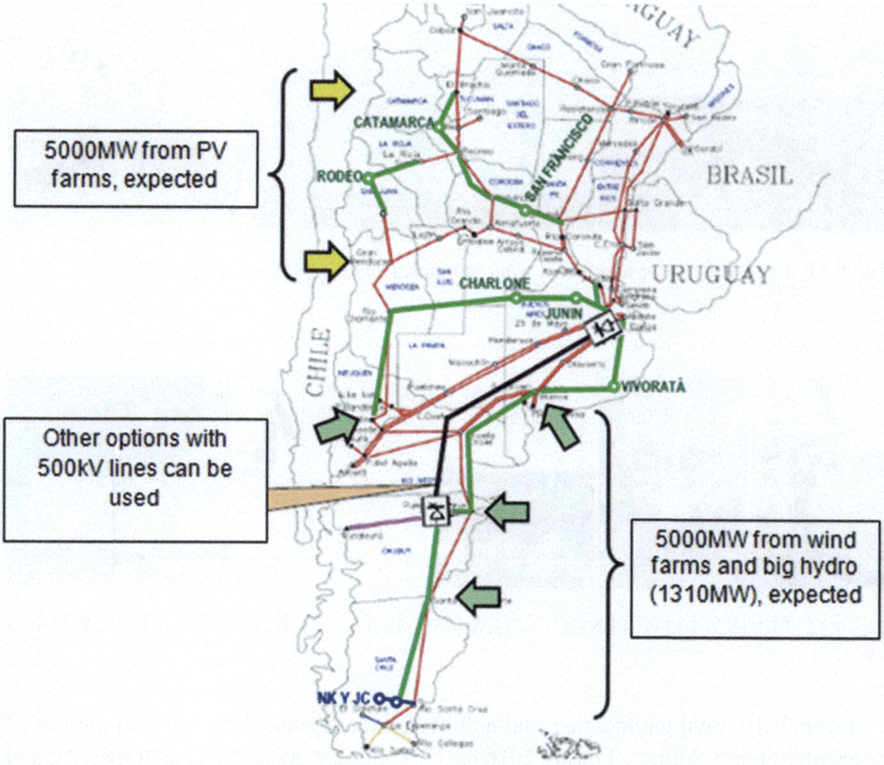

Fig. 9.10 AIES in medium and long term

9.8 Pending Applications of Wind Energy To Be Carried Out in Argentina

9.8.1 Small Wind Turbines in Urban Environments

During 2014 and 2015, laws were created in provincial governments promoting domestic electricity generation, where provinces of Neuquén [23], Santa Fé, Córdoba and San Juan were the pioneers. In addition, since 2008, the University of Córdoba has developed a PV Pairing Laboratory. Also the local market of small wind turbines needed to verify the characteristics of this devices, so consequently the National Institute of Industrial Technology (INTI) [24] developed a laboratory to verify them. Small wind turbines were mainly used to supply energy in isolated sites by means of provincial rural electrification plans (Chubut, Neuquén) and in remote communications stations, particularly in Patagonia. INTI installed in Cutral Có (Neuquén) a wind turbine testing plant, which is authorized to certify turbines up to 10 kW.

Fig. 9.11 Laboratory for testing small wind turbines, INTI, Cutral Co, Neuquen [24]

Fig. 9.12 FENUCo 100 W, FENUCo 1–3 kW [25]—EOLUX 1 kW, EOLOCAL 1 kW, INVAP 4 kW [24]

From 2010, civil engineering and architecture companies bear in mind the use of renewable energy sources. During 2016 the Government opens the market for solar and wind energy technologies. It made possible to integrate RES into homes and buildings, which are located in cities in the most critical areas of electricity demand.

Different companies and universities offer and design respectively several types of wind turbines with horizontal or vertical axis. This technology is shown in Fig. 9.12.

9.8.2 Wind Energy Applications Combined with Other Renewable Sources or Energy Accumulation

9.8.2.1 Pre-projects in Academic Institutions

The diversity of Renewable Energy Sources in Argentina motivates academic researchers to combine different RES in some places to obtain of electrical energy for remote or interconnected areas to AIES. Some researcher groups made since 2007 preliminary projects of hundreds to thousands of MW obtained from different RES. The following paragraphs and figures describe the relevant ones.

- During 2008/2009: Faculty of Engineering of National University of Comahue (FENUCo) presented a proposal for Offshore Windmills + Tidal Turbines (Hydrokinetic ones) for Santa Cruz (Puerto Deseado, Puerto San Julian and Rio

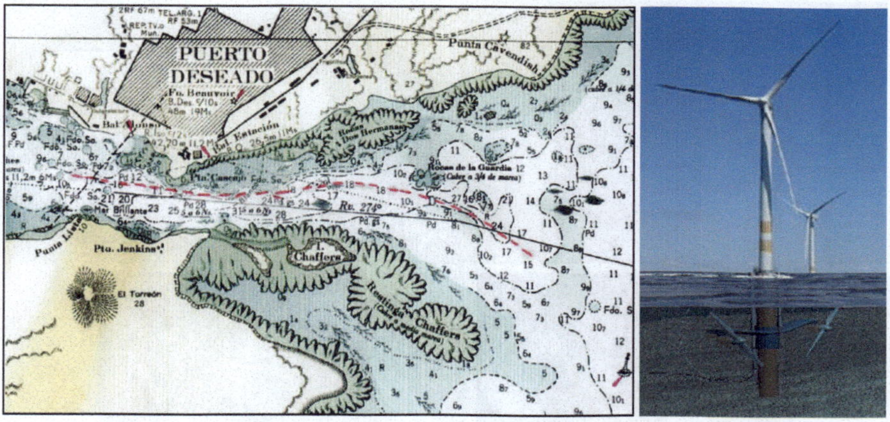

Fig. 9.13 Tidal Wind Combination (photomontage-right)—Example for Puerto Deseado (red dotted line for RE devices location)(left) [26]

Gallegos) [26] (Fig. 9.13). The expected installed power for these three harbors is 225 MW with a combined plant coefficient of 0.5. Santa Cruz is the southernmost continental province of Argentina and has three other places for this type of installation, but that places do not have access to AT/MT lines.

- During 2016–2018 FENUCo and Faculty of Human Sciences of the National University of Comahue, presented proposals of Wind–Tidal (Hydro-Kinetic Turbines + Bulb) application for Rio Negro Province [27]. It is proposed to flood Bajo del Gualicho (−71 m to NM) with Atlantic Sea water through tunnels and channels of 30–40 km long (see Fig. 9.14). This Hydro project is combined with a Wind Farm of 3.5 MW Type II wind turbines installation in several rows around the artificial albúfera (seawater lake) formed in Bajo del Gualicho. The installed capacity could be 12,000 MW (around 10,000 MW Hydro and 2000 MW Wind).

9.8.2.2 Provincial State Policies in Wind–Solar Developments

San Juan province is the only one that studies the full potential of the renewable energy sources and its applications as provincial state policy [28]. It have been investigated the potential of big Hydro (289 MW in operation and 255 MW in tender), Wind (Study of potential and combination with solar), Solar (7 MW in operation and 572 MW approved for installation RENOVAR I and II) and Geothermal (potential study). Particularly Wind–Solar Development near the town of Tocota is remarkable (Fig. 9.15).

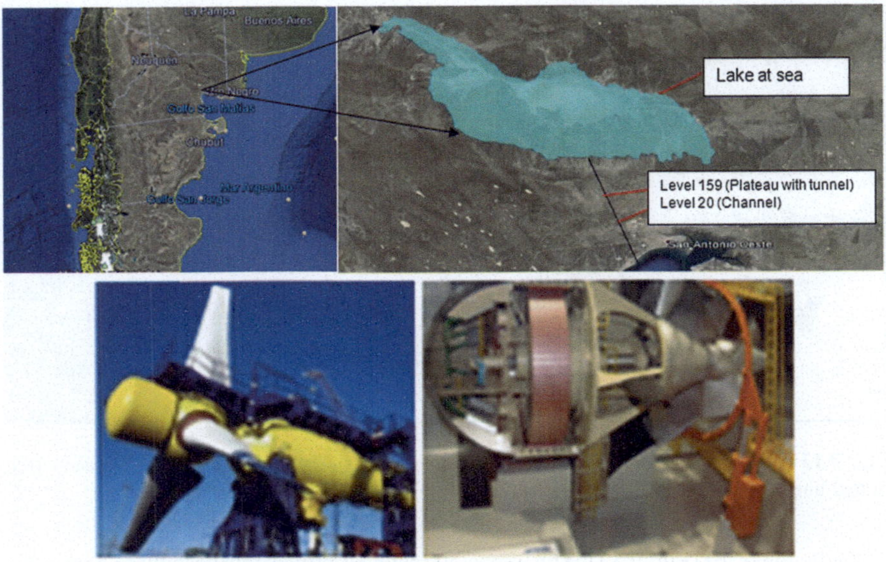

Fig. 9.14 Wind-Tidal proposal for Bajo del Gualicho as an artificial albúfera in Río Negro. Location of Bajo del Gualicho (Bigger pictures left and right)—Hydrokinetic and bulb turbines (next upper pictures) [27]

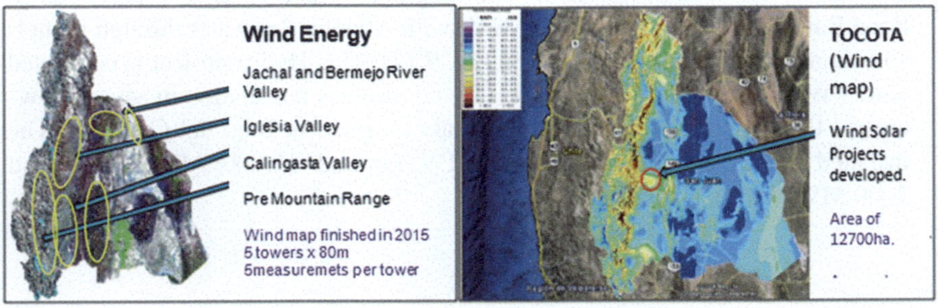

Fig. 9.15 Wind potential studies in San Juan (left); Wind–Solar plant in Tocota (right) [28]

9.8.2.3 Hydrogen as Energy Vector from Wind Energy [8]

During June 2002, the foundational stone of Pico Truncado Hydrogen Experimental Plant (HEP) was placed, and in 2003 it was built by the Municipality of Pico Truncado, through the Hydrogen Santa Cruz Foundation and the Argentine Hydrogen Association. It takes electric power from "Jorge Romanutti" Municipal Wind Farm (4× ENERCON 600 kW) and has approved prototypes of fuel cells, electrolysers, Hydrogen storages based on hydrides, conversion of OTO cycle engines and catalytic burners using H_2 as fuel.

On December 13, 1907, the first oil extraction was carried out in Comodoro Rivadavia. Almost a century later, on January 14, 2005, Hydrogen Experimental Plant (HEP) was inaugurated, initiating the first H_2 production by electrolysis of pure water whose electric power came from the wind. That year it becomes the most advanced installation of technology for obtaining H_2 in Latin America. This development is framed in Law No. 26123, promulgated on August 25, 2005, which declares "the development of technology, production, use and applications of hydrogen as fuel and energy vector is of national interest." This law also provided tax benefits to companies dedicated to obtaining H_2 and to those dedicated to generating energy to obtain H_2.

During 2005–2012, training of HHRRR in obtaining and using H_2 was a priority. It is carried out by means of international postgraduate training, through theoretical/practical courses of technologies in production and use of Hydrogen as a clean and renewable fuel. In this way HEP has a laboratory which is able to obtain H_2 with different types of Hydrolyzers and accumulate it in Hydride. It is also equipped with a compression system and pumps for loading H_2 in motor vehicles developed by the Argentinean company Galileo in stainless steel for H_2 use. The process of generation and use of H_2 is as follows:

$$\text{Water} + \text{Electrolyzer} + \text{Windmill} + \text{Good Winds(Patagonian)} = \text{Hydrogen} + \text{Oxygen}$$
$$= \text{Energy} + \text{Waste}\ldots\text{(again)Water} \Rightarrow \text{Pollution 0;Exhaustion of resources 0.}$$

During 2011, the HEP officially entered in a industrial phase. This achievement is due to the tenacious work of Dr. Juan Carlos Bolsich since the beginning of the twenty-first century. The advantages of H_2 as an energy vector are: once accumulated in special tanks, it can be injected both into a fuel cell to generate electricity and/or into a catalytic burner to obtain heating or cooking food. In addition, it is possible to use it in internal combustion engines to provide energy to all types of vehicles, generator sets or to accumulate it in the long term in hydrides using the waste heat from thermal power plants. The cells to obtain it are highly efficient, have a quiet operation, have a long life and require little maintenance.

Unfortunately, there are two inconvenient to solve: Law No. 26123 has not been regulated until now, for that reason it is waiting for the implementation of the facilities of the H_2 even its injection in gas pipelines up to 5% without danger of Creep. The other one in because of H_2 wide inflammation capacity with O_2, the user's mentality has not adapted to its acceptance as a safe fuel (Fig. 9.16).

9.9 Conclusions and Recommendations

9.9.1 Conclusions

It can be seen that the development of the applications of RES devices in Argentina has been very dependent on the economical and political situation in the country. But

Fig. 9.16 Hydrogen experimental plant of Pico Truncado [8]

Table 9.8 RES growth in the AIES up to 2025 according to MATER

Year	2017–2018	2019–2020	2021–2022	2023–2024	2025
%	8	12	16	18	20

in spite of this, during the last 40 years there has been a RES use, especially in the last decade, generating a legal and tariff framework that allows its constant growth of installed power, particularly in wind energy.

Argentina has been a pioneer and a leader in Latin America in the use of different sources of renewable energy: domestic photovoltaic applications (1980), geothermal (1985), commercial wind (1990), biomass (1983 ethanol–1995 soybean biodiesel), micro/mini hydro (1950). In the last three decades, also with more emphasis in the last 5 years, it is incorporating the generation of Wind energy, Biomass and Photovoltaic to the AIES, giving the basis in the next future to export electrical energy to Chile and Brazil.

Universities give the possibility of obtaining the necessary HHRR in short and medium terms for designing, construction, installation, operation and maintenance of RES energy systems.

This development of technologies and exchange of experience has been possible thanks to the European equipment installed in Argentina and the experience of this country in the development of biofuels.

There were some economical incentives as the proposal of Energy Plus since 2010–2012 and MATTER program since 2017. The former was not enough to solve the energy crisis 2015–2015, and the later is offered when Argentina has the worst recession since 2001. Argentina expects to raise economical index s during 2019.

In the case of MATER [29], the “Great Qualified User” is defined, which must have an average power of the previous year ≥300 kW, in order to make joint purchases with other users and have priority of dispatching the RES. It can also have dispatch priority and auto generate energy. It is only for projects after 2017, to stimulate the growth of the RES and tend to have 25% of them in 2025. Table 9.8 shows percentages of RES in AIES for each year until 2025.

9.9.2 Recommendations

The applications in buildings of RES (Micro-Hydro-Wind and Solar) would allow to save from 30 to 40% of the energy of common uses, particularly in the peak hour of consumption (18–22 h) [30, 31].

The GEN-REN and RENOVAR Plans have shown the planning deficit in HV sub-transmission (132 kV) in the short term and long-term UAT transmission, delaying works. So it is necessary to planning not only electrical generation to be installed, but also high and medium voltage lines to transport and distribute the electrical energy.

It is possible to say that Argentina has 200,000 MW of Wind Energy Technological Potential (80 m height, Wind Turbines of 2.5 MW Type II) [32], consequently Argentina can export to the rest of South America quotes of electrical power in medium term.

Taking into account all the gone and turns produced by political and economical crisis in Argentina to carried out Energy Projects, it is possible to make a list of future guidelines to bear in mind when a company wants to invest in RES application in Emerging country, such as:

- In the case of state electromechanical development to be sustainable on time, it is necessary to consider not only the economic, social and environmental aspects, but also:
  - Provision of spare parts and LV and MV lines.
  - Maintenance of the equipment.
  - HHRR for not only resource and design, but also for installation, operation and maintenance.
- For an electromechanical development to be sustainable in a emerging country, there must being short and medium terms:
  - Economic stability.
  - Predictable economy.
  - It is necessary to have a suitable agreed market tariff.
  - Technical and Environmental Planning.
- Agreement between Electrical Distributors and domiciliary user/generator about situation of the BT network.
- Professional Councils (Engineering, Architecture, Technicians, etc.) should provide a regulatory framework for the use of RE devices at home (houses and buildings), according to a duly proven RES.

The very relevant actions that emerge from the upper guidelines and to bear in mind when a company wants to carried out an electromechanical project in emerge country are:

- Agreement in tariff, between electrical distributors and users/generator.
- To regulate for development not for restricting energy production.
- Planning HHRR and location of Facilities (roads) before installing.
- Planning MV and HV networks to transport energy to urban centers.
- Economical predictability and legal security.
- To avoid gone and turns (economical and political ones).

References

1. Argentina, Map: Military Geographical Institute, maps and data of the Argentine Republic (2018)
2. *2010 Census: Population Census of the Inhabitants of the Argentine Republic* (National Institute of Statistics and Censuses, Argentina, 2010)
3. Schmukler M, Garrido S, Rural electrification in Argentina, social and technical adecuation of PERMER Program in Jujuy Province (ELECTRIFICACIÓN RURAL EN ARGENTINA ADECUACIÓN SOCIO-TÉCNICA DEL PROGRAMA PERMER EN LA PROVINCIA DE JUJUY. National University of Quilmes, ASADES Congress (2016)
4. C. Labriola, *Mediterranean Green Buildings & Renewable Energy Forum. Renewable Energy in Argentina. Part of World Renewable Energy Congress* (Faculty of Architecture, University of Florencia, Florence, Italy, 2017)
5. CAMMESA, 2002: Whole sale electricity market company (CAMMESA) report of wind energy, September 2002
6. COPELCO-MICON, *International Course on Wind Turbine Maintenance* (MICON, Electricity Cooperative of Cutral Co., Denmark, 1995)
7. TSERMA 2003–2005, *"Technicature" on Renewable Energy and Environment, First Course to Be Technicians in RE Applications* (National Technological University, 2003)
8. Bolsich, *Hydrogen Course. International Course of H_2 Production and Applications* (Hydrogen Experimental Plant of Pico Truncado, Santa Cruz, 2006)
9. Renewable Energy Trust Fund. Law No. 25.019: "National Regime od Wind Energy and Solar one" (REGIMEN NACIONAL DE ENERGIA EOLICA Y SOLAR). Target. Range, Cases and Authority of Application. Policies. Investments Regime. Beneficiaries. Benefits, Sanctions. Renewable Energy Trust Fund. Approved, 12 April 1998
10. Law No. 26.123: Hydrogen Promotion "Development of Technology, Production, Use and Applications of Hydrogen as Fuel and Energy Vector" (Desarrollo de la tecnología, la producción, el uso y aplicaciones del hidrógeno como combustible y vector de energía). Goals. Subjects, Application Authority, Infringements and sanctions. Creation of the National Fund for the Promotion of Hydrogen. Promotional Tax Regime. Supplementary provisions. Approved: 2 and 24 Aug 2006. Waiting for regulation, 2006
11. Law No. 26191: "National Promotion Regime for Use of Renewable Sources of Energy Destined to the Production of Electric Energy" (Régimen de Fomento Nacional para el uso de fuentes renovables de energía destinada a la producción de energía eléctrica). Object. Scope. Area of application. Application authority. Policies, Investment regime. Beneficiaries Benefits. Sanctions Renewable Energy Trust Fund. Approved: 6 Dec 2006. Regulated: 27 Dec 2006
12. GEN-REN, *Renewable Energy Plan 2010* (National Ministry of Energy, Argentina, 2010)
13. M.F. Gareis, *Review of the State of Wind Energy in Argentina, period 1994–2010 (Revisión del estado de la Energía Eólica en la Argentina, período 1994–2010)* (National University of Mar del Plata HYFUSEN Congress, 2011)
14. Provincial Ministry of Public Works, *Report of the Wind Installations in the Province of Chubut* (Provincial Ministry of Public Works, Chubut, 2017)

15. Provincial Ministry of Public Works and Environment, *Report of the Wind Installations in the Province of Santa Cruz* (Provincial Ministry of Public Works and Environment, Santa Cruz, 2018)
16. IMPSA, Pescarmona Industries S.A. First Wind Turbine of 2,5MW Manufactured in Argentina Presentation during Comodoro Rivadavia Energy Seminar, Sept 2014
17. NRG, NRG Industries, First Wind Turbine of 1,5MW Manufactured in Argentina. Presentation during Comodoro Rivadavia Energy Seminar, Sept 2014.
18. RENOVAR, National and International Announcement, Round 1, Ministry of Energy and Mining of the Argentine Nation, July, 2016
19. Renovar Plan, The Government Obtained Record Minimum Prices. Jorgelina do Rosario, INFOBAE, Septiember 30, 2016
20. RENOVAR 2, CAMMESA Report of RENOVAR Round 2, vinculation with Laws and Decree, Sept 2018
21. MINIREN, CAMMESA Report of MINIREN 2018 (RENOVAR Round 3), vinculation with Laws and Decree, Nov 2018
22. AIES growth 2025, CAMMESA report of how is growing AIES until 2025 because of GENREN, RENOVARs, Energy Plus and Mater Plans. Eng. Sinagra Chief of Power System Office on AIES High Voltage Grid. CAMMESA, Sept 2017
23. EPEN, *Rural Electrification Plan 1985–1995* (Provincial Development Council, Neuquen, 2017)
24. INTI 2018, National Institute of Industrial Technology, Mini Wind Turbine Test Laboratory, Cutral Co Plant, Neuquén. Presentation in Mini Eolo Seminar (CYTED REnet), University of General Sarmiento, Grat Buenos Aires, 12–15 Nov 2018
25. CV. Labriola, JL. Lassig, WEEF 2012: Renewable Energy Sources as a Link between Engineering and Social Change Work presented during World Engineerig Education Forum, Buenos Aires City, Sept 2012.
26. M. Biancucci, Carlos V.M. Labriola, WREC 2008: Wind-Tidal Combined Generation System for Patagonia. Argentina. Work presented during World Renewable Energy Congress, Scotland, United Kingdom, July 2008
27. L. Bertan, C. Labriola, A. Marchegiani' O. Audisio, S.T. Schenker, ICREPQ 2015: Electricity Generation from Tidal Power in Artificial "albúfera" in the Rio Negro Province, Argentina. International Conference on Renewable Energies and Power Quality, La Coruña (Spain), 25–27 March 2015
28. V. Doña, CYTED, 2018: Presentation of PhD. Ministry of Public work of San Juan Province, Mini Eolo Seminar (CYTED REnet), University of General Sarmiento, Grat Buenos Aires, 12–15 Nov 2018
29. Resolution MEyM 281-E2017: Term Market for Electricity from Renewable Sources (MATER). CAMMESA, 2017.
30. C. Labbriol, J. Lassig, P.b. Padilla, *ISCC 2014: 5th International Solar Cities Congress, "ENERGY IN CITIES: INNOVATION AGAINST CLIMATIC CHANGE". Work Title: Use of Intrinsic and External Renewable Energy Sources to a Multi-Floor Building* (EFNUCO, Buenos Aires, 2014)
31. J. Lassig, C. Palese, J.V. Sosa, U.J.Y. Carlos Labriola, *MED GREEN FORUM 2017: Mediterranean Green Buildings & Renewable Energy Forum –2017. Part of World Renewable Energy Congress Wind Energy Potential Research in a Low Building within an Urban Environment* (EFNUCO, Florence, 2017)
32. CVM. Labriola, WREC 2006 RTD: Wind Potential in Argentina, Situation and Prospects, Special Session of Renewable Technology Development office from EEC, World Renewable Energy Congress, Florence, United Italy, July 2006

Chapter 10
Advancements and Challenges Affecting Wind Turbine Implementation in the Member States of the Cooperation Council for the Arab States of the Gulf (GCC Countries)

Abdul Salam Darwish

10.1 Introduction

GCC countries are rapidly growing, driving the Middle East power demand to be doubled in 2040. The demand has been growing from 472 TWh in the year 2000 to 1132 TWh with an average growth of 5.6% a year [1]. This is due to the increase in population, growing economy, and high living standards. Renewable energy has come recently of interest and expected to contribute a share of more than 10% by 2040. The population of the GCC countries is 07% of the world population in 2014 [2]. The GCC contributes to a higher percentage of the global greenhouse gas emission than the European Union [2]. Total energy consumption is shown in Fig. 10.1

Renewable energy has always been the first-choice option for a modern power system. In the GCC area, wind and solar are now competing with conventional sources and command 90% of the investments in renewable power. The reasons behind this are the potential of these resources and the technological advancements that have made it easier to use and cheaper to invest in [4]. The cost of wind turbines has fallen by nearly one-third since 2009 and that of solar photovoltaic (PV) modules by 80% [5]. Increased number of countries held auctions to deploy renewables (from 6 in 2005 to 67 country in 2018). Global investment in renewables rose from less than $50 billion in 2004 to more than $360 billion in 2018 [6]. Figure 10.2 shows renewable power capacity installations have exceeded non-renewables by a rising margin.

Wind energy had a very good trend over the years and reached a worldwide capacity of 540 GW in 2017 [8]. It is expected to reach more than 610 GW (10% increase) by the end of 2018 as shown in Fig. 10.3.

A. S. Darwish (✉)
Phoenix Renewable Energy Centre, Manchester, UK

University of Bolton, Bolton, UK
e-mail: Darwish@phintc.com

A. Sayigh, D. Milborrow (eds.), *The Age of Wind Energy*, Innovative Renewable Energy, https://doi.org/10.1007/978-3-030-26446-8_10

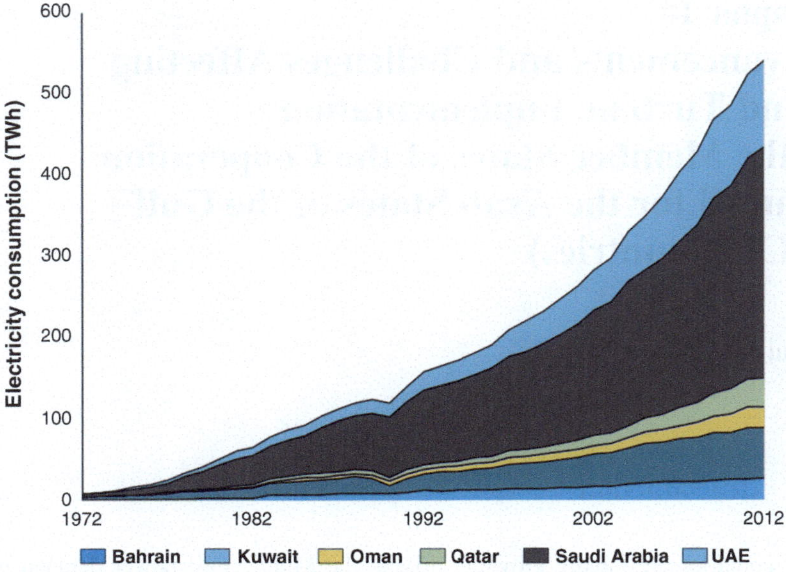

Fig. 10.1 Total primary energy consumption from 1972 to 2012, million tons of oil equivalent (Mtoe) [3]

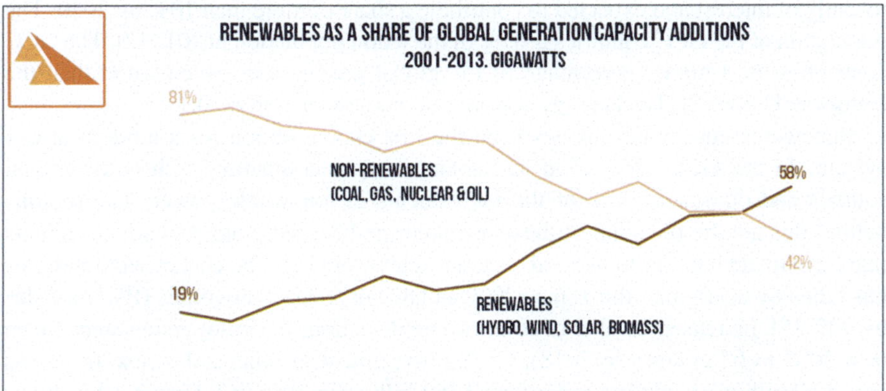

	2001	2002	2003	2004	2005	2006	2007	2008	2009	2010	2011	2012	2013
TOTAL NET CAPACITY ADDITIONS	104	134	150	153	134	180	200	174	185	232	234	233	207
RENEWABLES	20	22	34	36	41	47	56	67	85	94	115	116	120
NON-RENEWABLES	84	111	116	116	93	133	145	107	100	138	118	117	87

Fig. 10.2 Renewable and non-renewable power capacity additions, 2001–2013 [7]

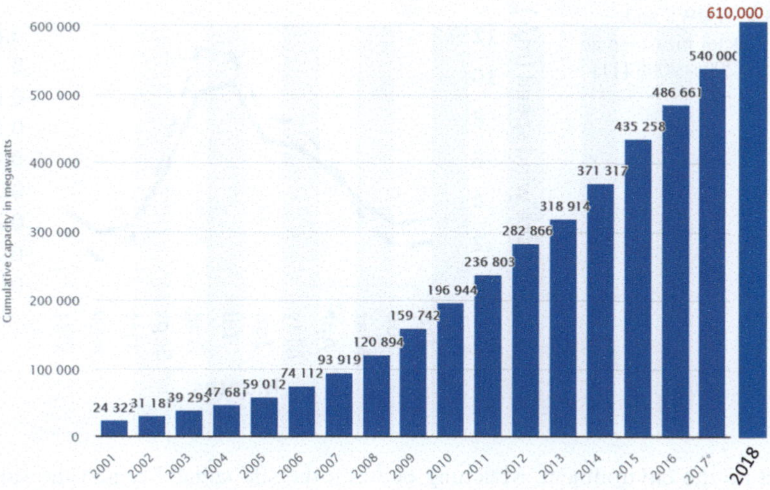

Fig. 10.3 Cumulative installation capacity

Table 10.1 Wind speeds at the GCC [9]

Country	Wind speed (m/s)	Hours of full load of wind per year
Bahrain	5–6	1360
Kuwait	5–5.5	1605
Oman	4–6	1463
Qatar	5–7	1421
Saudi Arabia	2.5–4.5	1789
United Arab Emirates	3.5–4.5	1176

Wind energy utilization in GGC countries has a moderate potential for the capacity of wind power generation. It can be considered a low speed region. However, the wind potential varies substantially from one country to the other. Table 10.1 [9] shows the recorded hours of the full load of wind per year for the GCC.

It can be seen that Saudi Arabia has the highest with 1789 h and UAE has the lowest with 1176 h. Higher speeds can be found in mountain terrains, coastal areas, and higher heights. According to the RES, countries with more than 1400 h per year are considered to have economically viable wind energy utility [10]. Some of the sites on the 2000 km coastline have great potential for offshore wind. An example of the average mean wind speed (AMWS) and capacity factor for Duqm, Oman is seen in Fig. 10.4 [11].

The GCC region is in need for this source of energy as solar can cover the daily demand only to a certain extent but wind can substitute the lack of supply in night-time. The wind energy utilization depends on several factors. These factors are the available wind speed frequency distribution over several years, social acceptance, economic benefits, technical support, distance from the main electricity grids and the

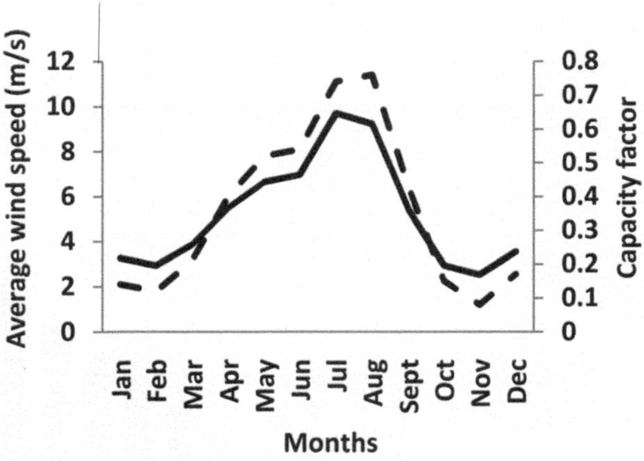

Fig. 10.4 AMWS and capacity factor measured at Duqm for 2004–2008 [11]

impact on the environment. Matching between the site suitability and the suitable wind turbine is an important issue before starting installing wind turbines in any country. Several studies were done to specify the most suitable sites and showed the feasibility of implementation (e.g. [12]). The lack of public awareness and detailed research has slowed down the implementation. This chapter aims to raise awareness and presents advancements in the technology and data analysis that could make the implementation possible.

Wind energy has become an important energy resource for many developed and developing countries. Technological advancement has supported the efficient harnessing of available wind energy and how to integrate it into energy transmission networks but is also due to growing recognition of the importance of renewable energy sources in climate change mitigation strategies [13, 14]. Many countries have relied on this source of energy to meet the total demand for electricity in a considerable amount. For example, Denmark is depending on wind for 40% of the domestic energy consumption [15] and is hoping to achieve 50% by 2020. In 2015, China reaches a total capacity of 148 GW [16].

High wind energy potential is available at different sites in GCC. For instant recent study showed that in Duqm, Oman has high wind speed region. Simulating a wind farm at that area showed the possibility of having a site with 25 MW with wind power density ranges between 200 and 800 W/m^2 at 80 m above the ground and can generate a total of about 75 GWh net AEP [17]. Wind availability across GCC is encouraging (e.g. KSA-central), peaks in the evening hours and decreases during daylight hours while it has an opposite behaviour in the coastal areas (e.g. KSA-east, Abu Dhabi) [18]. The highest wind speed of nearly 8 m/s winter time and 6.7 m/s at summer time. It is at higher values at higher heights.

Barriers to the implementation of wind power at the GCC region is merging within the barriers for overall renewable energy implementation which slowing down all efforts for energy utilization, such as [19–21]:

(a) Lack of renewables market. This is due to the subsidies of fossil fuel and energy use and non-liberalized market entry resulted in no renewable system can compete with the existing energy prices [20]
(b) Lack of related policies and regulations.
(c) Lack of appointed financial support.
(d) Reduced public awareness for implementing such schemes.
(e) No clear view on how to integrate the renewable energy systems with the national grid.

10.2 The Advancements of Wind Turbine Technology

Wind turbine technology had to fight for legitimacy among other renewables systems, which had an impact on the development of its technical improvements including aerodynamic parts, control, size, and other facilities. Advancements in those areas have also generated pains against the barriers that have always acted as obstacles. The recent improvements have enabled access to wind power and reduced the overall cost in spite of the existing industrial challenges [22]. The following advancements are enhancing the wind energy implementation.

10.2.1 Larger Turbines Improve Efficiency and Performance

Any percentage increase in average rotor diameter presided resulted in a large amount in energy generation and considered by many investors as attracted technology. Largest wind turbine ready for testing in the UK is the Haliade-X 12 MW [23]. Engineering giant GE Renewable Energy and the UK's Offshore Renewable Energy (ORE) signed a 5-year research programme to test the world's largest offshore wind turbine at a facility in northeast England. The turbine designed with 220 m rotor Diameter, 107 m blade length, 260 Height with 63% Capacity Factor [24]. Dong Energy has installed the world's largest wind turbines, Burbo Bank wind farm in the Irish Sea [25]. The turbine generates 8 megawatts (MW), with the largest rotor diameter of 164 m.

10.2.2 Clean Energy Meets Battery Storage

Most recent advancement in wind technology is the batteries. In this respect, Tesla here and the inroads it's made recently on the global stage to combine wind generation with battery storage in a way the world has not seen before. Batteries will support the power supply smoothness and help level out any trouble spots between the whims of the weather and the more predictable cycles of daily life [22].

10.2.3 Stable Way of Installation with Confidence

Nowadays wind turbine installations have been done with confidence as the whole process has been well modernized and experience been built up to its highest level.

10.2.4 Advanced and Efficient Site Selection Methodologies

Site selection has been well formulated with latest advanced models and programmes that have the ability to selecting and matching the right technology for the right site. Site selection is a critical part of the whole process and could have a direct effect on the success of a wind project. Advanced monitoring systems and surveying techniques have contributed to this selection process.

10.2.5 Specialised Blades Are Developed to Generate More Energy from Low-Wind Areas

The need for specialized wind rotor blades is essential particularly for the low speed regions to maximize the power extracted from the turbine over the hours of operation at any available wind speed [26].

10.2.6 Prices of Offshore Wind Is Falling Steeply and Will Keep Doing So

Figure 10.5 [27] shows that the offshore wind cost is falling steeply. This reflects on all types of wind technology and indicates a bright future for wind energy implementation, which will have an impact on the decision for investment.

10.3 Why Do We Need Wind Energy in GCC?

Several drivers are initiating the need for wind energy at the GCC countries in spite of the wealth of sunshine compared with the available wind energy potential. Those are:

- To secure power generation at night-time or through any unexpected climate change periods.
- Wind turbine normal costs dropped by more than 37%.

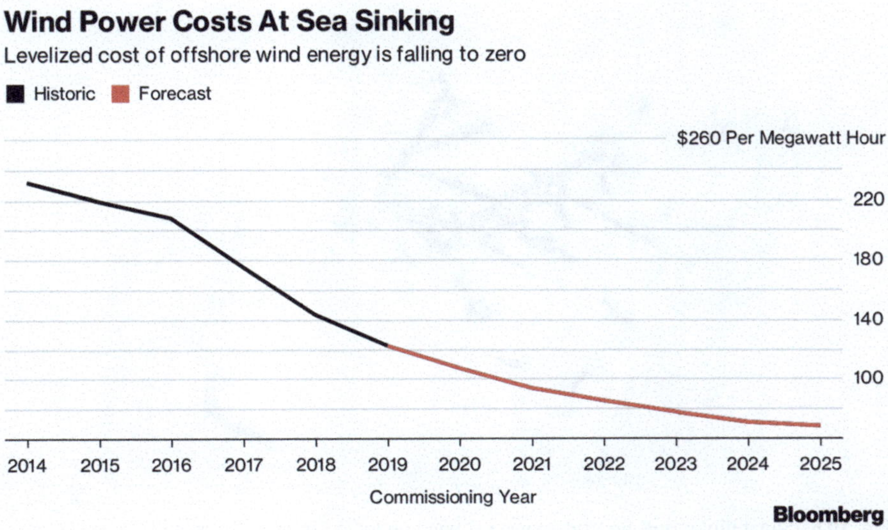

Fig. 10.5 Wind power costs falling steeply [27] (Source: Bloomberg New Energy Finance)

- By 2025, average electricity costs generated by the sun and wind could drop by as much as 59%.
- Technology to produce power from lower wind speeds is improving [26].
- Desalination is cheaper with the wind [28].
- More than 56% of the GCC's surface area has significant potential for wind deployment.
- Covering just 1% of this area could translate into an equivalent 60 GW of capacity.
- Figure 10.6 shows more than 70% (light blue) and up to 100% suitability (dark blue). Higher scores represent increased suitability [29].

10.4 Wind Energy Resources in the GCC Region Versus Oil

GCC regions exhibit significant untapped potential for wind power, driven by abundant wind resources and increasing power consumption [30]. The average wind speeds in GCC countries is comparable to Germany (3–7 m/s), wind potential has remained largely untapped so far with GCC cumulative capacities in 2012 being less than 2 MW (Germany: more than 31 GW) [30]. Wind power generation is significantly cheaper than oil-based power generation. Effectively, wind power frees up more locally produced oil available for export for all countries in this region. The average LCOE (Levelized Energy Cost) in favourable wind regions (i.e. >5 m/s) varies between 9 and 16 USD cents/kWh. This is well below the present cost of

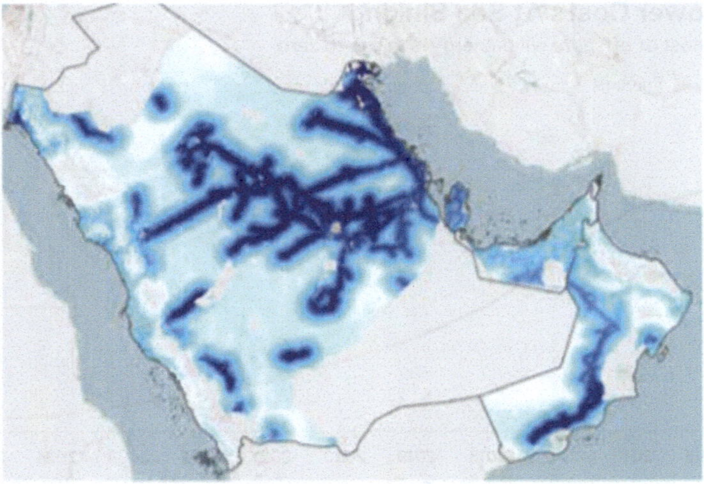

Fig. 10.6 Suitability scores for grid-connected wind up to 75 km from the grid [29]

oil-fired electricity generation of around 19 USD cents/kWh at an oil price of 100 USD/barrel. Figure 10.7 shows the Levelized cost of energy (inflation-adjusted) based on full load hours for different sites employing Vestas V100 turbine [30].

10.5 Regional Suitability for Onshore Wind Power Generation

Report by [31] indicated that site suitability is with a minimum wind speed of around 3.7 m/s at a height of 10 m in order for wind generation to be competitive with the LCOE of oil-fired electricity generation. Accordingly, Fig. 10.8 shows that there exist many suitable sites are available in GCC. For instance, in Saudi Arabia, Yanbu in the West and Juaymah and Dammam/Al Khobar/Dhahran in the East have a considerable potential for utilization and promising sites.

10.6 Wind Energy Availability Over View in the GCC Region

The GCC countries have started to raise awareness on the real need for alternative energy supply in spite of the wealth of fossil fuel. The following sections provide a brief overview for each country in this aspect.

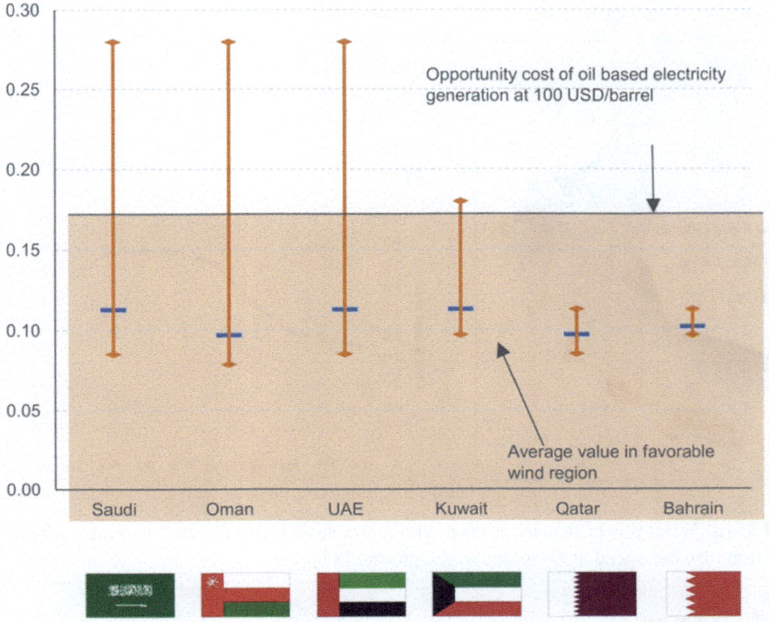

Fig. 10.7 Levelized cost of energy (inflation-adjusted) based on full load hours for different sites employing Vestas V100 turbine [30]

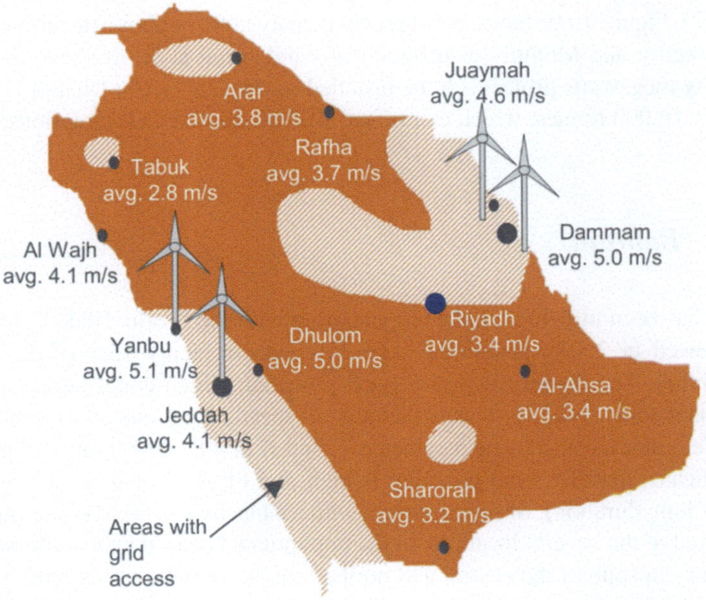

Fig. 10.8 Apricum analysis. Wind speeds measured at 10 m height [31]

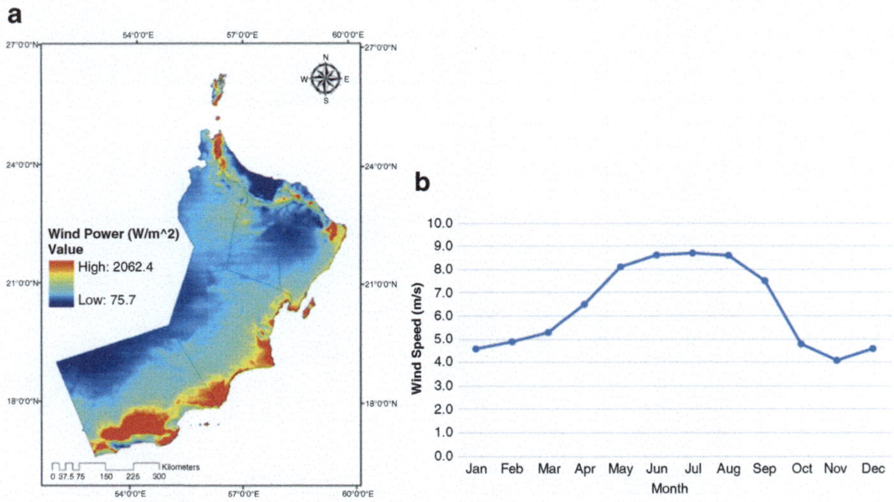

Fig. 10.9 (**a**) Wind power density at 80 m above the ground for the whole country. (**b**) Monthly distribution of wind speed at 80 m above the ground [17]

10.6.1 Oman

This country enjoys a promising amount of wind potential and started to implement the project for utilisation. Wind speed is more than 8 m/s during the summer months (June–September) (e.g. Duqm area) is shown and in many other areas between 5 and 7 m/s [17]. Figure 10.9 shows Wind power density at 80 m above the ground for the whole country and Monthly distribution of wind speed at 80 m above the ground [17]. Fifty megawatts project is to be installed by Masdar—Abu Dhabi [32]. This is to supply 16,000 homes, which can offset 110,000 tons of CO_2 emissions/year.

10.6.2 Bahrain

Bahrain has been utilizing wind energy starting with the World Trade Center (1950 and renewed in 2007), generating 0.66 MW which supply 15% of the building consumption. The country enjoys a good potential, which could attract investors. Assessment study by [33], reveals that a wind speed of 4.8 m/s at 10 m height and mean Weibull scale and shape parameters C and K of 4.8 and 1.74 m/s respectively. This indicates that the wind potential is high and of good quality (strong enough winds of long duration). Several projects with a capacity of 200 MW are possible to be invested at the several locations in the unpopulated central and southern parts of the country in spite of the central and northern parts are with higher wind potential, Fig. 10.10, [33]. Since wind energy can be used for reverse osmosis desalination plants (RO) [28], it is creating opportunities for investment in many areas in Bahrain.

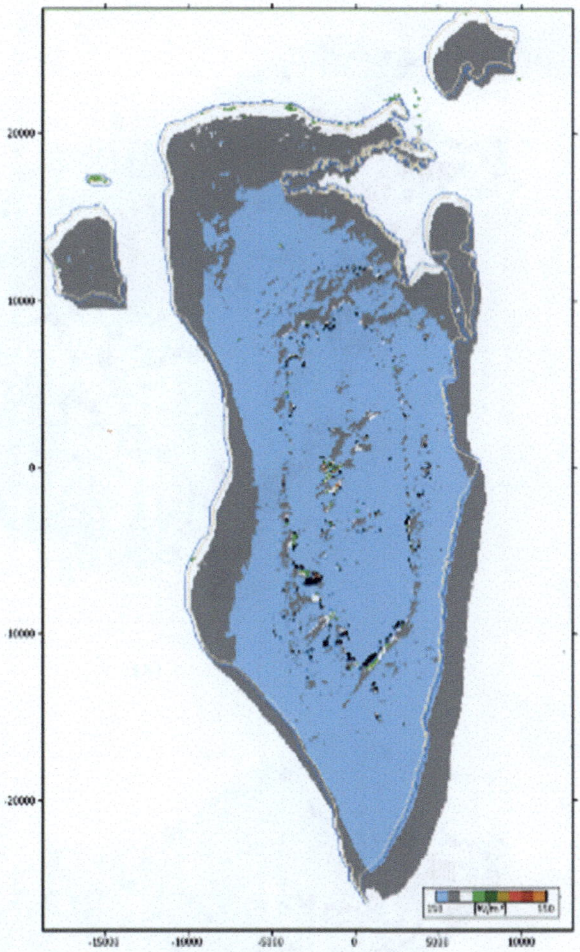

Fig. 10.10 The wind power geographical distribution at 10 m height [33]

10.6.3 *Saudi Arabia*

Within their energy vision for 2030, Saudi Arabia targets to generate 54 GW of renewable energy. Nine gigawatt target is from wind for 2040 which contributes to 6.3% of the total power capacity. Figure 10.11 shows the wind map for Saudi Arabia for February and August 2015 [34] which also shows three regions of wind availability in the coastal region, 7.5–8 m/s at Hafar Al-Batin, 7–7.5 m/s at Yanbu and 5.5–6 m/s T Riyadh in the central region [34]. Saudi Aramco commissioned the first wind turbine in Turaif [35], 2.75 MW can supply power to 250 households that saved 19,000 barrels of oil-equivalent. Keeping a similar investment will obviously enable the country to achieve their targets earlier.

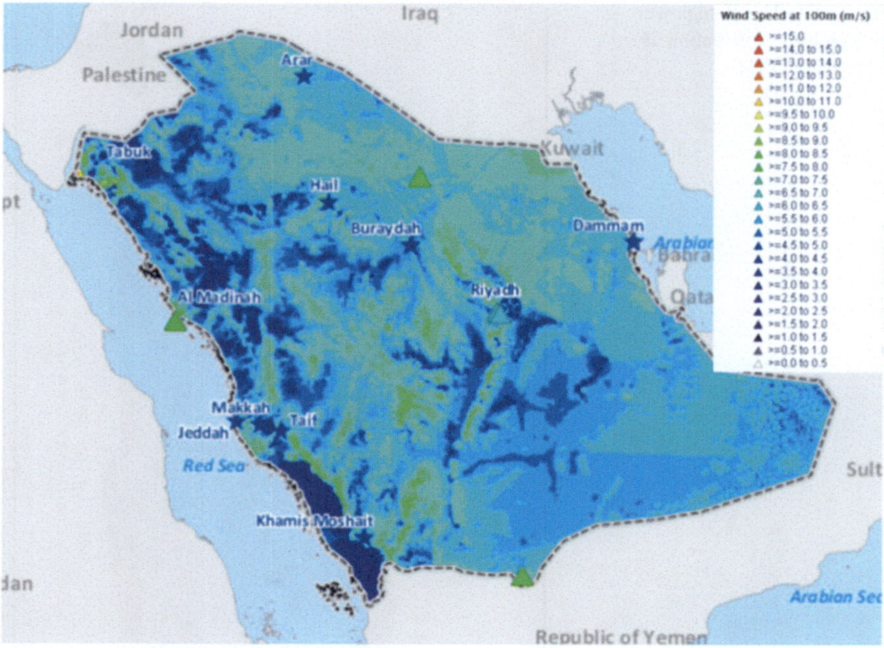

(a)

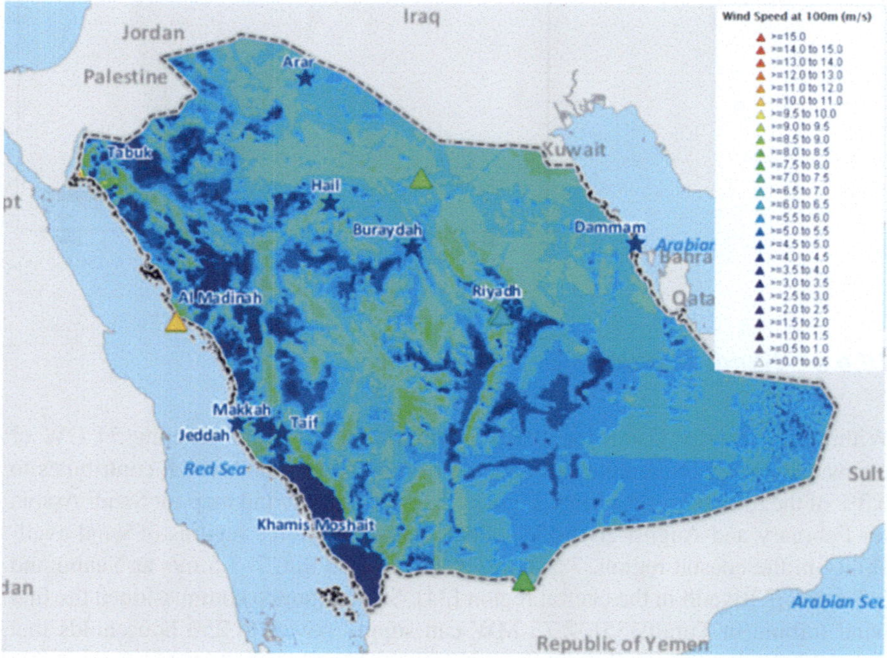

(b)

Fig. 10.11 Average wind speed in m/s for February 2015

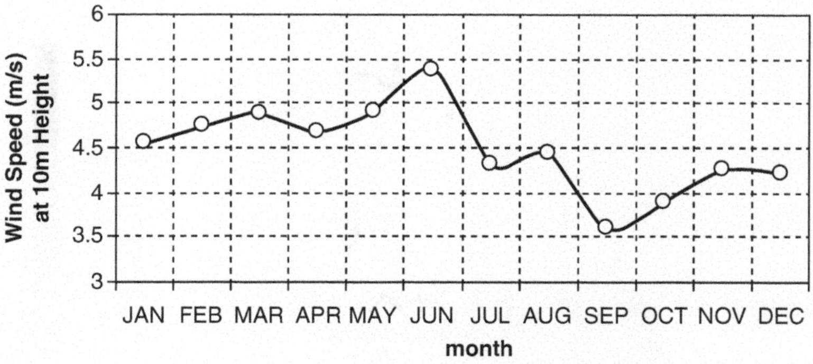

Fig. 10.12 Mean monthly variation of wind speed in Doha (1976–1989) [36]

10.6.4 Qatar

Wind energy potential in Qatar has also been studied [36]. Results showed that the average wind speed is 4.4 m/s. A good potential can be seen over the year between March and July, see Fig. 10.12. Utilization of available wind energy is possible through small wind energy systems, which can be both efficient and competitive. Wind pumping and small-scale electricity generation can be implemented. However, the present work recommends the offshore wind farms, as the nearby areas are promising.

10.6.5 Kuwait

Annual average wind velocity at the standard height of 10 m was found to be between 3.7 and 5.5 m/s [37]. This indicates a promising potential and many projects can generate a reasonable amount of energy. The Salmi Mini-Wind Farm of 2.4 MW was established to test and measure the performance of the equipment and located near the site of the first 70 MW Shagaya Renewable Energy Park [38]. The wind will have the capacity to generate 10 MW. The monthly variation analysis shows high WPD during the high electricity demand summer season than other months with maximum WPD of 555 at Al-Wafra see Fig. 10.13.

10.6.6 United Arab Emirates

As studies have shown [39], the possible occurrence of wind speeds greater than >5 m/s is monitored onshore, and wind speeds exceed >7.0 m/s near-shore. Moreover, energy utilization from wind at a height of 80 m is possible, to achieve a full load hour of 1176 h per year which refers to low capacity factor. Wind farms in the

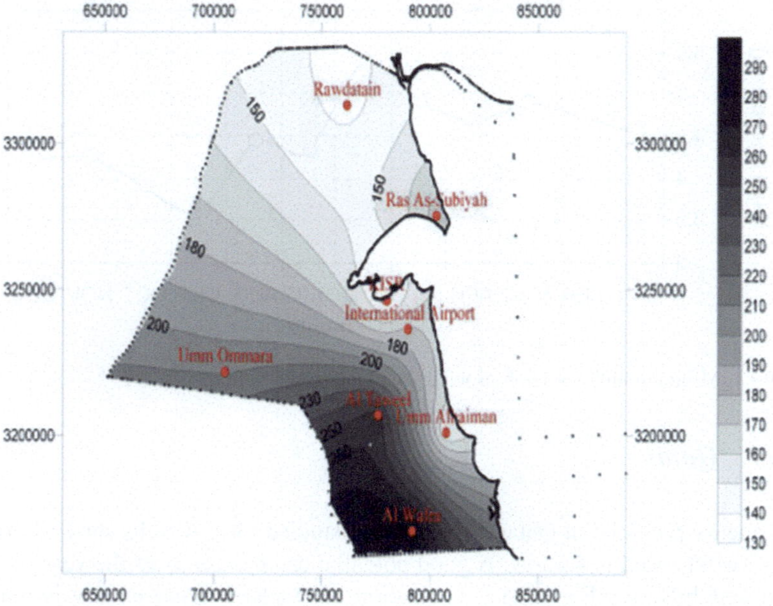

Fig. 10.13 Distribution of wind power density over Kuwait at 30 m height [38]

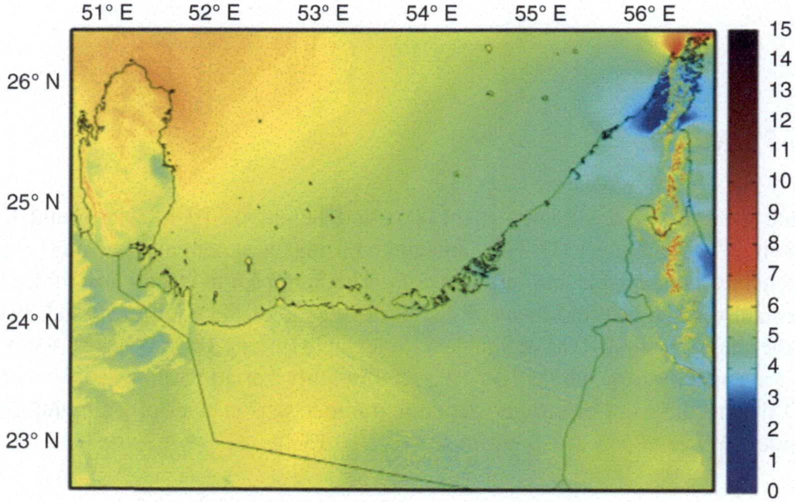

Fig. 10.14 The UAE wind resource map using GIS, WindPRO and WAsP programmes [40]

capacity of 1–2 MW are possible with this available wind speeds if careful attention is paid to the recent advances. Other low-speed turbines can well be implemented small to medium size. The maximum wind speed is generated in the north region in the UAE, whereas the maximum wind speed potential is mainly derived from coastal areas as seen in Fig. 10.14 [40]. High potential is also visible offshore.

10.7 Challenges to Implementation

The followings are some challenges that might form an obstacle to the implementation of wind energy systems at the GCC regions:

- Suitable wind sites are with low average wind speed.
  - Low-speed rotors have been developed for low wind speed regions [26].
- High initial investment.
  - Establishing a viable wind turbine site just cannot compete with electricity generated from more conventional means.
  - This is reduced over the years.
- Geography
  - Sites that best suit wind generation are not close to population centres; will require building lengthy and expensive transmission lines.
- Transportation
- Noise solved by:
  - Design changes (more air converted to power and less goes to acoustic).
  - Proper siting.
  - Insulating materials can be used to minimize noise impacts.
- *Visual Impacts* (Proper siting decisions can help to avoid any aesthetic impacts to the landscape and using today's larger and more efficient models of wind turbines).
- *Avian/Bat Mortality* (is minimised by proper site selection and use of latest technology to deter birds away from the turbine blades).
- Primary health and safety considerations.
- Interference with radar and telecommunication facilities.

10.8 Conclusion

The present wind energy business case is prevented by a number of barriers, but our survey results clearly show that the respondents believed that policies can make a major difference and allow for an expansion of wind systems installations at the GCC region. This work reviews the main advances with the design, implementation, and installation of the wind energy systems. It aims to highlight the fact that all barriers for the attempts to use this wind technology in this region is a myth, considering the available wind potential in each country and advancements in the technology. Hurdles still held back the deployment of more wind systems across many promising locations in the GCC. These hurdles have disappeared as the technology has developed for the low wind speed areas and there is a sharp reduction in the cost. The advancements in data analysis techniques have enabled accurate assessment of wind availability in these countries and have shown a wider range of promising sites for utilization, which formed a competitive advantage against the other forms of energy generation. Companies all over the world are much experienced nowadays to select and install and maintain which provides confidence to investors. This study highlights these advancements and the good potential of wind energy at the GCC region, thus recommending more investments in utilizing that potential in order to secure power generation at night-time or through any unexpected climate change periods. Wind energy resource for these countries will serve as the second source of clean energy after solar if implemented in the right way.

References

1. lbawaba, Business (2017). Retrieved from albawaba: https://www.albawaba.com
2. M. Indraganti, D. Boussaa, An adaptive relationship of thermal comfort for the Gulf Cooperation Council (GCC) countries: the case of offices in Qatar. Energy Build. **159**, 201–212 (2018)
3. Kapsarc.org. (2018), [online] https://www.kapsarc.org/wp-content/uploads/2017/11/KS-2017-MP04-GCC-Energy-Overview-2017.pdf. Accessed 10 July 2018
4. A. Darwish, Sustainable green smart buildings: future energy survivor. ISESCO. J. Sci. Technol. **12**(21) (2016)
5. IRENA, *Rethinking Energy 2017: Accelerating the Global Energy Transformation* (International Renewable Energy Agency, Abu Dhabi, 2017)
6. Centre/BNEF, r. S.-U, Global trends in renewable energy investment 2018 (2018), https://cleanenergysolutions.org/sites/default/files/.../christine-2018_e_short.pdf: Frankfurt School of Finance & Management gGmbH
7. J. Glennie, Wind/solar cost declines & global renewables growth (2014). Retrieved from SASK Wind www.saskwind.ca
8. S.A.M. Rehman, A review of wind-turbine structural stability, failure and alleviation, in *Advances in Civil, Environmental, & Materials Research (ACEM18)*, Songdo Convensia, Incheon, Korea: (ACEM18), 27–31, Aug 2018
9. R. Ferroukhi, N. Ghazal-Aswad, S. Androulaki, D. Hawila, T. Mezher, Renewable energy in the GCC: status and challenges. Int. J. Energy Sector Manag. **7**(1), 84–112 (2013). http://dx.doi.org.ezproxy.bolton.ac.uk/10.1108/17506221311316498
10. W.E. Alnaser, N.W. Alnaser, Solar and wind energy potential in GCC countries and some related projects. J. Renew. Sustain. Energy **1**(2), 1–28 (2009)
11. M.J. Shawon, L.E. Chaar, L.A. Lamont, The GCC: wind technology deployment potential, in *2011 IEEE GCC Conference and Exhibition (GCC)*, Dubai, 2011, pp. 174–177
12. A.M. Eltamaly, Design and implementation of wind energy system in Saudi Arabia. Renew. Energy **60**, 42–52 (2013)
13. T. Bruckner, I. Bashmakov, Y. Mulugetta, H. Chum, A. de la Vega Navarro, J. Edmonds, A. Faaij, B. Fungtammasan, A. Garg, E. Hertwich, et al., Climate change 2014: mitigation of climate change, in *Contribution of Working Group III to the Fifth Assessment Report of the Intergovernmental Panel on Climate Change*, Cambridge University Press, Cambridge, 2014
14. M.G. Genton, *Skew-Elliptical Distributions and Their Applications: a Journey Beyond Normality* (CRC Press, Boca Raton, 2004)
15. DEA, *The Danish Energy Model: Innovative, Efficient and Sustainable* (Danish Energy Agency, København, 2016)
16. WWEA, *WWEA Quarterly Bulleting Issue 1-206: Community Wind Special* (World Wind Energy Association, Bonn, 2016)

17. S. Al-Yahyai, Y. Charabi, Assessment of large-scale wind energy potential in the emerging city of Duqm (Oman). Renew. Sustain Energy Rev. **47**, 438–447 (2015)
18. S. Al Otaibi, Energy consumption in Kuwait: prospects and future approaches. Energy Policy **39**, 637–643 (2011)
19. A. Mokri, M.A. Ali, M. Emziane, Solar energy in the United Arab Emirates: a review. Renew. Sustain Energy Rev. **28**, 340–375 (2013)
20. A. Mondal, H.S. Khalil, *Renewable Energy Readiness Assessment Report: The GCC Countries* (Masdar Institute, Masdar, 2012)
21. J.P.A. Lilliestam, Barriers, risks and policies for renewables in the Gulf states, in *Energies*, ed. by V. Dovì, vol. 8, (2015), pp. 8263–8285. Available online: https://pdfs.semanticscholar.org/2c2d/0606dc6582592fd1f43dd8ddefd9562e9921.pdf
22. K. Matthews, BLOG POST (2017). Retrieved 7 Apr 2018, from planetizen. Inc. (US). www.planetizen.com
23. J. Murray, Haliade-X: World's largest wind turbine coming to the UK (2018). Retrieved 27 Apr 2018, from BusinessGreen: https://www.businessgreen.com/bg/news/3030821/haliade-x-worlds-largest-wind-turbine-coming-to-the-uk
24. G.R. Energy, Haliade-X offshore wind turbine platform (n.d.). Retrieved 24 June 2018, from ge.com: https://www.ge.com/renewableenergy/wind-energy/turbines/haliade-x-offshore-turbine
25. Environment, Dong installs world's largest wind turbines off UK coast (n.d.). Retrieved 27 June 2018, from Reuters: https://www.reuters.com/article/us-dong-energy-windfarm-britain-idUSKCN11E1F9
26. A. Salam Darwish et al., A methodology for improving wind energy production in low wind speed regions, with a case study application in Iraq. Comput. Indr. Eng. **127**, 89–102 (2019)
27. J. Shankleman, Subsidy-free wind farms risk ruining the industry's reputation (2017). Retrieved from Bloomberg NEF: https://about.bnef.com/blog/subsidy-free-wind-farms-risk-ruining-the-industrys-reputation/
28. J. Dodd, (2017). Retrieved 25 Apr 2018, from https://www.windpowermonthly.com/article/1436396/uae-desalination-cheaper-wind
29. IRENA, *Renewable Energy Market Analysis: the GCC Region* (IRENA, Abu Dhabi, 2016)
30. F. Mayr, Vast untapped growth potential exists for wind players in the Gulf region (2013). Retrieved from Apricum: https://www.apricum-group.com/hidden-champions-of-the-emerging-wind-markets/
31. Eltamaly et al. 2012, Rehman et al. 2003, Apricum analysis, Apricum, www.apricum-group.com
32. Grupotsk, TSK to build the largest wind farm in the Gulf (n.d.). Retrieved 25 Apr 2018, from en.grupotsk: http://en.grupotsk.com/noticias/tsk-construira-el-mayor-parque-eolico-del-golfo
33. H. Albuflasa, T. Murray, D. Infield, S. Watson, Feasibility study of wind power in the kingdom of Bahrain, in *European Wind Energy Conference and Exhibition 2007, EWEC 2007*, vol. 3 (2007)
34. M.A.M. Ramli et al., Analyzing the potential and progress of distributed generation applications in Saudi Arabia: the case of solar and wind resources. Renew. Sustain Energy Rev. **70**, 287–297 (2017)
35. M.M. Damoom et al., Adding sustainable sources to the Saudi Arabian electricity sector. Electr. J. **31**(4), 20–28 (2018)
36. A.H. Marafia, H.A. Ashour, *Feasibility of Wind Energy Utilization in Qatar* (Pergamon, Oxford, 2002)
37. W. Al-Nassar, , S. Alhajraf, A. Al-Enizi, L. Al-Awadhi, Potential wind power generation in the state of Kuwait, Renew. Energy, 30, 2005, pp. 2149-2161.
38. S.A. Fathi, All Arab states should follow renewable energy possibilities (2017). Retrieved from Gulf News: https://gulfnews.com/business/analysis/all-arab-states-should-follow-renewable-energy-possibilities-1.2026865

39. Areas with annual full load hours over 1400 h/y equivalent to a capacity factor of 16% are considered in this study as long-term economic potential; in German Aerospace Center (DLR). Concentrating solar power. Table 3–2. p. 57 and 68.
40. S. Sgouridis et al., RE-mapping the UAE's energy transition: an economy-wide assessment of renewable energy options and their policy implications. Renew. Sustain Energy Rev. **55**, 1166–1180 (2016)

Chapter 11
Wind Energy in the UK: Progress and Future Expectations

Abdul Salam Darwish

11.1 Introduction

The worldwide energy is evolving. The demand of conventional markets has surpassed supply due to fast-growing developing markets and almost doubled world GDP. The energy mix is moving, driven by innovative upgrades and ecological concerns. Like never before, energy supply industry needs to adjust to meet those demands in the close term. Consequently, the use of fossil fuels is to continue for more than a decade, as predicted by the German Advisory Council on Global Change (WBGU), and peak in 2030, then gradually fall off towards 2100, Fig. 11.1 [1].

Renewable energy technical innovations for utilization are frequently the first decision for growing modernizing power framework around the globe since the installation capacity of renewables has become greater than that of the non-renewables by a rising edge as shown in Fig. 11.2 [2].

This worldwide capacity reached 456 GW by the end of June 2016 [3] and was expected to approach 500 GW the first quarter of 2017. By this achieved capacity, the wind power will contribute with 5% to the global power supply, see Fig. 11.3 [3, 4].

Developing wind technology and generating power raises costs of electricity slightly. £18 added to the customers' average yearly electricity bill [5], in recent years in order to generate about 10% of the annual total. Offshore is more expensive than the onshore, which resulted in more rise in the cost of electricity, which has reached 2012–2014, had levelised cost of electricity of £131/MWh [5].

A. S. Darwish (✉)
Phoenix Renewable Energy Centre, Manchester, UK

University of Bolton, Bolton, UK
e-mail: Darwish@phintc.com

A. Sayigh, D. Milborrow (eds.), *The Age of Wind Energy*, Innovative Renewable Energy, https://doi.org/10.1007/978-3-030-26446-8_11

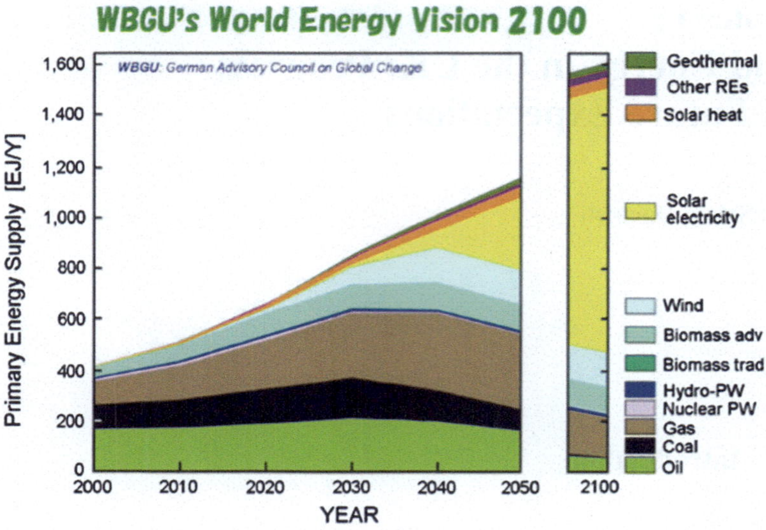

Fig. 11.1 Shows a forecast of the global energy supply until year 2100 [1]

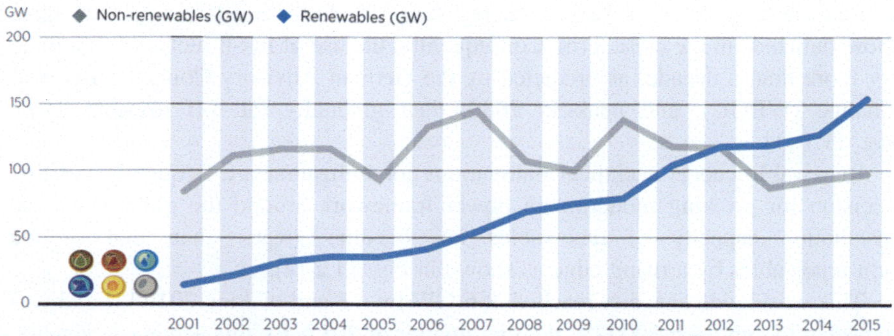

	2001	2002	2003	2004	2005	2006	2007	2008	2009	2010	2011	2012	2013	2014	2015
Renewables (GW)	14	21	31	35	35	41	54	69	75	79	104	118	119	127	154
Non-renewables (GW)	84	111	116	116	93	133	145	107	100	138	118	117	87	93	97

Source: IRENA, 2016b

Fig. 11.2 Renewables' and non-renewables' power capacity, 2001–2005 [2]

Electricity at the present and in the future of UK has a crucial role in economic and societal well-being. Developments in technologies are increasingly demanding electrical power including all types of transports, industry and all other sectors. Recent updates showed that renewables are the source for most of Europe's new power with a share of 90% of new EU power generation capacity recording an installed wind power capacity is 135 GW which is 17% of the available capacity 918.8 GW [6]. UK's electricity consumption December 2016 has increased slightly by 0.1% in 2016, from 302.7 TWh in 2015 to 302.9 TWh [7]. Figure 11.4 shows that total electricity generated In 2016 Q4 from renewables, decreased from 26.8% in 2015 Q4 to 22.2% 2016 Q4, due to fall in wind speeds and rainfall [7].

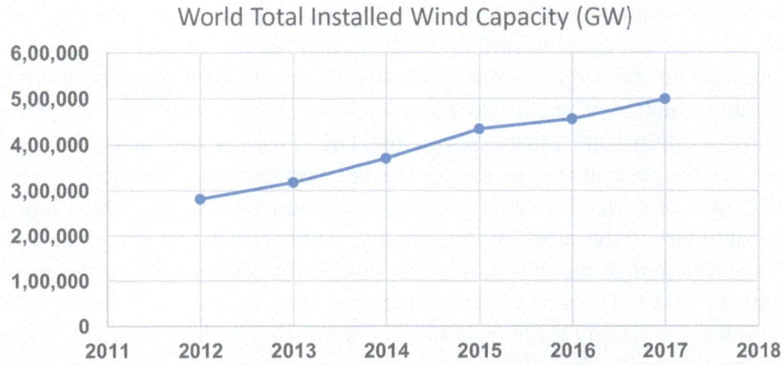

Fig. 11.3 World total installed wind capacity

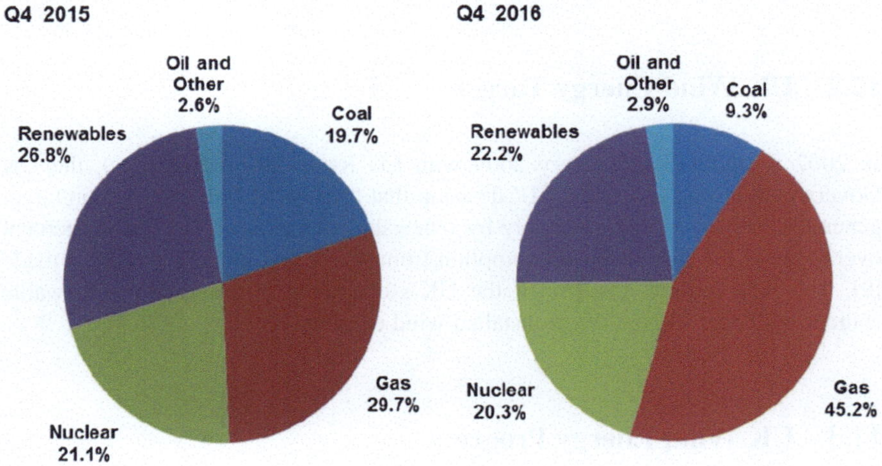

Fig. 11.4 Shares of electricity generation [7]

Onshore wind power generation has built up itself as a well-developed, clean and gainful innovation. It is currently the UK's biggest renewable and sustainable energy source. The onshore wind power generation contributed significantly to CO_2 reduction, added to the national economy and supplied a secured source of energy. The high onshore energy potential made the technology works very well and effectively. Projects are developed. The diverse range of people who have invested in onshore farms for development is getting wider such as large energy companies and independent developers and community groups or small businesses and farms [8].

Offshore wind capacity is getting greater over the recent years, which has contributed to an increasing amount of electricity every year and expected to stay at a higher trend. Offshore are leased in "rounds" [8] that made it easier than the onshore projects to be tracked and monitor its project developments. Rounds 1–3 were considered over the years 2001, 2003 and 2010, and in addition to these rounds, the Scottish government has overseen a further development programme. A survey of those projects discussed in further details in this paper.

In addition to the large onshore and offshore wind farm projects, there are a considerable number of small and medium wind size projects that are powering many homes, farms and businesses in the UK. Projects and incentives are put forward by the government to encourage the principle of "Generate your Own Power". Small and Medium Wind Strategy was launched by [8], which has introduced a summary to the scope of the industry, its benefit and its potential. A recent report suggested that the UK could be producing 85% of its own renewable electricity by 2030. The cost of renewables has been driven down by the government, resulting in having these technologies competing with other technologies. In addition, in this work, the effect of Brexit from the European Union on the Wind Energy development is reviewed with an emphasis on what will be expected to happen regarding the UK to meet its EU target.

11.2 UK Wind Energy Target

In 2007 (formalised in January 2009 with EU Renewables Directives), the UK Government has agreed to the EU' decision that the overall European Union target generates 20% of its energy supply by renewable energy in 2020. Fifteen percent overall target for UK energy consumption from renewable sources in 2020 is made [9]. This will require 35–40% of the UK's electricity to come from renewable sources by 2020, 33–35 GW of installed wind capacity [10].

11.3 UK Wind Energy Progress

In order for the UK to overcome this large challenge to meet the 2020 energy generation target, policies and practices considered. One of which is the Renewables Obligation (RO), through which the electricity supply companies must move in steps to source 10.4% of their electricity from renewable generators by 2011, or purchase equivalent Renewable Energy Certificates from companies who have excess, or pay a 3p/kWh "buy out" fine [11]. This has resulted in an increase in wind generation from 2 to 53% (Phased out in April 2017). Likewise, Contracts for difference (CfDs) which are being presented as a major aspect of Electricity Market Reform (EMR)—will supplant the Renewables Obligation (RO) for large-scale projects. "Newer build" [12] was introduced for onshore and offshore stations commissioned for the past 2 years to indicate the plant efficiency. Wind generation has since progressed significantly with a large amount of investment, which has had its impact on the generation of the non-renewable resources. Figure 11.5 (Schlissel, March 2017) shows the evolution of Britain's electricity supply and demand, which indicates a drop in combined gas and coal generation in recent years, which was balanced by a

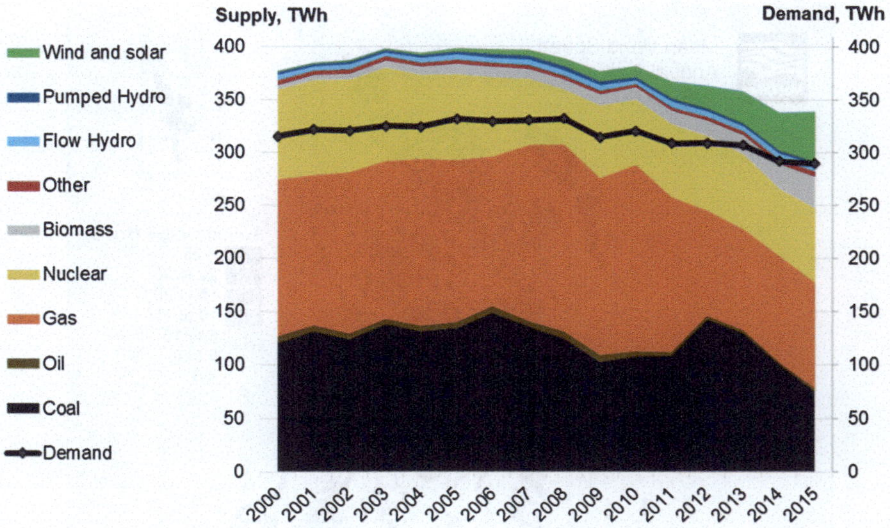

Fig. 11.5 UK generation and demand [13]

combination of falling demand due to energy efficiency practices, and growth in the amount of renewable technologies deployed with a considerable amount of wind energy technologies installed. There is a significant evidence that that greenhouse gas (GHG) emission is reduced by the development of the wind generation installation capacity [8]. Wind energy applications in the UK have been developed sharply for the last 8 years. The UK now is the world's SIXTH largest producer of wind power expected to approach 26 GW by 2020 [14, 15].

11.4 UK Wind Energy Potential and Electricity Generation from Wind Energy in the UK: 2017

The UK enjoys very high wind energy potential onshore and offshore. Figure 11.6 [16] shows that annual average wind speeds greater than 5 m/s are nicely spread all over the country and the northern regions (Scottish Islands and North Atlantic) are significantly windier than the south.

Ten windiest regions detected from this potential are shown in Table 11.1 and Fig. 11.7, Data from [16].

Offshore wind energy is in higher amount, Fig. 11.8 shows a highest energy potential as the distance from the coastal areas gets longer [17].

In general, wind farms categorised into four categories:

- Consented
- Planning
- Under construction
- Operational

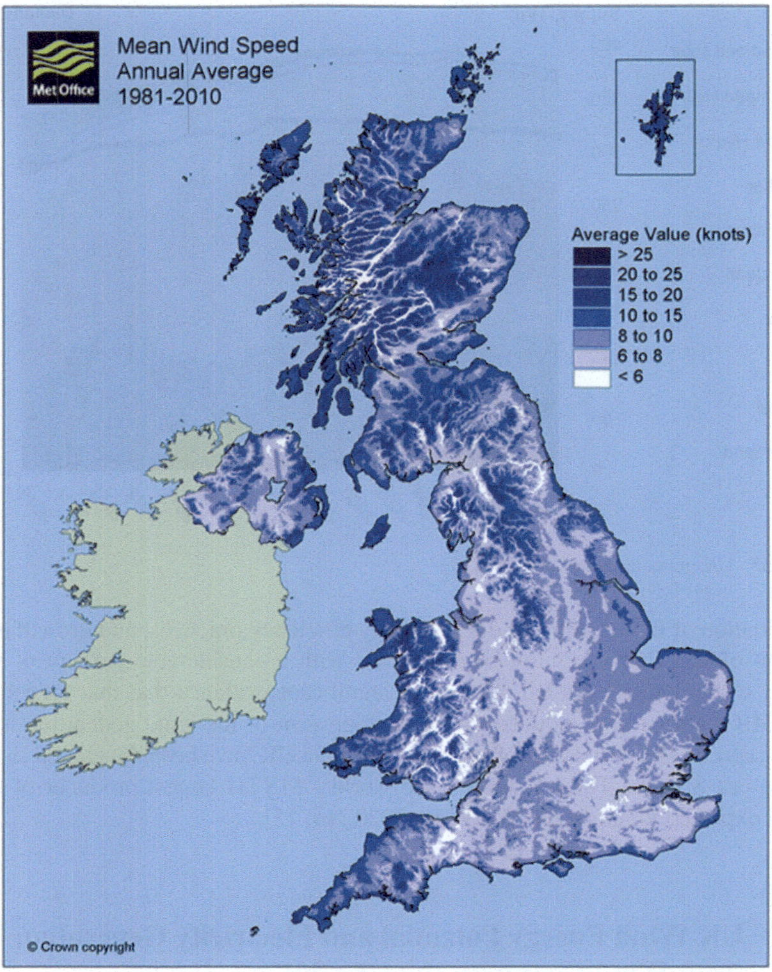

Fig. 11.6 Annual average mean wind speed, 1981–2010 [16]

Table 11.1 The ten windiest zones in the UK [16]

Station	Wind area	Wind speed in m/s
1	Shetland area	7.6
2	Buteshire	7.5
3	Orkney area	7.4
4	Caernarvonshire	6.7
5	Western Isles	6.5
6	Argyllshire	6.3
7	Anglesey	6.2
8	Inverness	6.1
9	Peeblesshire	5.8
10	Ross and Cromarty	5.8

Fig. 11.7 Ten windiest zones in the UK

Table 11.2 Onshore and offshore wind energy installed capacity, 2017

	UK 2016 Q3–2017 Q1 [8]		UK 2020 target		Gap 2020	
	GW	GWh	GW	GWh	GW	GWh
Onshore wind	9.5	22,700	14.9	34,150	5.4	11,450
Offshore wind	5.1	16,207	13	44,120	7.9	26,490
Total	14.6	38,907	27.9	78,270	13.3	37,940

Up to date statistics showed the amount of wind energy installed capacity for both onshore and offshore with a significant progress achieved by March 2017. This is shown in Table 11.2, which also indicated the amount of the wind capacity required to meet the European 2020 target and what would be the gap between what actually been achieved and what must be developed in order to achieve that target.

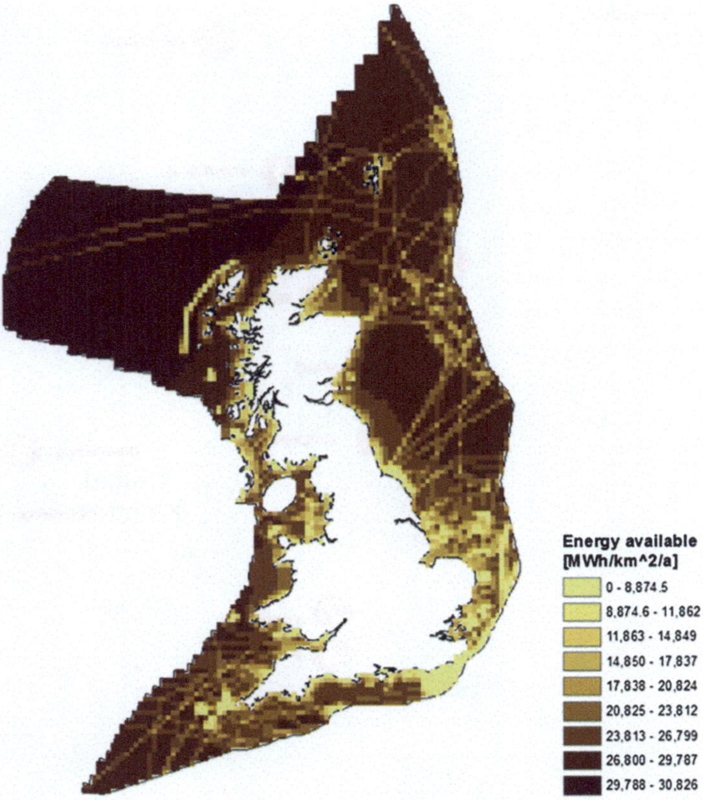

Fig. 11.8 Energy available for exploitation (MWh/km^2/a) [17]

Evolution of electricity generation from the wind technology 2008–2017 is summarised in Fig. 11.9, which shows an increased capacity up to date.

11.5 Onshore largest Wind Farms Projects

There are about 1465 onshore projects in operation [8], from which are many at a large capacity. Whitelee wind farm (539 MW) is one of the largest onshore located on Eaglesham Moor (Scotland). It consists of 215 Siemens and Alstom wind turbines each turbine with a hub height of 65 m and rotor diameter of 90 m with an overall cost around £600 M. The farm is powering 300,000 UK homes. It occupies 55 sq. km and being in operation since 2009. Second extension installed in 2011 [21, 22]. Second interesting wind farm is CLYDE wind farm (350 MW), located on South Lanarkshire (Scotland) and situated between Biggar and Moffat occupying around 47 square kilometres. It consists of 152 turbines are split into three sections and being in operation summer 2011 [22]. Table 11.3 shows some of these farms.

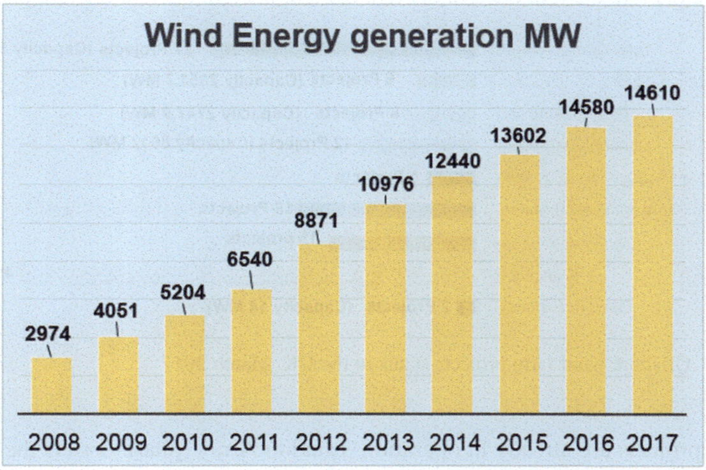

Fig. 11.9 Electricity generation from wind energy in UK (2008–2017)—data collected by author and from [8, 18–20]

Table 11.3 Six largest onshore wind farms in the UK

Wind farm	Location	Turbine	Power per turbine (MW)	Number of turbines	Total capacity (MW)	Date in operation
Whitelee	Eaglesham Moor	Siemens and Alstom	2.3	215	539	2009, 2011
CLYDE	South Lanarkshire	Siemens	2.3	152	350	2011
Crystal Rig	Scottish borders	Nordex N80/2500 Siemens	2.5	85	200.5	2004, 2007, 2010
Griffin	Perthshire	Siemens	2.3	65	156.4	2012
Hadyard Hill	South Ayrshire	Bonus	2.3	52	130	2006
Black Law	Lanarkshire	Siemens	2.3	54	124.2	2005

11.6 Offshore largest Wind Farms in Operation

Offshore wind farms are classified as fully commissioned, partial generation (under construction), pre-construction, consent authorised, consent submitted, concept planning, and developed [23]. Twenty-seven projects are in operation up to date with Capacity 5154.3 MW with stated cost of £16 bn, and around six large projects are under construction with partial operation with Capacity 2553.7 MW. Another six projects are in pre-construction phase with capacity of 2747.9 MW and 12 projects have had an authorised consent with capacity of 8632 MW. This makes it about 14 GW of offshore wind generation is coming in the near future (Fig. 11.10).

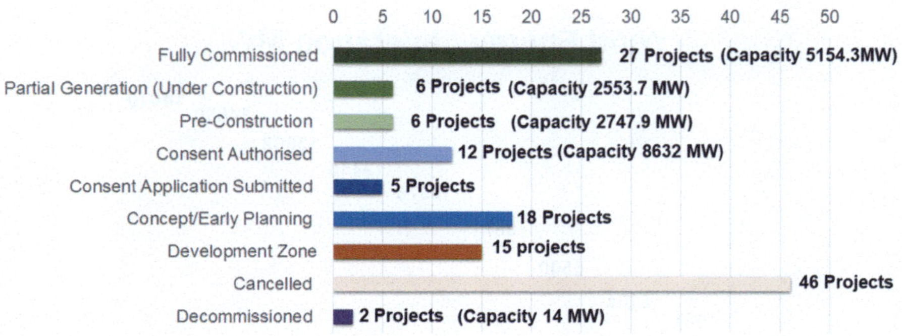

Fig. 11.10 Offshore wind farm projects status in the UK, March 2017

Large projects are already in operation. London Array (phase one) is the world's largest project with 630 MW capacity located 20 km off the Kent coast in the outer Thames Estuary in the UK. One hundred seventy-seven foundations were started on 2011 with 175 turbines with 120 m rotor diameter, each turbine generating 3.6 MW. It occupies an area of 100 square km [24]. The project is supported with two offshore substations and one onshore substation and used nearly 450 km of offshore cabling. The project generates enough power for nearly 500,000 UK homes a year equivalent to two-thirds of the homes in Kent. The project achieves 925,000 tons per year in CO_2 reduction. Four major investors are behind the London Array. They are [24] as follows:

- E.ON, which is one of the main green generators, with 20 wind farms, found both on and offshore has 30% share of the London Array project.
- La Caisse is one of Canada's leading institutional fund managers, has 25% share of the London Array Project.
- Denmark-based DONG Energy is one of the leading European energy groups, has 25% share of this London Array Project.
- Masdar is Abu Dhabi's renewable energy company advancing the development, commercialisation and deployment of clean energy technologies and solutions, has 20% share of this London Array project.

11.7 What Impact Is Brexit Likely to Have on the UK's Wind Energy Industry and Targets?

The UK decision to leave the European Union (EU) has created a misunderstanding to the future of renewable energy targets and interrelations between investors of these technologies. Following the referendum June 2016, ministers have announced on different occasions that the UK is continuing to invest at the same momentum on clean technology and will work hard to achieve all targets. It is obvious as a consequent to Brexit; the UK will not be under the pressure of the EU's targets under the EU Renewable Energy Directive. It is likely that the electricity generated from renewables will still be in the future energy mix for the reason of the UK is still to comply with its national emission reduction targets under the Climate Change Act 2008 and its international climate change obligations [25]. This chapter gathers the following thoughts for this issue:

- The UK would follow an easier route on heat and transport policies in the short term.
- Most renewable electricity projects that are needed to achieve the 30% goal have already consented and many are under construction.
- The UK will introduce tougher requirements to make sure that CO_2 emissions are cut with the promised amount in 2050.

11.8 Conclusion

The UK enjoys a large amount of wind energy potential; if utilized appropriately, it will provide six times its future electricity demands. Onshore wind energy installed capacity is still higher than the offshore capacity. The cost of wind turbines has fallen by nearly one-third since 2009, encouraging an increased number of countries to hold auctions to deploy renewables. Offshore wind has the potential to be a significant part of an economic and low-carbon UK energy system. Floating offshore wind technology can provide access to high-quality wind resources discarding the limitation of depth. Challenges require a key move in society's attitude to and proper use of energy and will only be met with the support of both household and business clients. It is concluded that the UK will not be able to meet its EU target by 2020. However, by considering the large number of projects under construction and already consented, a good outcome of installed capacity will be seen and the total energy generated is expected to reach 26 GW by 2020.

References

1. I. Leung, Australia's solar industry: threats and opportunities (2011), http://www.electronicsnews.com.au/Features/Australia-s-solar-industry-threats-and-opportuniti. Accessed 9 Dec 2015
2. P. Koekoek, 7 powerful charts that prove renewable energy is not "hippie talk" (2017). Retrieved from dailyplanet.climate-kic.org: https://dailyplanet.climate-kic.org/7-powerful-charts-that-prove-renewable-energy-is-not-hippie-talk/
3. L. Tyler, WWEA: global wind expected to reach 500 GW by year-end (2016). Retrieved from North American Wind Power: http://nawindpower.com/wwea-global-wind-expected-to-reach-500-gw-by-year-end
4. A.S. Darwish, *Wind Energy Potential in the UK. Lecture Notes and Private Contributions* (Phoenix Renewable Energy Centre, Manchester, 2017)

5. en.m.wikipedia, Wind power in the United Kingdom (2017). Retrieved from Wikipedia: https://en.m.wikipedia.org/wiki/Wind_power_in_the_United_Kingdom
6. Vaughan, Almost 90% of new power in Europe from renewable sources in 2016 (2017). Retrieved from the Guardian: www.theguardian.com/environment/2017/feb/09/new-energy-europe-renewable-sources-2016
7. Government, Electricity statistics (2016), Gov.uk, https://www.gov.uk/government/uploads/system/uploads/attachment_data/file/604090/Electricity.pdf. Accessed 10 Feb 2017
8. RenewableUK, Onshore Wind (2017). Retrieved from renewableUK: http://www.renewableuk.com/page/OnshoreWind
9. S. Ashworth-Hayes, The UK's renewable energy target (2017). Retrieved from Full Fact: https://fullfact.org/economy/uks-renewable-energy-target/
10. en.wikipedia, Renewable energy in the United Kingdom (2017). Retrieved from en.wikipedia.org: https://en.wikipedia.org/wiki/Renewable_energy_in_the_United_Kingdom
11. D. Elliott, REFIT 'Feed in-Tariffs', the UK renewables obligation and the new EU directive (2008). Retrieved from THE ROBERT GORDON UNIVERSITY: www2.rgu.ac.uk/abs/ESRC%20Seminars/Seminar%202/D.Elliott.ppt
12. J. Atkins, Closure of the renewables obligation (2017). Retrieved from Energy UK: https://www.energy-uk.org.uk/press-releases/energy-uk-blogs/6091-energy-uk-blog-on-the-closure-of-the-renewables-obligation.html
13. G.W. Schlissel, *Electricity Grid Transition in the U.K.* (Institute for Energy Economics and Financial Analysis, Cleveland, 2017)
14. G. Dolphin, 2020 signposts (2013). Retrieved from National Grid: http://nationalgridconnecting.com/2020-signposts/
15. reve, 8 countries that produce the most wind energy in the world (2016). Retrieved from Evwind: http://www.evwind.es/2016/06/23/8-countries-that-produce-the-most-wind-energy-in-the-world/56665
16. Metoffice, Where are the windiest parts of the UK (2015). Retrieved from Met Office: http://www.metoffice.gov.uk/learning/wind/windiest-place-in-uk
17. S.D.A. Cavazzi, An Offshore Wind Energy Geographic Information System (OWE-GIS). Renew. Energy **87**, 212–228 (2016)
18. D. o. DECC, Energy trends (2014), Gov.uk. Retrieved from https://www.gov.uk/government/uploads/system/uploads/attachment_data/file/295362/ET_March_2014.PDF
19. D. F. Government, Offshore wind: part of the UK's energy mix (2013), Gov.uk
20. RenewableUK, New records set in best ever year for British wind energy generation (2016). Retrieved from http://web.archive.org/web/20160321164419/http://www.renewableuk.com/en/news/press-releases.cfm/05-01-2016-new-records-set-in-best-ever-year-for-british-wind-energy-generation
21. S. Shankleman, Updated: top 10 UK onshore wind farms (2012). Retrieved from Business Green: http://www.businessgreen.com/bg/analysis/2205704/top-10-uk-onshore-wind-farms
22. The wind power (2017). Retrieved from The wind Power: http://www.thewindpower.net/store_windfarms_view_all_en.php
23. Wind Farms (2017). Retrieved from 4C Offshore: http://www.4coffshore.com/windfarms/rampion-united-kingdom-uk36.html
24. londonarray, The project (2017). Retrieved from London Array: http://www.londonarray.com/the-project-3/
25. N. Pincott, UK offshore wind: what impact is Brexit likely to have on the UK's offshore wind industry? (2016). Retrieved from Norton Rose Fulbright: http://www.nortonrosefulbright.com/knowledge/publications/141735/uk-offshore-wind-what-impact-is-brexit-likely-to-have-on-the-uks-offshore-wind-industry

Chapter 12
Urban Environment: Characterization of the Wind in Flat Roofs

J. Lassig, U. Jara, J. J. Valle Sosa, and C. Palese

12.1 Introduction

Wind is a resource that exists throughout the world, and that under certain orographic conditions can significantly increase its energy potential. The so-called "concentrating effects" can significantly improve the average speed of wind flow.

These effects not only occur in nature but also in urban environments, which makes it advisable to study them as an energy resource (Fig. 12.1).

The integration of wind turbines in buildings, is becoming a new possibility, which has begun to be studied in Academic Research Centers in wind energy, such as the Technical University of Delft in Holland which begins to be explored on a small scale in several Dutch cities, such as Amsterdam, The Hague, Tilbug, and Twente, and also in the UK [1].

So to locate a wind turbine above a building at the moment, it is a real possibility to obtain electricity, taking advantage of the effect of wind acceleration on the turbine. Wang et al. [2] analyze this possibility. On the other hand Grant et al. [3] they study to locate turbines of vertical and horizontal axis inside ducts located in the edges of the terraces of tall buildings, taking advantage of the suction that is produced there. Lin Lu and KaYan Ip [4] carry out simulation studies (in CFD: Computational Fluid Dynamics) with different sets of buildings, and assess that the increase in speed between these buildings can be increased between 1.5 and 2 times the average wind speed, producing increases in available wind power between 3 and 8 times. On the other hand, Mithraratne [5] evaluates the possibility of installing small wind turbines on the roofs of houses in New Zealand, and concludes that it could reduce between 26 and 81% the carbon emissions in that country.

Urban wind has a number of problems to be solved: from the prediction of winds to the type of wind turbine to be used. Dr. Cruz [6] of the CIEMAT of Spain, in his

J. Lassig (✉) · U. Jara · J. J. V. Sosa · C. Palese
Laboratory of Environmental Fluid Dynamics (LaDiFA), Faculty of Engineering, National University of Comahue (UNCo), Neuquén, Argentina

A. Sayigh, D. Milborrow (eds.), *The Age of Wind Energy*, Innovative Renewable Energy, https://doi.org/10.1007/978-3-030-26446-8_12

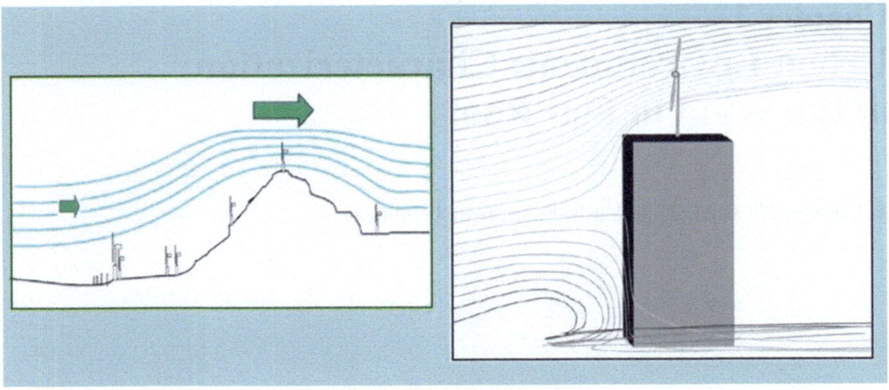

Fig. 12.1 Illustration of how orography modifies the wind intensity (left) and also the buildings (right)

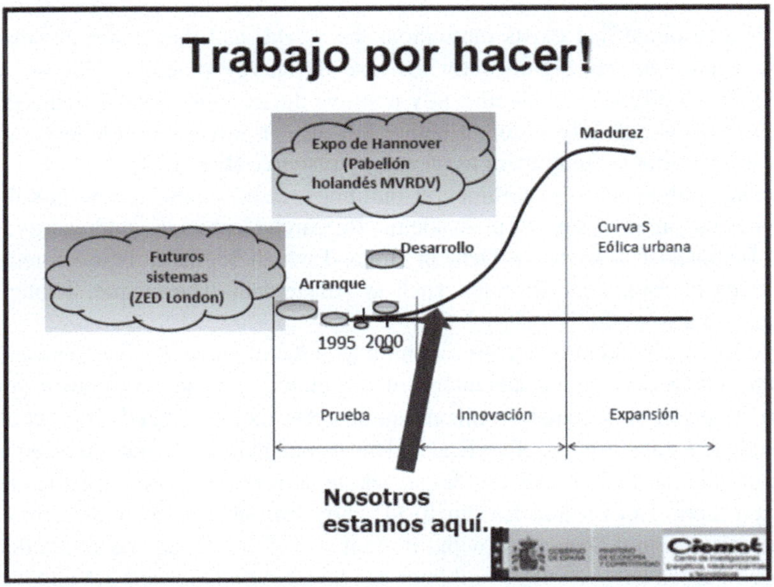

Fig. 12.2 Extracted from the CIEMAT website, https://o2o3.files.wordpress.com/2012/03/cruz_ciemat_2012-01-26_getafe_minieolica-entorno-urbano.pdf

2012 conference held in Getafe, presents a very interesting graph (Fig. 12.2) on the state of development of urban wind energy, where it indicates that it is in the beginning. For what still do much research is missing in this regard.

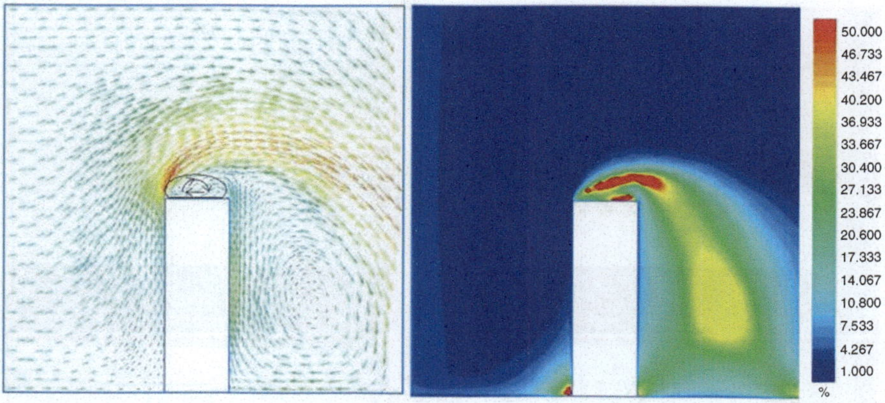

Fig. 12.3 (Left) Wind speed generated in the environment of the building (2D model), with the recirculation area on the flat roof, (right) turbulence intensity

12.2 Buildings with Flat Roofs

All buildings produce an acceleration of free flow of the wind in certain places near each building, and as we move away, the wind speed approaches the wind speed of the free current.

In a building with sharp edges windward of the wind, the boundary layer separates at these edges [7] and the separation bubbles are formed on the sides and top of the building. The main current deviates around it, and a large wake is formed to leeward. Figure 12.3 shows a partial scheme of the above description, since it is a 2D CFD output which partially reproducing what happens.

The results of this separation are a region with low speeds, a high level of turbulence, and the recirculation of flow in the roof. This recirculation region should be avoided for the location of wind turbines, and the location of the measuring towers. Therefore, it is important to know the size of the recirculation region.

On the other hand there is a zone of wind acceleration on the roof, which in Fig. 12.3 (left) is indicated with the vectors of colors from yellow to red. This would be the area with the highest wind power is available.

The determination of these suitable and unsuitable areas for the installation of wind turbines, can be established by means of physical simulations (wind tunnels) or computational simulations (CFD).

12.3 The Role of Roof Geometry

The geometrical shape of the ceiling changes the flow pattern on it. In Fig. 12.4, the flow pattern visualized with smoke from wind tunnel tests is illustrated with four flat roof shapes.

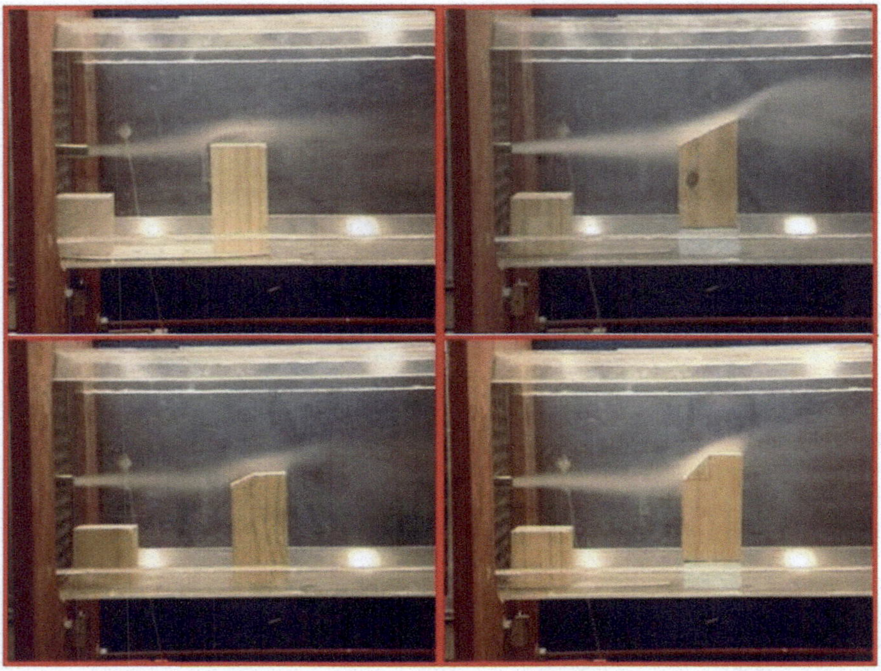

Fig. 12.4 (Left above) Horizontal roof, where the recirculation area is observed attached to the roof. In the other three figures, the inclination of the lee roof eliminates the recirculation zone

For the horizontal flat roof you can see the recirculation area that marks the smoke on it. For inclined and semi-inclined roofs there are practically no recirculation zones, so the maximum wind speeds are closer to the roof.

Figure 12.5 shows the output in 2D CFD for 45° inclined roof (left) and semi-inclined roof (right), where the quality of the flow is reproduced, indicating what was observed in the wind tunnel tests (Fig. 12.4) where they have been reduced the recirculation bubble.

Taking into account the intensity of turbulence, in both cases it is lower than the horizontal ceiling as illustrated in Fig. 12.6.

If the wind impacts from the opposite direction, the flow patterns are different and more damaging place to locate the wind turbines, so Figs. 12.7 and 12.8 plot the distribution of velocities and turbulence intensity, respectively.

For the explained above, we can conclude that the geometry of the roof and its location with respect to the wind, is relevant for the good location of wind turbines on them.

What has been shown so far, is for an isolated building. If we "submerge" this building within a city, the pattern of the incident flow will be more complex due to the interaction between nearby buildings, and the increase in the turbulence intensity of the urban environment.

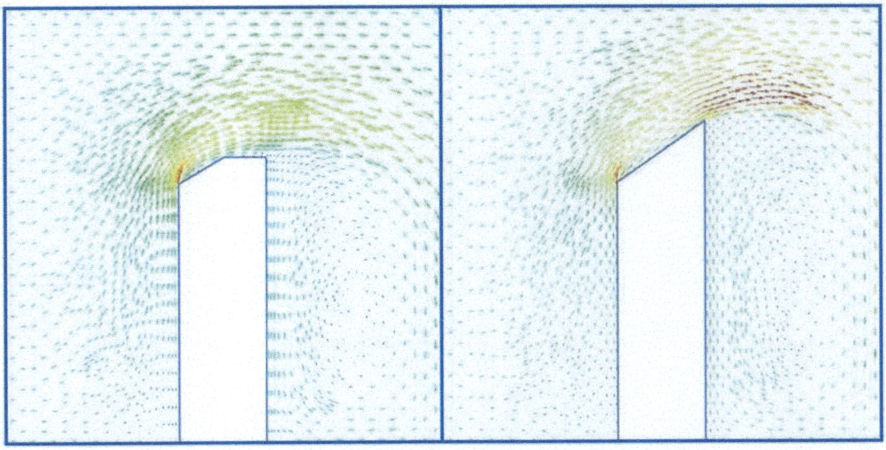

Fig. 12.5 Flow pattern on the inclined roofs simulated by 2D CFD, where it is observed that the recirculation zone is very small

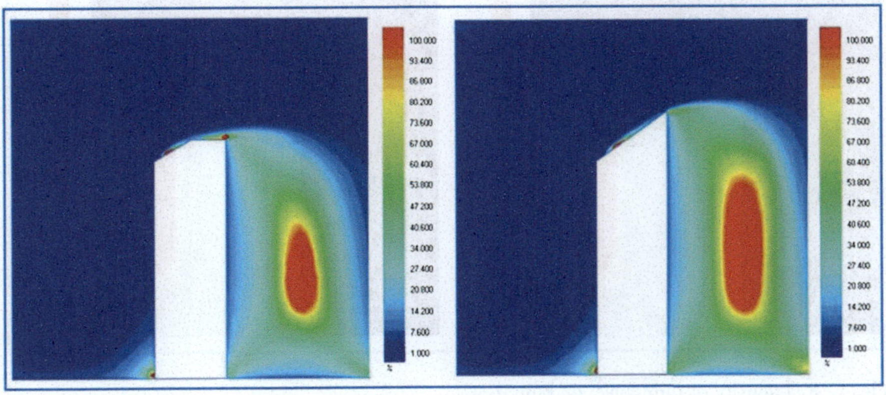

Fig. 12.6 Low turbulence intensity on both inclined roofs

12.4 Wind Tunnels Tests Versus CFD

Installation of wind turbines in buildings requires a particular study of how the wind is modified in its environment. This study can be done in scale models in wind tunnels, or by CFD simulations.

Actually, the data obtained from the wind tunnel tests are more reliable because they not only describe the flow pattern but also quantify it adequately. On the contrary, the CFDs adequately qualify the flow pattern, but the numerical values (speed, pressure, intensity of turbulence) still differ from reality. This situation is due to the fact that even these programs are not calibrated for the fluid dynamic characteristics of an environment urban, especially in personal computers. In large

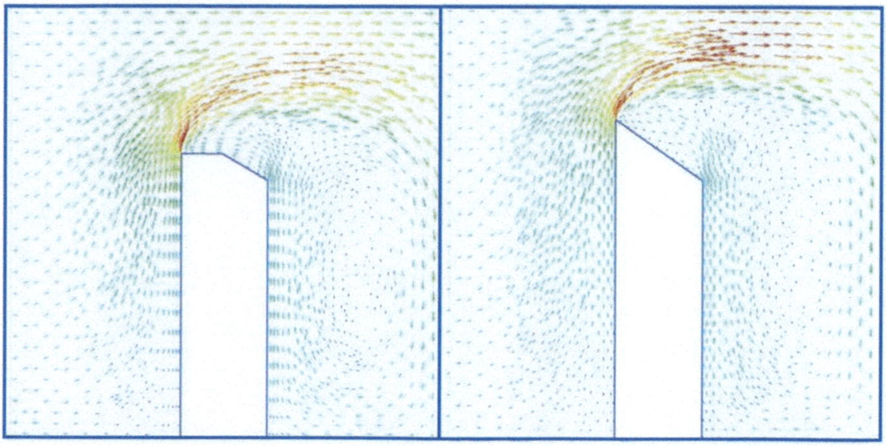

Fig. 12.7 Wind speed distribution

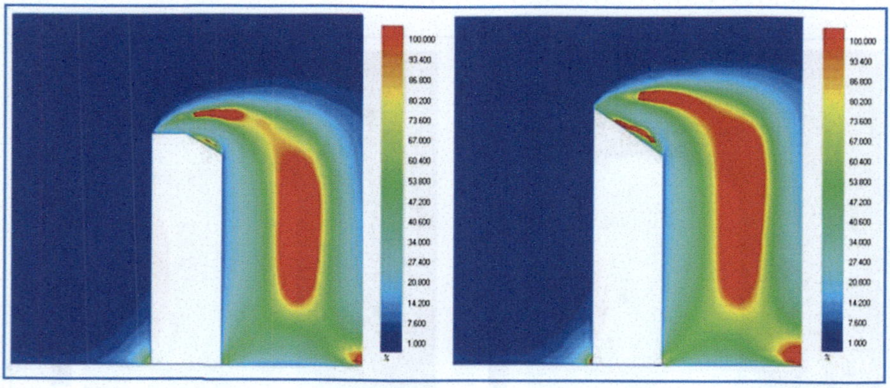

Fig. 12.8 Intensity of turbulence on both roofs

computing centers with thousands of microprocessors in parallel it is possible to simulate better turbulence models and there the numerical values converge more to reality, but the cost is high compared to the possible wind investment in a building.

12.5 A Case of Study

Homes with flat roofs are very common in Latin American and European countries. In particular the tall buildings dedicate their roof to use it as a terrace where they add additional constructions such as gyms, swimming pool with solarium, concierge department, and there are also water tanks, and machine rooms (such as elevators, and air conditioning systems). The latter are more bulky and can standing out from

Fig. 12.9 Examples of buildings with flat

Fig. 12.10 Buildings with flat roofs in the city of Buenos Aires, Argentina

the roof, causing it to lose its "flat" shape and become the "stepped" roof. Examples of these roofs of tall buildings are indicated in Fig. 12.9.

So we will divide the analysis into two parts: (A) in which the additional constructions are low and we will consider them as flat roofs, and (B) where the constructions standing out above the roof and we will consider them as stepped flat roofs.

12.5.1 *Case A: High Building with Flat Roof*

These types of ceilings are idealized, or have water tanks under a roof, in particular this type of roofing we have already analyzed in Sect. 12.3, so we will not abound in more details. An example of these types of ceilings can be seen in Fig. 12.10.

Fig. 12.11 EMA indicates the location of the building under study, in the north of the city of Rosario

12.5.2 Case B: High Building with Stepped Ceiling

The studied building is located in the city of Rosario (Argentina), with a height of 43 m on the terrace, and 6 m more in the water tanks. In Fig. 12.11, a graph with the location of the building is presented, in which the proximity to the Paraná River can be appreciated. This location permit expect a very favorable condition for the exploitation of the wind resource, since the winds coming from the quadrants East and North will not be affected by large obstacles generating turbulence, given the extensive area of islands with low altitude vegetation and flat relief. Figure 12.12 shows comparatively the low height of the building surrounding the building under study, and the location of the Automatic Meteorological Station (EMA) 1 m above the water tank, so wind measurements were made at 50 m height.

The last floor is smaller than the rest, producing a kind of base at the top, and on top of it, the water tank is located. This configuration is typical of buildings between 30 and 50 m high that are built in Argentina. Locating wind turbines there makes it very restricted.

With the meteorological station installed on the roof of the building, the characteristics of the wind were determined. Figure 12.13 indicates the curve of Weibull and the Rose of Winds obtained there [8]; the average winds have an intensity of 3.5 m/s, and the most frequent directions are North, East and South.

In the analysis of the wind tunnel, an attempt was made to determine the increase in speed over different parts of the roof of the building, and the region of greatest turbulence, for the three directions of main winds that exist in the city of Rosario (Argentina).

The model was built with a scale of 1:100. Figure 12.16 shows that the front of the building faces North, and in the upper part of it there is North-South symmetry,

Fig. 12.12 (Left) Location of the anemometer on the water tank of the building; (right) the building and its surroundings where lower buildings can be seen

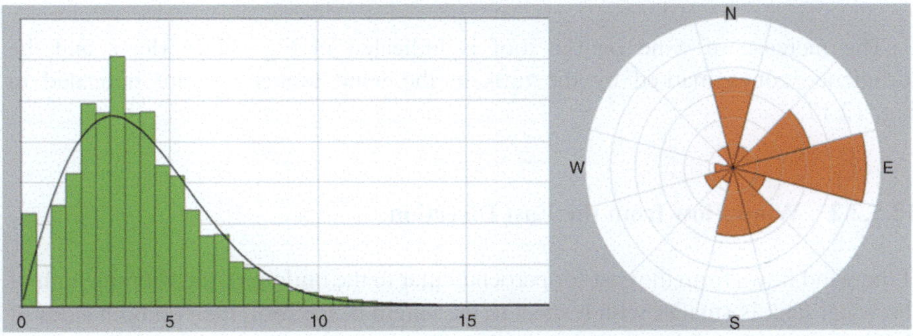

Fig. 12.13 Weibull curve and wind rose obtained at 50 m height, and 1 m above the water tank

so the tests in both directions coincide in the flow pattern. Thus, tests were carried out with winds from the North, South, East and North East directions.

12.5.2.1 Wind Flow from North and South Directions

In Fig. 12.14 (left) it can be seen that the wind to the windward rises and accelerates to a height of 5 m above the water tank bypassing it, but to the leeward of the water tank the created depression produces vortices reducing the intensity of the wind and increasing the intensity of turbulence. So this area is not suitable for the installation of wind turbines.

To windward of the water tank (Fig. 12.14 center) a vortex appears determining another zone of great intensity of turbulence.

On both sides of the water tank the wind passes and accelerates over the roof of the top floor, and there are two windward and leeward vortices of that department in the base of the terrace.

Fig. 12.14 (Left) wind flow passing above the water tank, (center) flow that occurs in the step formed by the roof of the last department and the water tank; (right) wind flow passing lateral to the water tank

The increase of wind on the roof is indicated in Fig. 12.16 (left), and the turbulence zones marked by the tufts in the wind tunnel test are indicated in Fig. 12.17.

12.5.2.2 Wind Flow from the East Direction

If the wind flows from the east (or perpendicular to the building), the lateral length of the water tank is smaller with respect to the length of the roof, then on both sides of the water tank there is a larger area available by the wind. Figure 12.15 indicate this: (left) the wind passing over the water tank accelerates and generates vortices to leeward and windward thereof; (right) the wind passing in front and/or behind the water tank accelerates to a height greater than 5 m and close to the floor of the roof the recirculation bubble occurs.

The increase of the wind speed on the roof is indicated in Fig. 12.16 (right), and the turbulence zones marked by the tufts in the wind tunnel test are indicated in Fig. 12.17 (center).

12.5.2.3 Wind Increase in the Ceiling

Figure 12.16 shows the dimensional wind increments on the roof for three directions.

The reference speed is at the height of the water tank before reaching the building, so the values indicated are small.

$$V_{\text{Adimensionalized}} = \frac{V_{\text{average}}}{V_{\text{reference}}}$$

It is observed that according to the direction of the wind the speed increase is different, and that obeys the geometry of the building.

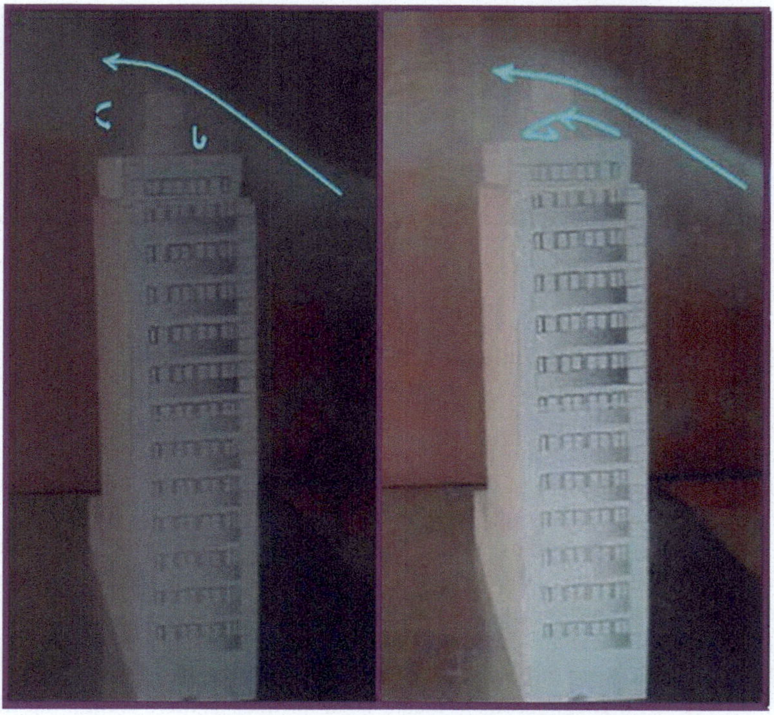

Fig. 12.15 Flow pattern on the water tank (left) and on both sides of it (right)

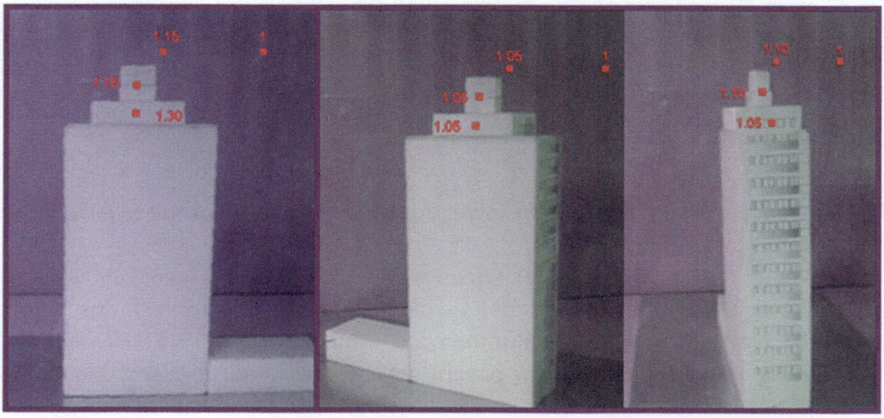

Fig. 12.16 Adimensionalized speeds for three wind directions: (left) North and South; (center) Northeast; and (right) East

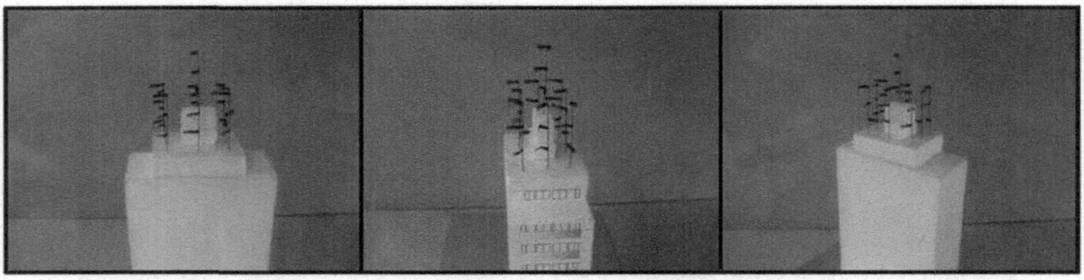

Fig. 12.17 Tufts indicating the turbulence zones, for wind directions from left to right of the North, East and Northeast

12.5.2.4 Qualitative Determination of the Areas with Highest Turbulence Intensity

In the analysis with the wind tunnel, we try to determine the region of greatest turbulence, and the increase of the wind speed over the building at a height outside said zone.

In Fig. 12.17, the windings on the building indicate the degree of turbulence, according to the direction of the wind. Each tufts equals a height above the ceiling of 2.50 m. The greatest turbulence is indicated by the tufts.

We can interpret the tufts information in three levels of risk: red, it is very dangerous to locate wind turbines due to the very high intensity of turbulence higher than 0.40; yellow, it is dangerous to locate wind turbines as there is high turbulence intensity between 0.30 and 0.40; and green, it is the place to locate wind turbines with turbulence intensities lower than 0.30. Figure 12.18 illustrates the aforementioned, and Fig. 12.19 the numbering for each plane studied according to the location of the windscreen shown in Fig. 12.17.

12.6 Conclusion

Urban wind has several problems to solve, from the prediction of winds to the type of wind turbine to be used.

In a building with sharp edges windward of the wind, the boundary layer is separated in the edges and on the roof a recirculation bubble is formed, this region should be avoided for the location of the wind turbines.

Within a city, the incident flow pattern will be even more complex due to the interaction between nearby buildings, and the increase in turbulence intensity. As a result of the complex urban environment, the determination of suitable and unsuitable areas for the installation of wind turbines must be established by means of physical simulations (wind tunnels) or computational simulations (CFD).

Using the wind tunnel as a tool to determine the location of the wind turbines, the method presented for the study of the stepped flat roof (Sect. 12.5.2) would be the simplest.

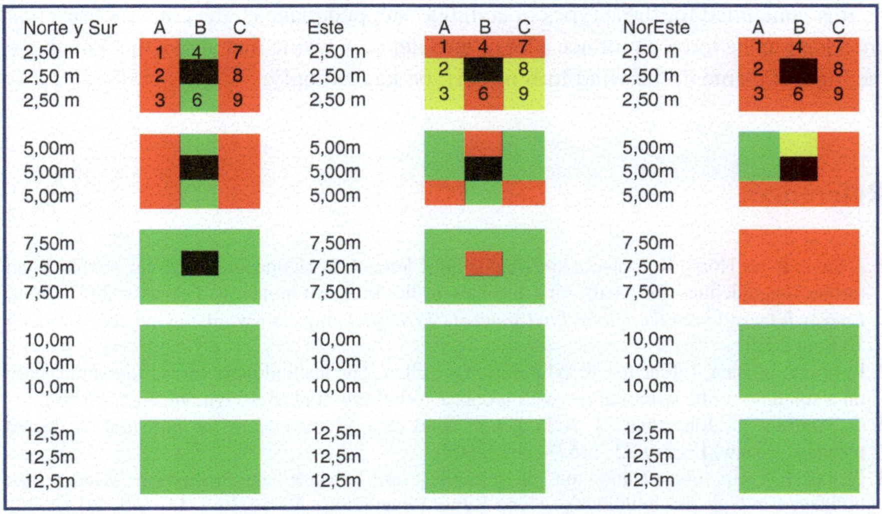

Fig. 12.18 For three wind directions, the level of risk to locate wind turbines in the studied building is indicated. Black color indicates the solid structure of the water tank, very high red turbulence, large yellow turbulence and moderate green turbulence

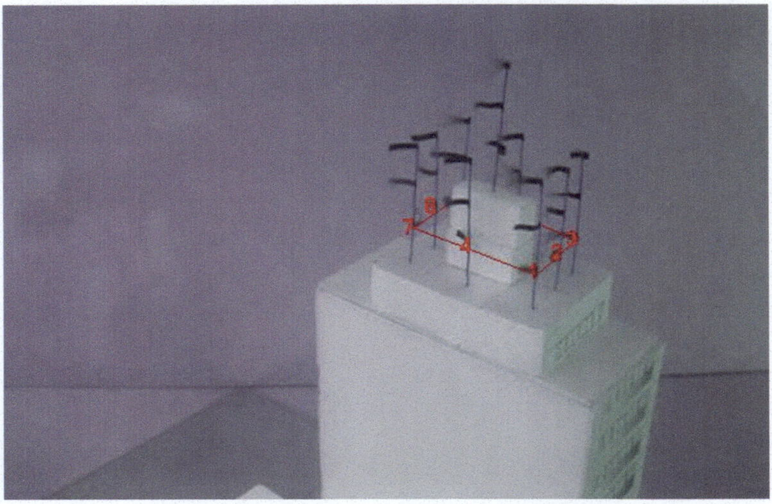

Fig. 12.19 Location of each point in the five levels where the tuff have been located

In the case studied, the location of wind turbines should be performed as shown in Fig. 12.18, where it should be satisfied in all directions with turbulence intensity less than 0.30 (represented in green). This is only achieved from the 10 m levels above the water tank.

It is concluded for these types of ceilings "stepped planes," the constructions that are added in the terrace produce a lot of turbulence, causing that above the 5 m above the highest element, the wind turbines can be located only.

References

1. J. Cace, E. ter Horst, K. Syngellakis, M. Niel, P. Clement, R. Heppener, E. Peirano. Urban wind turbines—guidelines for small wind turbines in the built environment; February 2007; *Wind Energy Integration in the Urban Environment (WINEUR)*. http://www.urbanwind.net/ Accessed 10 Sept 2009
2. F. Wang, L. Baia, J. Fletcher, J. Whiteford, D. Cullen, The methodology for aerodynamic study on a small domestic wind turbine with scoop. J. Wind Eng. Ind. Aerodyn. **96**, 1–24 (2008)
3. G. Andrew, C. Johnstone, N. Kell, Urban wind energy conversion: the potential of ducted turbines. Renew. Energy **33**, 1157–1163 (2008)
4. L. Lu, K.Y. Ip, Investigation on the feasibility and enhancement methods of wind power utilization in high-rise buildings of Hong Kong. Renew. Sust. Energ. Rev. **13**, 450–461 (2009)
5. N. Mithraratne, Roof-top wind turbines for microgeneration in urban houses in New Zealand. Energ. Build. **41**, 1013–1018 (2009)
6. I. Cruz., https://o2o3.files.wordpress.com/2012/03/cruz_ciemat_2012-01-26_getafe_minieolica-entorno-urbano.pdf
7. H. Schlichting, *Boundary Layer Theory*, 7th edn. (McGraw-Hill, New York, 1979), p. 817
8. F. Estavillo, Determinación del Aprovechamiento Eolico para un Edificio Sustentable en la Ciudad de Rosario. *Tesis de Maestría*, FCEyN-UNR, 2014

Chapter 13
Wind Energy in Morocco: Resources, Potential, and Progress

Hassan Nfaoui

13.1 Introduction

As a result of the increase in oil prices in the 1970s, the supply of energy consumption became increasingly important factors for the balance of the Moroccan economy. Thus, energy from oil accounted for about 54% of all energy consumed in Morocco in 1981, 74% of imported petroleum products.

The Moroccan government has taken steps to deal with this situation, trying to develop renewable energies, especially solar and wind energy. For this reason, there was the creation of two institutions, one academic: Solar Energy Laboratory (SEL) of the Science Faculty, Rabat, in 1980, and the other governmental: Center for the Development of Renewable Energies (CDER), in 1982. Their mission is to direct and coordinate research and development efforts of the study of the solar radiation and the wind speed in Morocco, as well as the photovoltaic and wind conversion and applications.

To this end, in 1986, the CDER published the Moroccan Wind Atlas on the basis of data from 17 synoptic stations, using 5 years of hourly average of wind speed (1978–1982). Most of the them are located in airport, which in turn established in less windy areas for the security of Air Navigation The document proposes the first statistical analysis of existing anemometric data, in order to assess the Moroccan wind potential. The results presented in this document can be considered preliminary. Morocco's wind map published has remained incomplete due to the lack of wind data from areas where there are no airports.

In 1990, for the detection of the windiest sites, the CDER launched a program of assessment of the Moroccan wind potential. This program was reserved for the evaluation of the wind potential in the coastal areas from Dakhla to Tetouan and the regions of Taza and Midelt. Measurements for several stations have shown that

H. Nfaoui (✉)
Laboratory of Solar Energy and Environment—Sciences Faculty,
Mohammed V University, Rabat, Morocco

A. Sayigh, D. Milborrow (eds.), *The Age of Wind Energy*, Innovative Renewable Energy, https://doi.org/10.1007/978-3-030-26446-8_13

Morocco has a significant wind potential for the production of electricity on a large scale.

For the development of wind energy in Morocco and to evaluate projects implementing wind energy, in 1991, CDER began a feasibility program of assessment of the economic wind farms by the installation of automatic anemometers in the windy sites through Morocco, which allows for the acquisition of the characteristics of the wind resource.

The research on wind potential assessment and its exploitation for the electricity production on a large scale in the windiest sites in the Sahara provinces in south of Morocco remain limited until the early twenty-first century, despite the existence of wind speed data dating back to 1978 which are available at Directorate of National Meteorology (DNM). A statistical analysis of these data is necessary and important, to assess quantitatively and qualitatively the wind potential available in this region.

In addition to its location in an arid zone, south of Morocco enjoys a significant wind potential. This allows this region to be a favorable area and economic for the installation of wind farms to produce electricity on a large scale. Thereafter, we present the characteristics and the assessment of wind potential in this region, mainly in Tan-Tan, Laâyoune, and Dakhla sites.

In order to evaluate and exploit wind energy, statistical and dynamic analysis of wind, related to the climatic data of the concerned region, consists essentially of determining the force of the wind, frequency as well as the periods during which it blows. So to that effect, it can be measured at different times of the day, the year, and year over year.

On the other hand, the evaluation of wind energy necessitates a statistical assessment of the measurements of the wind speed available on the site over a long measurement period, taken at a small step, for example an hour, or by using reconstruction models using a limited number of parameters, such as Weibull Hybrid function.

To contribute to the improvement of the wind Atlas in Morocco, Nfaoui took up again the research work of Knidiri and Laâouina for the site of Tangier. The choice of this windy zone is due to the fact that, for the Tangier site, we have available 12 year of HAWS of data. In addition, Tangier is located in a promising region for large scale wind power electricity through the installation of wind farms. Measurements are taken by DMN at 10 m height.

Measuring wind speed every hour needs the use of measuring instruments (anemometer, acquisition chain, etc.). They operate all the time in addition to their permanent control and maintenance, in order to minimize errors due to breakdowns, as well as their calibration in order to increase the reliability of the measurements. This requires an investment and a skilled staff working continuously, which is not always easy to achieve. To facilitate taking measurements in an isolated site in order to evaluate its wind potential, it is more practical and economic when we optimized taking measurements by selecting only one to four measurements daily over a short period so as to evaluate the wind potential with a good accuracy for a given site.

Morocco imports more than 90% of its fossil fuels. This energy dependence has a negative impact on state budget in terms of foreign currencies. The electricity consumption increases annually by 6%. In 2016, the contribution of wind electricity to net national electricity consumption reached 9%.

Fortunately, Morocco has a strategic location in the North West of Africa and is only 14 km from Europe, across the Strait of Gibraltar. In addition, it is characterized by a varied hilly relief (the Atlas and Rif Mountains, the Sahara, etc.) and the length of its coasts (the Atlantic Ocean and the Mediterranean Sea) constitute 3500 km of coastline. This allows Morocco to have a significant potential of wind resources that can be exploited for the production of large electricity power connection to the national grid. There are several pathways for the generation of electricity power from renewable energy sources, among the most developed techniques are wind farms. So it has good solar and wind potential.

The launch of wind projects in Morocco allows local resources to contribute, in addition to its energy independence, to the economic and social sustainable development efforts. Thus, the high energy demand, the need to create jobs and reduce oil imports, and the environmental protection of the country support the installation of wind farms. For these reasons, in 2009, a large-scale solar energy use strategy was launched by the creation of the Moroccan Agency for Solar Energy (MASEN) to develop renewable energy projects. This policy permits Morocco to promote regional integration between sub-Saharan African and European countries and to strengthen the links with the main players in this sector on a worldwide basis. Thanks to its model of sustainable development, Morocco will be able to identify the latest innovations, energy efficiency, and clean technologies. Morocco has undertaken an oath to ensure its energy independence and become a leader in renewable energy in Africa. Thus, plans for development of renewable energy have been launched. The Moroccan project of solar energy, controlled by MASEN and its wind counterpart, led by the National Electricity Office (ONE), comes within this framework. The aim is to bring to 42% the share of electricity produced from renewable energy by 2020, 14%, respectively, from hydraulic, solar, and wind power each.

In this perspective, a full renewable energy option will be fully discussed based on what is feasible and appropriate for Morocco. The new strategy adopted is based on increasing the contribution of renewable energies to the national electric power installed at 52% by 2030 and on the development of a clean energy economy. This will enable Morocco to reduce its greenhouse gas (GHG) emissions and to fight against climate change (CC) and thus contribute to the preservation of the environment. The twenty first century will be a century of less and less water, sustainable development and environmental emigration. Morocco is known for its know-how of dam construction and its renewable energy models. The sharing of these skills with African countries will contribute to its development and consequently to the reduction of emigration to Europe.

13.2 Wind Characteristics and Resources

13.2.1 General Characteristics of the Wind Resource

13.2.1.1 Fundamental Causes of the Winds

The original source of the renewable energy contained in the earth's wind resource is the sun. Global winds are caused by pressure differences across the earth's surface due to the uneven heating of the earth by solar radiation. For example, the amount of solar radiation absorbed at the earth's surface is greater at the equator than at the poles. The variations in incoming energy set up convective cells in the lower layers of the atmosphere (the troposphere). In a simple flow model, air rises at the equator and sinks at the poles. The circulation of the atmosphere that results from uneven heating is greatly influenced by the effects of the rotation of the earth (at a speed of about 1670 km/h at the equator, decreasing to zero at the poles). In addition, seasonal variations in the distribution of solar radiation give rise to variations in the circulation [1].

The spatial variations in heat transfer to the earth's atmosphere create variations in the atmospheric pressure field that cause air to move high pressure to low pressure. There is a pressure gradient force in the vertical direction, but this is usually cancelled by the downward gravitational force. Thus, the winds blow predominately in the horizontal plane, responding to horizontal pressure gradients. At the same time, there are forces that strive to mix the different temperatures and pressure air masses distributed across the earth's surface. In addition to the pressure gradient and gravitational forces, inertia of the air, the earth's rotation and friction with the earth's surface (resulting in turbulence), affect the atmospheric winds. The influence of each of these forces on atmospheric wind systems differs depending on the scale of motion considered.

As shown in Fig. 13.1 worldwide wind circulation involves large-scale wind patterns which cover the entire planet. These affect prevailing near surface winds. It should be noted that this model is an oversimplification because it does not reflect the effect that land masses have on the wind distribution [1].

13.2.1.2 Temporal Characteristics of Wind

A more accurate assessment of wind potential depends on the measurement step and the wind speed measurement period. In this section, we will review some of the compilation and calculation steps required by this method of data analysis, to evaluate the wind potential in a given site with good accuracy.

(a) Interannual

The interannual variation is a temporal variation that corresponds to the difference observed from 1 year to another. It gives information on the periodicity and irregularities of the wind. Nfaoui concluded that it takes 10 years of hourly data to determine long-term values of weather or climate and to arrive at a reliable assessment of wind potential at Tangier location [2].

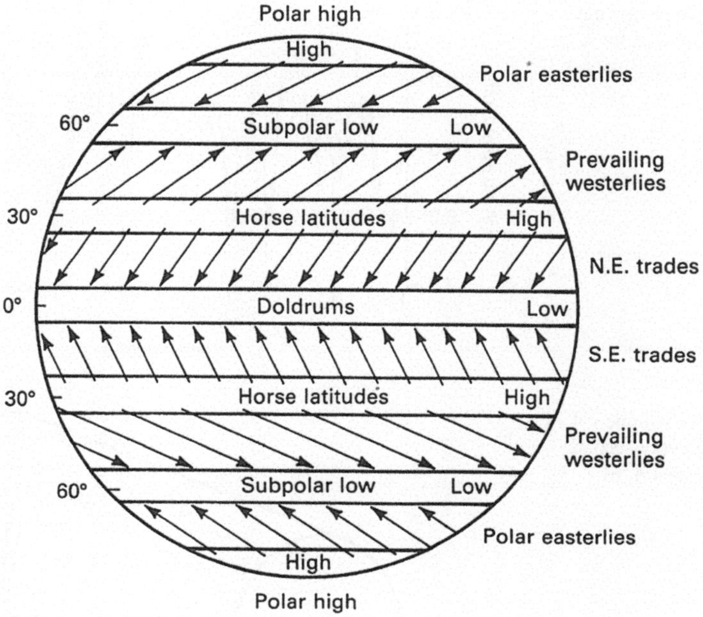

Fig. 13.1 Simplified general circulation [1]

(b) Diurnal (Time of day)
Large wind variations also can occur on a diurnal or daily time scale. This type of wind speed variation is due to differential heating of the earth's surface during the daily variation in solar radiation cycle. A typical diurnal variation is an increase in wind speed during the day with the wind speed lowest during the hours from midnight to sunrise.

(c) Short-term
Short-term variations usually mean variations over time intervals of 10 min or less (Fig. 13.2). Ten-minute average is typically determined using a sampling rate of about 1 s. It is generally accepted that variations in wind speed with periods from less than a second to 10 min and that have a stochastic character are considered to represent turbulence. For wind applications, turbulent fluctuations in the flow need to be quantified. For example, turbine design considerations can include power quality. Turbulence can be thought of as random wind speed fluctuations imposed on the mean wind [1].

13.2.1.3 Available Potential of the Wind Resource

Through a rotor disc of area A, the mass flow of air is $\dot{m} = \frac{dm}{dt}$ (Fig. 13.3). From the continuity equation of fluid mechanics, the mass flow rate is a function of air density, ρ, and air velocity (assumed uniform) v. It is given by

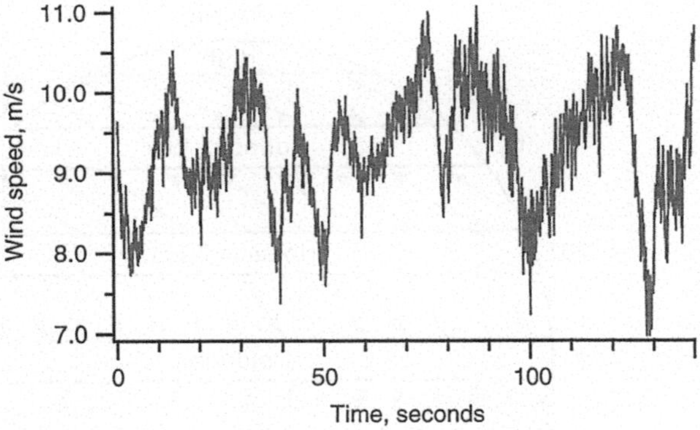

Fig. 13.2 Typical wind speed variation as function of time for a short period

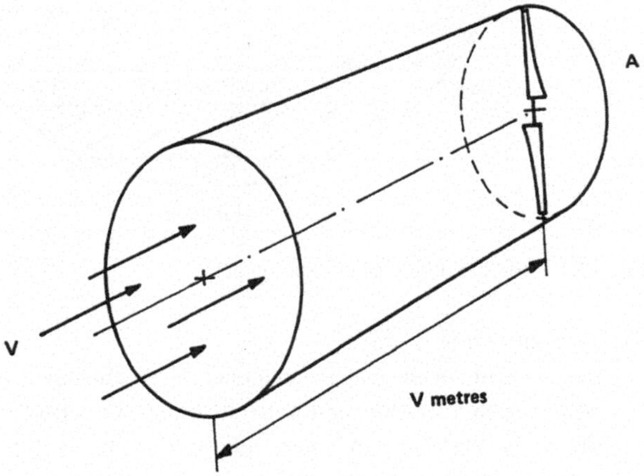

Fig. 13.3 A volume $v{\cdot}A$ of air flowing every second through an area A, $\dot{m}$

$$\dot{m} = \frac{\mathrm{d}m}{\mathrm{d}t} = \rho A v\,(\mathrm{kg/s}) \tag{13.1}$$

and thus a flow of kinetic energy per second or kinetic power P_{kin} of the flow is given by

$$P_{\mathrm{kin}} = \frac{1}{2}(\rho A v)v^2 = \frac{1}{2}\rho A v^3\,(\mathrm{W}) \tag{13.2}$$

where

$A =$ area swept by the rotor blades (m^2),
$v =$ undisturbed wind speed (m/s),
$\rho =$ air density (kg/m^3).

The wind power density is proportional to the cube of the wind velocity. The wind velocity is an important parameter and significantly influences the power per unit area available from the wind. But more accurate estimate of P_{kin} can be made if hourly average, v_i, is available for a year rather than using annual average wind speed. Then, the average of power estimated for each hour can be determined. The average wind power density, based on hourly average is

$$\frac{\overline{P}}{A} = \frac{1}{2}\rho\overline{v}^3 K_e \tag{13.3}$$

where $\overline{v}$ is the annual average wind speed and K_e is called the energy pattern factor given by

$$K_e = \frac{1}{N\overline{v}^3}\sum_{i=1}^{N} v_i^3 \tag{13.4}$$

where N is the number of hours in a year, 8760. Some sample qualitative magnitude evaluations of the wind resource are as follows [1]:

$$\begin{aligned} &\frac{\overline{P}}{A} \prec 100\ \text{W/m}^2 - \text{Low} \\ &\frac{\overline{P}}{A} \approx 400\ \text{W/m}^2 - \text{Good} \\ &\frac{\overline{P}}{A} \succ 700\ \text{W/m}^2 - \text{Excellent} \end{aligned} \tag{13.5}$$

Since P varies proportionally to the cube of the wind speed. Therefore, an assessment of the available wind energy at a given site requires the knowledge of the wind speed taken at smaller measuring steps, 1 h, for example, for a sufficiently long period. So it pays to carefully select a good site for a wind turbine: 10% more wind gives 30% more power [1].

13.2.1.4 Maximum Power Extraction: The Betz Limit

A wind rotor extracts power from the wind by slowing down the wind. Too much slowing causes the air to flow around the rotor area instead of passing through it. It is clearly not possible to extract all the energy from the air which passes through the windmill disc. If this were so then the air would cease to move and pile up behind the windmill. In practice, the airflow through the disc is as shown in Fig. 13.4. It was shown by Betz that under conditions of maximum power extraction the air velocity at the windmill disc will have fallen to 2/3 of the value of the upstream (undisturbed) wind velocity v_∞ and further decreases to a final value of 1/3 the initial velocity well downstream of the windmill. This means that we can only extract a fraction 24/27 of the initial wind energy. It is more convenient to express the performance of a windmill as the power output divided by the power in the wind passing through an area equal to the swept area of the mill. From Fig. 13.4 this is seen to be a fraction

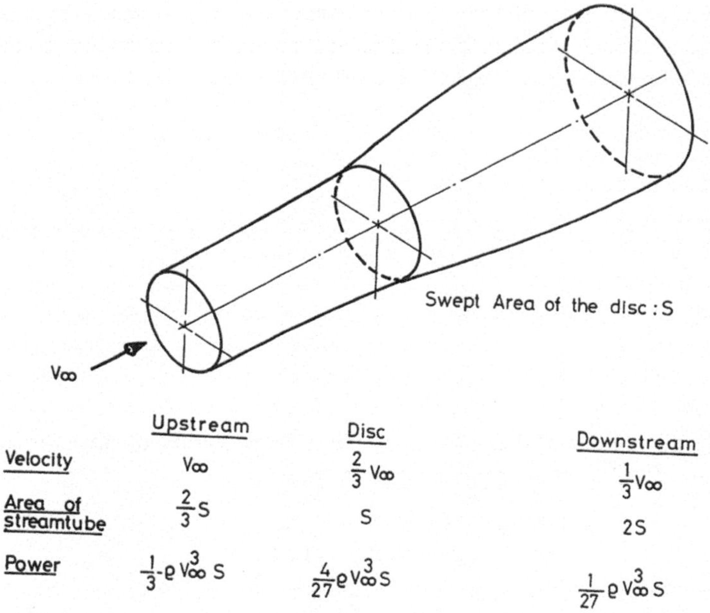

Fig. 13.4 Condition for maximum power extraction, Betz limit [5]

16/27. This fraction works out to be 59% and is called the Betz limit. Sometimes windmill performance is given as a fraction of the Betz limit. In that case the rotor itself receives a velocity $2/3v_{\infty}$. So the effective mass flow is only $2/3\rho A v_{\infty}$. If this mass flow is slowed down from v_{∞} to $1/3v_{\infty}$ then [3–5]:

$$P_{\max} = \frac{16}{27}\rho A \frac{v_{\infty}^{3}}{2} \tag{13.6}$$

Practical windmills will not achieve the Betz limit; thus, the wind velocity at the disc will be somewhat greater than the value of $2/3v_{\infty}$. The value of the wind velocity at the disc is expressed in terms of an inflow factor "a" such that the wind velocity is given by $v_{\infty}(1 - a)$. The inflow factor will lie between 0 when the wind mill is not operating and 1/3 when the mill is operating at maximum efficiency. In windmills the blades rotate, giving a further limitation to efficiency. In order that the rotor should rotate it is necessary for the wind to exert a torque about the rotor axis; by the principal of angular momentum the rotor will exert an equal and opposite torque on the air stream causing it to rotate in the opposite direction to that of the rotor. This rotating wake is particularly important at low rotor speeds.

The fraction of extracted power, which we call power coefficient C_p, in practice seldom exceeds 40% if measured as the mechanical power of a real wind rotor. The subsequent conversion into electrical power, for example, gives a reduction in available power, depending on the efficiency η of transmission and generator.

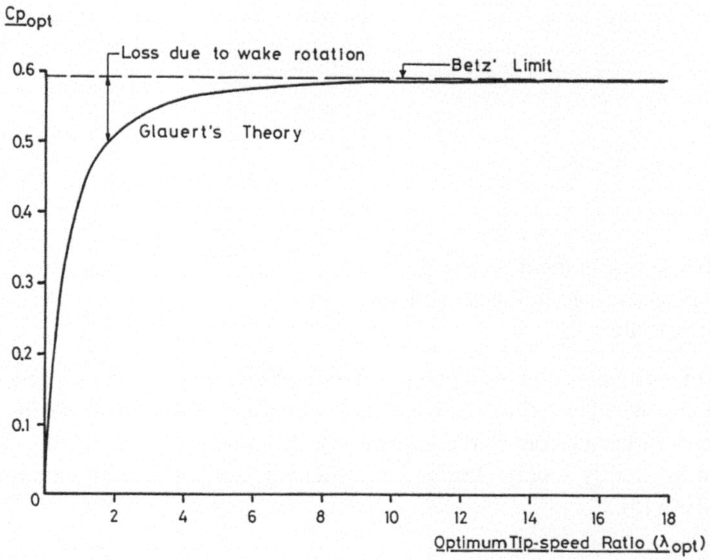

Fig. 13.5 Effect of rotating wake on maximum efficiency [5]

A further reduction of the available power is caused by the fluctuation in speed and direction which an actual wind turbine experiences in the field [1].

In order to evaluate the performance of a rotor it is necessary to satisfy simultaneously the equations which define the air flow through the rotor, together with the equations which define the forces on the blades due to that air flow. An aerofoil of known characteristics is selected and a value of the tip speed to wind speed ratio λ chosen.

$$\lambda = \frac{\Omega R}{v_\infty}$$

where λ is the tip speed ration, Ω is the angular velocity, R is the rotor radius, and v_∞ is the undisturbed wind speed.

The rotating wake will contain kinetic energy which will need to be subtracted from the energy available to the rotor. The effect of the rotating wake on maximum efficiency is shown in Fig. 13.5. The efficiency is plotted in terms of a dimensionless speed λ which is defined as tip speed/wind speed. The effect of wake rotation is seen to be less important at higher values of λ where efficiency approaches to the Betz limit, but is a severe limitation at low speeds. In a similar manner to that for the axial reduction of wind speed the increase in angular velocity of the wake is expressed in terms of the rotation speed of the rotor Ω and a radial inflow factor a_1 such that the average wake angular speed at the rotor is $a_1\Omega$. It should be noted that because of the axial and radial inflow effects, the magnitude and direction of the wind nearby the rotor blade is not the same as the undisturbed wind velocity v_∞.

The wind power density available in the wind can be estimated by the following equation:

$$P = \rho \sum_i \frac{f_i v_i^3}{2} \tag{13.7}$$

where

f_i: wind frequency in the *i*th interval,
v_i: wind speed average in the *i*th interval,
N: number of intervals.

Since the instantaneous wind power varies as the cube of the wind speed, it might be questioned whether estimating wind energy on the basis of hourly means is likely to lead to significant error. Studies shown that the structure of the wind is such that the use of the hourly average leads to an underestimation of useful wind energy by about 5–10% [5].

13.2.2 Characteristics of the Atmospheric Boundary Layer

The atmospheric boundary layer is the lowest part of the atmospheric and its characteristics are directly influenced by contact with the earth's surface. Here, physical quantities such as velocity, temperature, and relative humidity can change rapidly in space and time. For example, an important parameter in the characterization of the wind resources is the variation of horizontal wind speed with height above the ground.

One would expect the horizontal wind speed to be zero at the earth's surface and to increase with height in the atmospheric boundary layer. This variation of wind speed with height is called the vertical profile of the wind speed. There are at least two basic problems of interest with the determination of variation in wind speed as a function of height for wind energy applications, namely, instantaneous variation and seasonal variation. In addition to variations due to the atmospheric stability, the variation of wind speed with height depends on surface roughness on terrain.

13.2.2.1 Atmospheric Density and Pressure

It is of interest to locate the way of calculating the air density. As a function of moisture content and derived from Boyle–Gay-Lussacs' law:

$$\rho = \frac{Mp}{RT} \tag{13.8}$$

where,

M: mol weight (kg/mol),
p: pressure (N/m),
R: universal gas constant = 8.31434 (J/mol K),
T: temperature (K).

Moist air is slightly less dense then dry air. Knowing that a standard mol of air weights 28.966 g, or $M = 0.028966$ kg/mol, we can calculate the density of (dry) air at various temperatures and pressures. But the presence of water vapor decreases the density. The decrease depends upon the ratio of vapor pressure e and atmospheric pressure p[3]:

$$\rho_{\text{wet}} = \rho_{\text{dry}}\left(1 - 0.3783\left(\frac{e}{p}\right)\right) \tag{13.9}$$

The effect of the water vapor is rather small, even for completely saturated air. But using the copula theory, Bahraoui et al. [6] demonstrated that the estimated wind potential is higher if the air density is not constant. ρ is dependent mainly on temperature T and the pressure P, both of which vary with height. In most studies, the air density is taken constant, and it is replaced with the standard air density, taking the average sea temperature 15 °C and 1 atmospheric pressure, that is, 1.225 kg/m^3. Now employing the ideal gas law, the air density could be expressed as [5]:

$$\rho_{(T,p)} = 1.225\left[\frac{288.15}{T}\right]\left[\frac{p}{1013.3}\right] \tag{13.10}$$

The wind power density, Eq. (13.2), can be calculated by two ways. The first way is considering the air density $\rho_{(T,p)}$ constant. In this case the mean power produced until an observation z is:

$$\frac{P(z)}{A} = \frac{\rho_{(T,p)}}{2}\int_0^z f(v)v^3\,\mathrm{d}v \tag{13.11}$$

where f is the probability density function (pdf) of the wind speed. When the value $z \to \infty$, we have the average of the power energy. The second case if the air density $\rho_{(T,P)}$ is not constant, for n registration of the data $\rho_{(T,p)} = \left(\rho_{(T_1,p_1)}, \rho_{(T_2,p_2)}, \ldots, \rho_{(T_n,p_n)}\right)$, the mean wind power energy can be calculated until the observation z_k:

$$\frac{P(z_k)}{A} = \frac{1}{2}\sum_{i=1}^{k}\rho_{(T_i,p_i)}z_i^3 \tag{13.12}$$

Now to simulate the wind power energy density we come back to simulate the wind speed variable coupled with the temperature and the pressure.

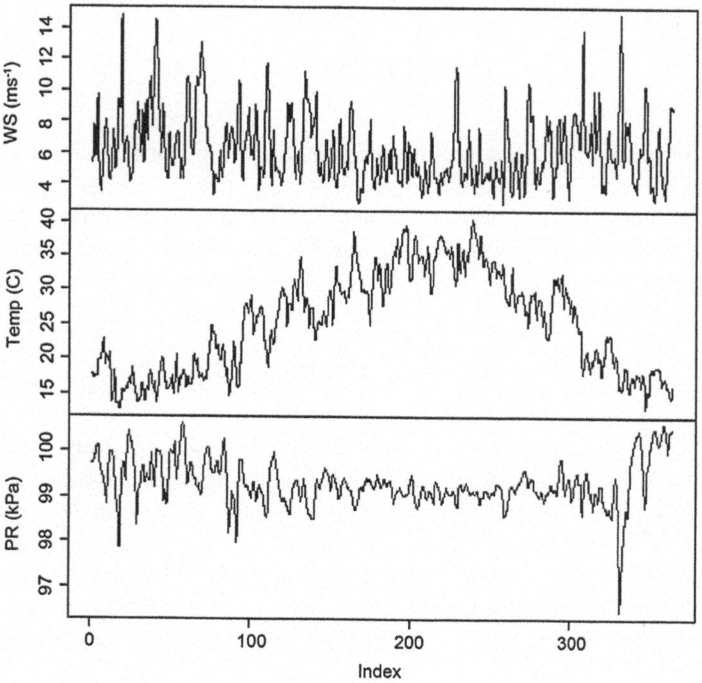

Fig. 13.6 Daily wind speed, temperature and pressure in Hrarza, west-north of Morocco [5]

Figure 13.6 represents the registration of daily wind speed, temperature and pressure for 1 year collected in the region Hrarza, situated in the north of Morocco. Close to the straits of Gibraltar and surrounded by two seas, the Mediterranean and the Atlantic, this region suffers a gusty wind. The registration covers 365-day, maximal wind speed and their correspondent temperature and pressure. The first lecture of the graphical behaviour of these variables is that the temperature variable is more predictable than the wind speed and the pressure. It is probably due the seasonal comportment.

13.2.2.2 Stability of the Atmospheric Boundary Layer

A particularly important characteristic of the atmospheric is its stability—the tendency to resist vertical motion or to suppress existing turbulence. The stability of the earth's atmosphere is governed by the vertical temperature distribution resulting from the radiative heating or cooling of its surface and the subsequent convection mixing of the adjacent to the surface. If the atmosphere is approximated as a dry ideal gas, using conventional thermodynamic relationship for a fluid element in a gravitational field is given by

$$dp = -\rho g\,dz \tag{13.13}$$

where$g$: local gravitational acceleration.

The first law of thermodynamics for an ideal gas closed system of unit mass undergoing a quasi-static change of state is given by

$$dq = du + pdv = dh - vdp = c_p\,dT - \frac{1}{\rho}dp \tag{13.14}$$

q: heat transferred,
u: internal energy,
h: enthalpy,
v: specific volume and
c_p: constant pressure specific heat.

For an adiabatic process (no heat transfer) $dq = 0$, and Eq. 13.4 becomes

$$c_p\,dT = \frac{1}{\rho}dp \tag{13.15}$$

Substitution for dp in Eq. (13.13) gives

$$\left(\frac{dT}{dz}\right)_{\text{adiabatic}} = \frac{1}{c_p}g \tag{13.16}$$

If the changes in g and c_p with elevation are assumed negligible, the change in temperature, under adiabatic conditions, is a constant. Using $g = 9.81\ \text{m/s}^2$ and $c_p = 1.005$ kJ/kg/K yields to the dry adiabatic Lapse rate, with no heat transfer is about 1 °C per 100 m [1].

13.2.2.3 Turbulence

Turbulence in the wind caused by dissipation of the wind's kinetic into thermal energy via the creation and destruction of progressively smaller guest. Turbulent wind may have a relatively constant mean over time periods of an hour or more, but over shorter times (minutes or less) it may be variable.

Turbulent wind consists of longitudinal, lateral, and vertical components. The longitudinal component, in the prevailing wind direction, for example, is designed $u(z, t)$. Each component is frequently conceived of as consisting of a short-term mean wind U, for example for the longitudinal component, with a superimposed fluctuating wind of zero mean, u', added to it, thus [1]:

$$u = U + u' \tag{13.17}$$

where u = instantaneous longitudinal wind speed, z = height above ground and t = time.

Note that the short-term wind speed, in this case U, refers to mean wind speed averaged over some (short) time period, Δt, longer than the characteristic time of the fluctuations in the turbulence. This time period is usually taken to be 10 min but can be as longer as an hour. In equation form it is as follows [1]:

$$U = \frac{1}{\Delta t}\int_0^{\Delta t} u\mathrm{d}t \tag{13.18}$$

Assuming that the sample interval is δt, such that $\Delta t = N_s \delta t$, where N_s = number of sample during each short-term interval, then turbulent wind can be expressed as a sequence u_i. The short-term mean wind speed can then be expressed in simpled form as

$$U = \frac{1}{N_s}\sum_{i=1}^{N_s} u_i \tag{13.19}$$

The short-term average longitudinal wind speed, U, is the one most often used in times series observations.

The most basic measure of turbulence is the turbulence intensity. It is defined by the ratio of the standard deviation of the wind speed to the mean wind speed. In this calculation both the mean and standard deviation are calculated over a time period longer than that of the turbulent fluctuations, but shorter than periods associated with other types of wind speed variations (such as diurnal effects). The length of this time period is normally no more than an hour, and by convention in wind energy engineering it is usually equal to 10 min. The sample rate is normally at least once per second (1 Hz). The turbulence intensity, TI, is defined by [1, 7]:

$$\mathrm{TI} = \frac{\sigma_u}{U} \tag{13.20}$$

where σ_u is the standard deviation, given in sampled from by

$$\sigma_u = \sqrt{\frac{1}{N_s - 1}\sum_{i=1}^{N_s}(u_i - U)^2} \tag{13.21}$$

Turbulence intensity is frequently in the range of 0.1–0.4; in general, the highest turbulence intensities occur at the lowest wind speeds, but the lower limiting value at a given location will depend on the specific terrain features and surface conditions at the site. Figure 13.7 illustrated a graph of a typical segment of wind data sampled at

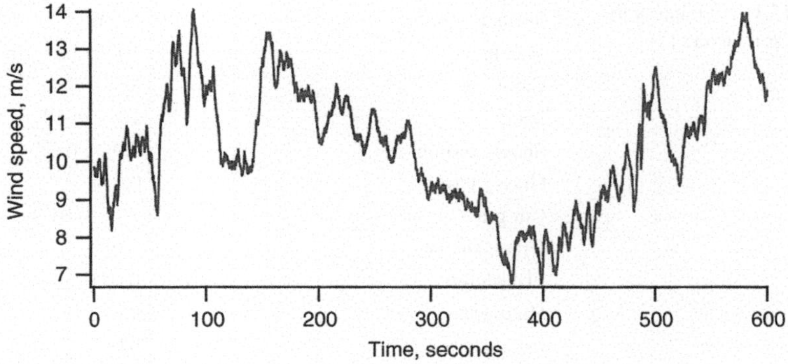

Fig. 13.7 Sample wind data [1]

8 Hz. The data has a mean of 10.4 m/s and a standard of 1.63 m/s. Thus, the turbulence intensity, over the 10 min period, is 0.16 [1].

13.2.2.4 Wind Speed Variation with Height

It has been shown that information on wind speeds at heights of 10–150 m above the ground level is very desirable for any decision about selection of a wind turbine. Often, these data are not available and some estimated must be made from wind speeds measured at a height z_a. This requires an equation that predicts the wind speed at one height z_1 in terms of the measured speed at another z lower height. The common of these simpler expressions is the power law expressed as [6]:

$$\overline{v}_z = \overline{v}_a \left(\frac{z}{z_a}\right)^{\alpha} \tag{13.22}$$

where

v_z: wind speed average at height $z(m)$,
v_a: wind speed average at height $z_a(m)$,

The exponent α is determined empirically. It varies with height, time of day, season of the year, nature of terrain, wind speeds and temperature [7]. It depends also on the average meteorological stability and surface roughness of the soil z_0. The z_0 values for different terrain types are summarized in Table 13.1.

Early work on this subject showed that under certain conditions α is equal to 1/7, indicating a correspondence between wind profiles and flow over flat plates [1]. In practice, the exponent α is a highly variable quantity. It has an important effect on estimates of wind power.

If long-term wind data are available, the empirical results of Justus make it possible to evaluate the value of α (Fig. 13.8). For a usual range of roughness of

Table 13.1 Roughness for various soil covers [7]

Surface type	z_0 (m)
Marsh, ice	10^{-5} to $3 \cdot 10^{-5}$
Calm sea	$2 \cdot 10^{-4}$ to $3 \cdot 10^{-4}$
Sand	10^{-4} to 10^{-3}
Snowy terrain	4.9×10^{-3}
Grassy terrain	0.017
Cut grass	10^{-3}–0.01
Grass, steppe	0.032
Flat region	0.021
High grass	0.039
Corn	0.045
Beet	0.064
Dwarf palms	0.1–0.3
Shrubs	0.05–0.1
Trees	0.2–0.9
Suburbs	1–2
City	1.4

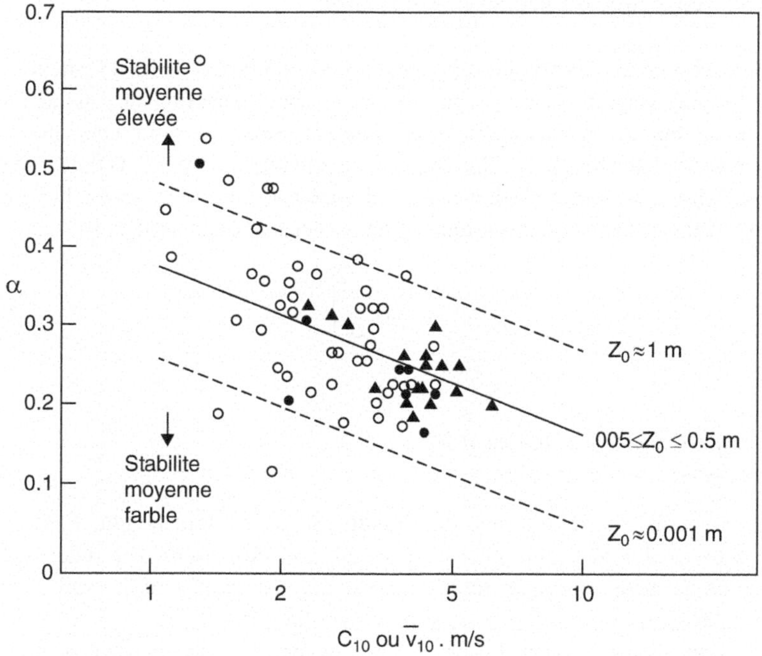

Fig. 13.8 Variation of the exponent α as a function of the average wind speed, measured at a height of 10 m [6]

the soil surface, $0.05 < z_0 < 0.50$, the exponent α is given by the following expression [7]:

$$\alpha = \frac{0.37 - 0.088 \ \ln(v_a)}{1 - 0.088 \ \ln\left(\frac{z_a}{10}\right)} \tag{13.23}$$

where v_a is given in m/s and z_a in *m*.

13.2.3 Mathematical Modeling of the Distribution of Wind Speed Frequencies

It is a form of data presentation that favors threshold phenomena rather linearity of the wind conversion systems. The histogram represents a number of hours, for example, for which the wind speed is equal to 0, 1, 2 m/s etc. (on the abscissa). It informs about the hour distribution in different classes. Such a presentation is useful to know the operation threshold of a wind turbine installed in a given site and to evaluate its generated energy output. On the other hand, if projection of measured data from one location to another is required, or when only summary data are available, then there are distinct advantages to the use of analytical representations for the probability distribution of wind speed [1–8].

The most complex, but probably the most important, aspect of wind data analysis is the compilation and characterization of wind frequency distributions. Indeed, the determination of the energy available at a given site requires a detailed and precise knowledge of the wind data and their variations over time. Suppose, for example, that in a certain place the winds are 5 m/s during a whole day and, in another place, 10 m/s for half a day and zero (0 m/s) the rest of the day. The average speed is 5 m/s in both cases, but we will have much more wind energy in the second case. Therefore, to evaluate available wind energy, the average speed alone is rarely sufficient.

13.2.3.1 Weibull Function

In order to calculate the output of a wind turbine at a particular site one must know the distribution of wind speed frequencies there, but this is equally difficult to manipulate. A simplified mathematical representation would be useful to characterize frequency distributions, to eliminate the influences both of random errors in wind measurements and the calculation of power potential. This would become particularly useful if we want to estimate wind speeds at different heights above the ground. As much as possible, we use different distribution functions for the occurrence of wind speeds. The application of the Weibull function and log-normal distributions to data from more than of hundred stations in the United States of America (USA) shows that it gives the best results [4]. The basic form of the two-parameter of Weibull function has the following expression:

$$f(v) = \left(\frac{k}{c}\right)\left(\frac{v}{c}\right)^{k-1} exp\left[-\left(\frac{v}{c}\right)^{k}\right] \quad (13.24)$$

where k is the shape factor and c is the scale factor of the considered probability density function (pdf). It is defined only for values greater than 0. As the value of k increases, the corresponding curve has a sharper peak, indicating that there is less wind speed variation. The energy pattern factor K_e (defined as the total calculated amount of power available in the wind divided by the power from cubing the average wind speed) is given by:

$$K_e = \frac{\overline{v^3}}{(\overline{v})3} = \frac{\Gamma\left(1+\frac{3}{k}\right)}{\Gamma^3\left(1+\frac{1}{k}\right)} \quad (13.25)$$

Examples of some parameters of interest are given in Table 13.2.

The mean and variance are, respectively,

$$\overline{v} = c\Gamma\left(1+\frac{1}{k}\right) \quad \text{and} \quad \sigma_v = c^2\left[\Gamma\left(1+\frac{2}{k}\right) + \Gamma^2\left(1+\frac{1}{k}\right)\right] \quad (13.26)$$

such that $\Gamma(x) = \int_0^{+\infty} t^{x-1}e^{-t}dt$ is the Euler's Gama function.

The cumulative distribution function (CDF) obtained by integration $f(v)$ has the form of

$$F(v) = 1 - exp\left(-\left(\frac{v}{c}\right)^{-k}\right). \quad (13.27)$$

The Weibull function—its necessity and application—is discussed in [2–5]. In these references a complete methodology was presented with example applications. In references, a simple method is available for fitting Weibull distributions to data by a linear regression method where values of ln[−ln(1 − $F(v)$)] are plotted versus ln (v), and a straight line is fitted to the points. The slope of the line is k, and the intercept on the ln[−ln(1 − $F(v)$)] axis is $-k$.ln(c), using the Eq. (13.26).

Table 13.2 Variation of parameters with Weibull k shape factor [1]

K	$\sigma_U/\overline{U}$	K_e
1.2	0.837	3.99
2	0.523	1.91
3	0.363	1.4
5	0.229	1.15

This same approach can be used to calculate the parameters k and c as well as the average power at heights other than the measurement height, using the following formulas [2, 7]:

$$k_z = k_a \frac{1 - 0.088 \ln\left(\frac{z_a}{10}\right)}{1 - 0.088 \ln\left(\frac{z}{10}\right)} \tag{13.28}$$

$$c_z = \frac{\overline{v_z}}{(1 - f_0)\Gamma\left(1 + \frac{1}{k_z}\right)} \tag{13.29}$$

$$\overline{P}_z = \frac{1}{2}\rho \int_0^{\infty} v^3 f_z(v)\, \mathrm{d}v \tag{13.30}$$

The value of the exponent α, given by the Eq. (13.22) is only valid for a certain roughness interval. To take into account the different values of the roughness of the ground surface, this expression can be written in the following form [7]:

$$\alpha = \frac{x - 0.088 \ln(v_a)}{1 - 0.088 \ln\left(\frac{z_a}{10}\right)} \tag{13.31}$$

where

z_0 (m)	x
0.000–0.005	0.25
0.005–0.050	0.31
0.05–0.50	0.37
0.50–4.00	0.48

In addition, when using the Weibull function, the frequency of zero winds relative to new heights must be estimated. This is relatively difficult without real data, but the hypothesis of identical frequencies of zero winds seems reasonable.

13.2.3.2 Weibull Hybrid Function

Nfaoui and Knidiri et al. have shown that for the areas where frequencies of calm wind are greater than 15–20%, the classic Weibull function did not fit the observed distribution [2, 7]. On the top of that, calm and/or strong winds were not estimated correctly. The Weibull hybrid method is a particular case of Weibull where the frequency of calm winds is temporarily neglected. The Eq. (13.24) becomes

$$f_{\mathrm{H}}(v) = (1-f_0)\left(\frac{k}{c}\right)\left(\frac{v}{c}\right)^{k-1}\exp\left[-\left(\frac{v}{c}\right)^{k}\right] \quad \text{for } v > 0$$
$$f_{\mathrm{H}}(v) = f_0 \quad \text{for } v = 0 \tag{13.32}$$

where f_0 is the probability of observing calm wind speeds. The Weibull hybrid function is then a function of three parameters (f_0, k, and c). This is a little complex to use, but the results are better. The value of f_0 is known directly from the wind data.

A method which is relatively long but extremely precise for determining the Weibull parameters is called the maximum-likelihood method given by Jonson and Kotz [10] to calculate the parameters k and c as follows [2, 6]:

$$k = \left[\left[\frac{\sum_{i=1}^{N} V_i^k \ln(V_i)}{\sum_{i=1}^{N} V_i^k}\right] - \left[\frac{\sum_{i=1}^{N} n_i \ln(V_i)}{N}\right]\right]^{-1} \tag{13.33}$$

$$C = \left[\frac{\sum_{i=1}^{N} n_i V_i^k}{N}\right]^{\frac{1}{k}} \tag{13.34}$$

where

N: total number of observations of nonzero wind speed,
n_i: number of observations of wind speed v_i,
v_i: midpoint wind speed interval i.

It can be seen that the parameter k appears on both sides of Eq. (13.33), so it must be solved by an iterative method. Once the value of k is determined, we can then calculate c from Eq. (13.34).

13.2.4 Wind Turbine Energy Production

Even if we manage to collect a large quantity of wind measurements data, it is not easy to exploit them in their raw state. There a number of ways to summarize them in a compact form so that one may evaluate the wind resource or wind power production potential of a particular site. These include both direct and statistical techniques.

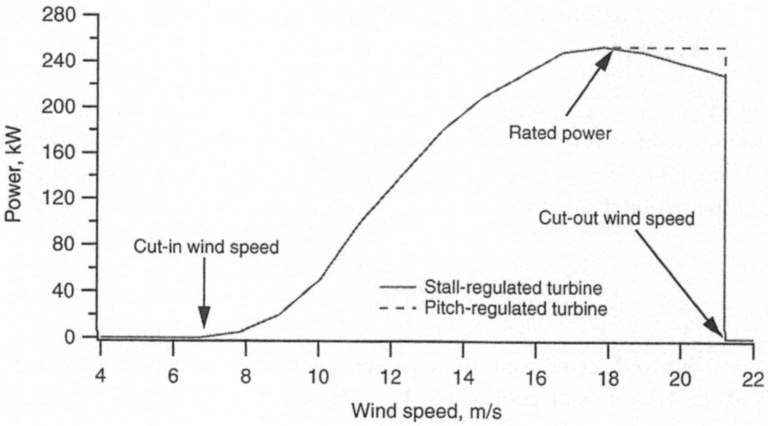

Fig. 13.9 Power output curve for wind turbine [1]

13.2.4.1 General Aspects of Wind Turbine Power

In this section we will determine the productivity (both maximum energy potential and machine power output) of a given wind turbine at a given site in which wind speed information is available in either time series format or in a summary format (average wind speed, standard deviation, etc.). In practice, the power available from a wind turbine P_w, can be shown by a machine power curve (Fig. 13.9). Two typical curves, $P_w(v)$, simplified for purposes of illustration, are shown in Fig. 13.7. Later sections of this chapter describe how such curves can be estimated from analytical models of the wind turbine system [1].

13.2.4.2 Wind Turbine Energy Production Using Direct Method

Suppose one is given a series of N wind speed observations, v_i, each averaged over the time interval Δt. The average wind machine power, $\overline{P_w}$, is

$$\overline{P_w} = \frac{1}{N}\sum_{i=1}^{N} P_w(v_i) \tag{13.35}$$

where $P_w(v_i)$ is the power output defined by a wind machine power curve.

The energy from a wind machine, E_w, is

$$E_w = \sum_{i=1}^{N} P_w(v_i)(\Delta t) \tag{13.36}$$

13.2.4.3 Wind Turbine Energy Production Estimates Using Statistical Techniques

The related performance parameter, the capacity factor, CF, of a wind turbine at a given sites is defined as the ratio of the energy actually produced by the turbine to the energy that could have been produced if the machine ran at its rated power, P_R, over a given time period Thus [1],

$$CF = \frac{\overline{P_w}}{P_R}. \quad (13.37)$$

It is possible to determine a turbine power curve based on power available in the wind and the rotor power coefficient, C_p. The result is the follow expression for $P_w(v)$ [1]:

$$P_w(v) = \frac{1}{2}\rho A C_p \eta v^3 \quad (13.38)$$

where η is the drive train efficiency (generator power/rotor power). The rotor power coefficient is defined by [1]

$$C_p = \frac{\text{Power extracted}}{\text{Power in the wind}} = \frac{P_{rotor}}{\frac{1}{2}\rho A v^3}. \quad (13.39)$$

13.2.4.4 Production Calculations for a Real Wind Turbine Using Weibull Distribution

A measure of the maximum possible average power from a given rotor diameter can be calculated assuming an ideal wind turbine and using a Weibull probability function. The average wind machine power $\overline{P_w}$ is calculated using:

$$\overline{P_w} = \int_0^\infty P_w(v)p(v)dv \quad (13.40)$$

Or the derivative of the cumulative distribution function is equal to the probability density function, that is,

$$p(v) = \frac{dF(v)}{dv}. \quad (13.41)$$

So it is also possible to rewrite this Eq. (13.40) using the cumulative distribution function; thus,

$$\overline{P_{\mathrm{w}}} = \int_0^\infty P_{\mathrm{w}}(v)\mathrm{d}F(v). \tag{13.42}$$

Based on Eq. (13.41) and replacing the integral in Eq. (13.42) with a summation over N_{B} bins, the following expression can be used to find the average wind machine power [1]:

$$\overline{P_{\mathrm{w}}} = \sum_{j=1}^{N_{\mathrm{B}}} \left\{ \exp\left(-\left(\frac{v_{j-1}}{c}\right)^k\right) - \exp\left(-\left(\frac{v_j}{c}\right)^k\right) \right\} P_{\mathrm{w}}\left(\frac{v_{j-1}+v_j}{2}\right) \tag{13.43}$$

13.2.4.5 Assessment of Capacity Factor of a Wind Turbine

The capacity factor is one of the performance parameters of a wind turbine which both the user and the manufacturer need to know. It is defined as the ratio between average power output and rated power. Due to lack of wind speeds during fabrication of wind turbines and its wide spectrum specification in the international market, manufacturers cannot give exact information about their capacity factors. Hence, they are not able to estimate their energy productions. Therefore, they tend to use approximate formulas to estimate the capacity factor C_{f} for any turbine at any site. A comparison made by El-Mallah and Soltan between three methods leads to the following formula more realistic for estimating the capacity factor of a turbine [9]:

$$C_{\mathrm{f}} = \frac{\overline{P}}{P_{\mathrm{r}}} = \frac{\exp\left(-\left(\frac{v_i}{c}\right)^k\right) - \exp\left(-\left(\frac{v_{\mathrm{r}}}{c}\right)^k\right)}{\left(\frac{v_{\mathrm{r}}}{c}\right)^k - \left(\frac{v_i}{c}\right)^k} - \exp\left(-\left(\frac{v_0}{c}\right)^k\right) \tag{13.44}$$

where

- k and c are related to average wind speed at the site as: $\overline{v} = c.\Gamma\left(1 + \frac{1}{k}\right)$,
- v_i, v_{r}, and v_0 are the cut-in, rated, and cut-off wind speeds, and $\overline{P}$ and P_{r} are the average and rated powers.

From Darwish and Sayigh's research work [10], it can be concluded that c is usually about 12% larger than the mean wind speed. So the optimum design for energy production is a rated speed of about twice the mean speed. If the mean wind speed at a site is v_{m}, the rated speed falls at $(1.8–2.2)v_{\mathrm{m}}$. The cut-in and cut-off wind speed are estimated as $v_{\mathrm{i}} = 0.5v_{\mathrm{r}}$ and $v_0 = 2v_{\mathrm{r}}$, respectively [11].

13.3 Wind Potential in Morocco: Bibliographic Study

13.3.1 Introduction

As a result of the increase in oil prices in the seventies, the supply of energy consumption became increasingly important factors for the balance of the Moroccan economy. Thus, energy from oil accounted for about 54% of all energy consumed in Morocco in 1981, 74% of imported petroleum products.

The Moroccan government has taken steps to deal with this situation, trying to develop renewable energies, especially solar and wind energy. For this reason, there was the creation of two institutions, one academic: Solar Energy Laboratory (SEL) of the Science Faculty, Rabat, in 1980 [12, 13], and the other governmental: Center for the Development of Renewable Energies (CDER), in 1982 [7]. Their mission is to direct and coordinate research and development efforts of the study of the solar radiation and the wind speed in Morocco, as well as the photovoltaic and wind conversion and applications.

To this end, in 1986, the CDER published the Moroccan Wind Atlas on the basis of data from 17 synoptic stations, using 5 years of hourly average of wind speed (1978–1982). Most of the them are located in airport, which in turn established in less windy areas for the security of Air Navigation The document proposes the first statistical analysis of existing anemometric data, in order to assess the Moroccan wind potential. The results presented in this document can be considered preliminary. Morocco's wind map published has remained incomplete due to the lack of wind data from areas where there are no airports.

For the development of wind energy in Morocco and to evaluate projects implementing wind energy, in 1991, CDER began a feasibility program of assessment of the economic wind farms by the installation of automatic anemometers in the windy sites through Morocco, which allows for the acquisition of the characteristics of the wind resource [14].

13.3.2 Wind in Morocco

Winds are produced by a multitude of complex meteorological forces. In general, it is the earth-scale pressure systems that are responsible for the prevailing winds in such a region.

Morocco is subject to several dominant wind systems, including mid-latitude west winds, northeastern trade winds, and calm winds associated with the transition between these two systems. The Azores anticyclone creates a strong circulation of trade winds, which is reinforced in summer. Migratory depressions sweep the area, especially in winter. Several local winds, such as a simple thermal circulation, sea/land and valley/mountain breezes, are also important.

In southern Morocco and the Saharan provinces, meteorological weather is subject throughout the year to the domination of a dry continental air mass. In addition, the dry continental air is subject to strong diurnal variations in temperature, due to rapid heating or cooling. In the North, the conditions are quite different, especially in winter, when colder polar air brings clouds and rain, the local thermal circulation is much less important.

In Morocco, mountains have all kinds of influences on the winds, and are likely to play a role in creating local winds (e.g., Midelt site). The ranges of Atlas and Rif, in many areas, are perpendicular to the prevailing winds. They intercept humid winds coming from the Atlantic and the Mediterranean, forcing the air to rise and to get cooler, causing clouds and rain or snow.

South of the Rif Mountains, there is the Atlas, a mountainous ensemble formed by three chains superimposed, oriented along a west-south and north-east axis. The northernmost range, the Middle Atlas is separated from the Rif by the narrow corridor of Taza where the mountains canalize and accelerate the winds. The position of the Taza site is appropriate for the generation of electricity by the installation of wind farms.

13.3.3 Available Wind Speed Measurement Data

The Directorate of National Meteorology (DMN) used to manage all the Moroccan meteorological stations, and supplies data summaries. At the time, the DMN takes care of 32 synoptic sites with hourly wind records. In certain cases the data goes back several years to 1948. A great number of sites in Morocco were very well chosen so that a good exposure of the instruments is achieved (17 sites), but some were poorly exposed due to the proximity of buildings [7]. Several of them are located in the airports, are equipped with anemometers and wind vanes to measure wind speed and its direction. By 1978, the number had risen to about thirty. For most of these stations, we have available the hourly average wind speed (HAWS) and wind direction data, measured at a height of 10 m above sea level with the exception of Laâyoune where it is 15 m. The measurement data are archived on files, but their exploitation has remained very limited, apart from some research carried out at the universities and some institutions for research and development of renewable energies, such as the Development Agency of Renewable Energies and Energy Efficiency (ADEREE), former CDER [7, 14].

13.3.4 Quantitative Assessment of Wind Potential

In 1986, CDER published the "Wind Atlas of Morocco" on the basis of data from 17 synoptic stations of the DMN (Table 13.3). These stations are generally installed in airports that are located in the less windy sites for the safety of the aerial

Table 13.3 Synoptic stations selected [7]

Station	North latitude (°)	Longitude	Altitude (m)	Starting measurements	Anemometor type	Anemometor height (m)
Agadir	30.23	9.34	18	1949	CA/JR	10
Béni-Mellal	32.22	6.24	468	1970	CA	10
Casablanca	33.34	7.4	56	1947	CA/JR	10
Dakhla	23.46	15.56	11	1980	JR	10
Fes	33.55	4.58	571	1961	CA	10
Ifrane	33.3	5.1	1664	1961	CA	10
Kenitra	34.18	6.36	5	1951	CA/JR	10
Laâyoune	27.1	13.13	63	1976	CA	15
Marrakech	31.37	8.02	464	1947	CA/JR	10
Midelt	32.41	4.44	1508	1951	CA	10
Nador	35.9	2.55	7	1976	JR	10
Ouarzazate	32.56	6.54	1136	1950	CA	10
Oujda	34.47	1.56	465	1947	CA/JR	10
Rabat	34.3	6.46	75	1948	CA/JR	10
Errachidia	31.56	4.24	1037	1973	CA	10
Safi	32.17	9.14	43	1955	CA	10
Tanger	35.43	5.54	15	1949	CA/JR	10

CA Chauvin Arnoux, *JR* Jules Richard

navigation. This book is about wind characteristics and resources, statistical analysis of wind measurements data, the assessment of the wind potential and the mapping of the annual mean wind regimes, based on 5 years of wind speed measurements (1978–1982), limiting into four measurements per day, responding to the standards defined by the World Meteorological Organization (WMO [7, 14].

The preliminary analysis of these data carried out by CDER not only makes it possible to assess the Moroccan wind potential, but also to provide the parameters useful for the dimensioning of wind farms. These parameters are the frequencies of absence of wind f_0, the monthly wind power received per unit of surface and the parameters k and c of the Weibull density function. The results are presented in the form of hourly frequency data, monthly and annual averages of wind speed, as well as the characteristic parameters of the Weibull density function (Tables 13.4, 13.5, 13.6, 13.7, and 13.8). Using annual averages of wind speed, a map of the annual average of wind in Morocco was plotted (Fig. 13.10). Given the short duration of the data used (5 years), this map gives only qualitative indications, with the recapitulatory Tables 13.4, 13.5, 13.6, 13.7, and 13.8 will serve as a guide and basis for this research work.

The series of wind speed data used for the establishment of the Moroccan Wind Atlas, published by Knidiri and Laâouina [7], are insufficient despite the interest of the subject being treated. The map of the average wind speed of the winds in Morocco is interesting, but does not take into account corrections of the local

Table 13.4 Monthly and annual averages of measured wind speed, V (m/s) [7]

Stations	J	F	M	A	M	J	Jt	A	S	O	N	D	Annuals	Periods
Agadir	2.5	2.9	2.4	3.1	3	2.7	2.2	2.2	2.3	2.1	2.6	1.7	2.5	1978–1982
Béni-Mellal	0.7	1	1	1.1	1.3	1.6	1.6	1.5	1.2	0.9	0.6	0.6	1.1	1978–1982
Casa	2.9	3.9	3.6	3.9	4.5	3.1	3.7	3.3	3.1	3.2	2.5	3.3	3.5	1978–1982
Dakhla	5.8	7	7.9	9.1	10.6	11.2	11.2	10.3	8.5	6.6	5.5	6.4	8.4	1980–1983
Fes	2.9	2.8	2.8	2.6	2.7	2.4	2.8	2.9	3	3.3	3.2	3.4	2.9	1978–1982
Ifrane	2.6	3.1	3	2.8	2.8	2.1	2.6	2.7	2.8	3.1	2.8	2.8	2.8	1978–1982
Kenitra	2.2	3.1	3.2	3.8	3.9	3.9	3.8	3.7	3.4	3	2.4	2.9	3.3	1978–1982
Laâyoune	4.8	5	5.7	5.2	6.4	6.7	7.3	7.5	5.7	4.4	4.5	4.7	5.7	1978–1982
Marrakech	2.1	2.5	2.9	2.9	3.2	3.5	3.1	2.8	2.7	2.3	1.9	1.5	2.6	1978–1982
Midelt	4.4	5.7	5.3	5.6	4.8	3.3	3.3	3	2.7	3.7	3.9	4.6	4.2	1978–1982
Nador	3.3	3.9	3.9	3.9	4	3.9	3.9	3.5	3.4	3.3	2.9	4.3	3.7	1979–1983
Ouarzazate	1.7	2.1	2.9	4	3.6	3.5	3.5	3.2	2.4	2.4	1.8	2.2	2.8	1978, 1980–1983
Oujda	3.7	3.9	3.6	3.7	3.9	3.9	3.8	3.6	3.3	3.5	3.3	4.5	3.7	1978–1982
Rabat	3.2	3.3	2.9	3.3	3.3	2.9	2.7	2.6	2.7	2.8	2.4	2.8	2.9	1978–1982
Errachidia	2.7	3.2	3.9	4.4	4.6	5.5	4.1	4	3.4	3	2.3	2.4	3.6	1978–1982
Safi	3.9	3.9	4.4	4	4.3	4.2	4.3	4.2	3.9	3.9	3.5	3.6	4	1978–1982
Tanger	4.7	5.2	4.9	5.1	5.4	4.9	5.9	5.5	6.1	5.6	6.3	5.2	5.4	1978–1982

Table 13.5 Monthly and annual averages of estimated wind power, P (W/m^2) [7]

Stations	J	F	M	A	M	J	Jt	A	S	O	N	D	Annuals
Agadir	56	109	29	56	53	35	21	27	27	37	69	23	45
Béni-Mellal	6	8	5	5	22	8	11	12	7	4	4	4	8
Casablanca	73	171	71	77	108	71	61	49	43	57	42	70	74
Dakhla	235	390	530	787	989	1205	1247	925	517	283	183	284	640
Fes	45	45	40	34	33	24	32	34	31	48	53	84	42
Ifrane	38	44	39	36	32	16	24	29	26	40	43	49	35
Kenitra	56	91	78	94	111	91	88	85	76	74	76	90	84
Laâyoune	125	160	185	137	221	227	273	288	155	86	120	116	175
Marrakech	25	39	53	40	55	62	51	41	46	34	25	18	41
Midelt	269	343	344	299	231	110	102	87	75	170	222	324	208
Nador	107	143	126	105	118	107	105	73	69	88	83	257	115
Ouarzazate	52	57	89	162	115	93	89	82	45	61	34	87	81
Oujda	107	121	78	90	95	93	93	89	69	87	117	194	103
Rabat	44	51	31	38	40	26	21	18	22	24	26	30	31
Errachidia	20	93	81	129	141	178	89	79	61	40	24	24	80
Safi	71	67	83	61	73	65	62	62	57	58	49	55	64
Tanger	161	218	159	170	182	192	340	291	298	200	278	225	226

Table 13.6 Monthly and annual averages of the estimated form factor, *K* (without unit) [7]

Stations	J	F	M	A	M	J	Jt	A	S	O	N	D	Annuals
Agadir	1.48	1.32	1.96	1.81	1.87	2.1	2.05	1.84	2	1.57	1.37	1.52	1.63
Béni-Mellal	1.58	1.51	1.76	1.84	1.43	1.96	1.85	1.76	1.76	1.81	1.59	1.58	1.67
Casablanca	1.59	1.6	2.22	2.31	2.45	2.56	2.44	2.5	2.3	2.03	1.76	1.68	1.99
Dakhla	2.11	2.23	2.27	2.26	2.91	2.82	2.69	2.88	3.05	2.52	2.45	2.3	2.22
Fes	1.63	1.66	1.78	1.78	1.9	1.98	2.05	2	2.26	1.96	1.74	1.49	1.8
Ifrane	1.5	1.67	1.8	1.74	1.81	1.89	1.96	1.96	2.09	1.86	1.58	1.61	1.74
Kenitra	1.47	1.64	1.87	2.23	2.04	2.26	2.29	2.21	2.09	1.93	1.57	1.72	1.9
Laâyoune	2.18	2.06	2.43	2.48	2.94	3.51	4.27	4.59	3	2.5	1.99	2.3	2.52
Marrakech	1.8	1.83	1.85	2.07	1.99	2.05	2.05	1.98	1.97	1.84	1.78	1.74	1.89
Midelt	2.01	2.17	1.9	2.27	2.21	2.22	2.07	2.21	2.21	2.19	1.97	2.04	2
Nador	1.53	1.6	1.67	2.09	1.91	2.2	2.07	2.24	2.36	1.82	1.65	1.39	1.77
Ouarzazate	1.22	1.34	1.39	1.47	1.43	1.53	1.55	1.53	1.54	1.38	1.39	1.13	1.39
Oujda	1.78	1.72	2.01	2.07	2.11	2.23	2.14	2.06	2	1.85	1.55	1.75	1.88
Rabat	1.94	1.92	2.01	2.38	2.16	2.35	2.46	2.54	2.45	2.23	1.8	1.9	2.08
Errachidia	2.28	1.53	1.91	1.77	1.77	1.97	1.95	2.04	1.88	1.87	1.79	1.82	1.71
Safi	2.11	2.2	2.5	2.46	2.78	2.99	3.21	3.12	2.63	2.51	2.15	2.06	2.46
Tanger	1.79	1.77	1.99	2.05	2.21	1.82	1.7	1.77	1.97	2.21	2.15	1.71	1.88

Table 13.7 Monthly and annual averages of the estimated scale factor, *C* (m/s) [7]

Stations	J	F	M	A	M	J	Jt	A	S	O	N	D	Annuals
Agadir	3.73	3.96	3.49	4.23	4.2	3.84	3.27	3.2	3.56	3.23	3.61	2.71	3.61
Béni-Mellal	2.3	2.29	2.16	2.21	2.43	2.37	2.57	2.54	2.28	2.11	2.02	1.99	2.31
Casablanca	4.26	5.21	4.94	5.05	5.68	5.03	4.73	4.52	4.25	4.46	3.76	4.62	4.74
Dakhla	6.88	8.33	9.16	10.4	11.8	12.6	12.6	11.6	9.61	7.66	6.7	7.5	9.71
Fes	3.52	3.64	3.67	3.61	3.61	3.41	3.71	3.67	3.7	3.97	3.84	4.01	3.71
Ifrane	3.45	3.91	3.91	3.75	3.7	3.16	3.52	3.78	3.64	3.98	3.79	4.17	3.74
Kenitra	4.03	4.87	4.99	5.57	5.78	5.47	5.44	5.34	5.17	5.08	4.87	5.29	5.2
Laâyoune	5.56	5.96	6.58	6.05	7.25	7.44	8	8.21	6.43	5.19	5.4	5.64	6.54
Marrakech	3.39	3.83	4.17	4.03	4.39	4.55	4.34	4.1	4.25	3.78	3.32	3.08	3.98
Midelt	9.17	9.41	9.1	8.96	8.47	7.04	6.6	6.54	6.42	8.21	8.73	9.8	8.21
Nador	4.8	5.5	5.32	5.57	5.61	5.82	5.57	5.13	5.14	5.08	4.71	5.9	5.37
Ouarzazate	3.9	4.09	4.63	5.66	4.81	4.54	4.7	4.63	3.95	3.91	3.34	3.66	4.38
Oujda	5.32	5.36	4.98	5.38	5.5	5.65	5.55	5.47	4.91	5.02	5.13	6.42	5.4
Rabat	3.73	3.86	3.39	3.86	3.79	3.43	3.22	3.11	3.28	3.21	2.95	3.26	3.43
Errachidia	3.26	3.87	4.69	5.27	5.61	6.24	5.09	4.99	4.44	3.67	2.8	3.05	4.42
Safi	4.55	4.58	5.07	4.53	4.97	4.84	4.78	4.77	4.51	4.53	3.99	4.14	4.62
Tanger	5.69	6.3	5.95	6.11	6.45	6.15	7.14	6.95	7.39	6.67	7.39	6.18	6.53

Table 13.8 Monthly and annual averages of observed wind frequency, f_0 (%) [7]

Stations	J	F	M	A	M	J	Jt	A	S	O	N	D	Annuals
Agadir	25	18	21	18	20	20	25	24	27	26	22	31	23
Béni-Mellal	64	52	49	45	42	25	31	32	40	51	64	63	47
Casablanca	24	15	17	12	10	13	12	18	18	19	24	19	17
Dakhla	4	5	3	1	0	0	0	0	0	2	7	3	2
Fes	7	12	13	18	15	21	14	12	8	6	7	6	11
Ifrane	14	12	13	14	16	25	16	20	14	13	17	24	17
Kenitra	40	29	27	23	24	20	20	22	26	33	44	38	29
Laâyoune	2	4	3	4	2	1	0	1	1	4	6	7	3
Marrakech	31	26	22	18	18	14	19	23	28	31	35	46	26
Midelt	46	32	35	29	36	47	44	48	52	49	50	47	43
Nador	24	20	18	21	20	24	20	23	24	26	31	20	23
Ouarzazate	53	43	32	21	17	14	18	24	32	33	39	36	30
Oujda	21	19	18	22	20	22	22	25	23	22	27	21	22
Rabat	2	2	5	3	3	5	6	6	7	2	9	4	4
Errachidia	8	8	6	6	7	1	8	10	12	8	8	10	8
Safi	4	4	2	1	2	3	1	1	1	4	2	3	2
Tanger	6	7	6	5	6	10	7	10	7	6	5	5	7

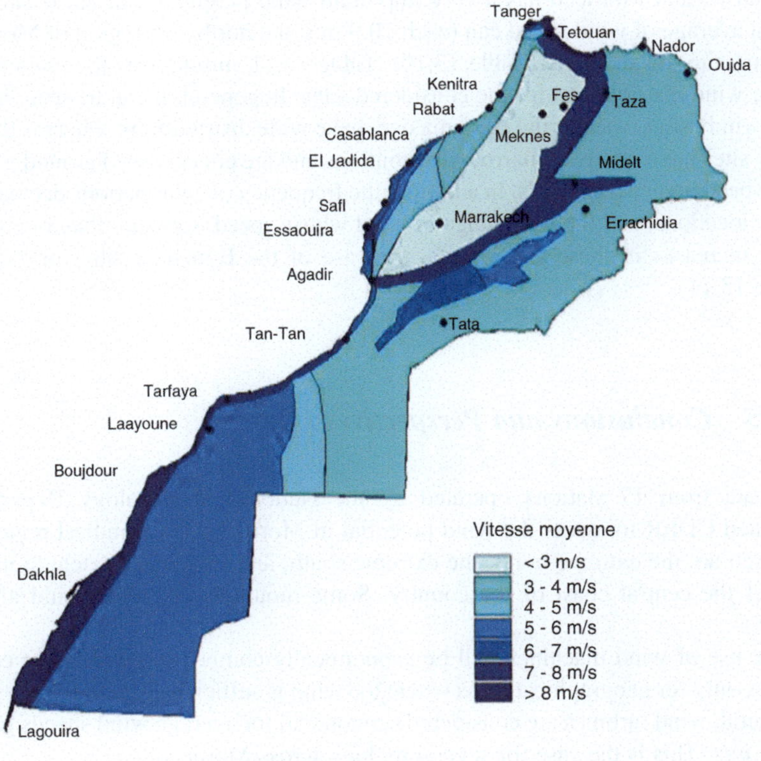

Fig. 13.10 Annual average of wind speed [7]

environment (topography, obstacles, etc.) as described, for example, in the "European Wind Atlas" [15].

One of the most detailed research work carried out by Nfaoui [2, 16] concerned a detailed study of the statistical and dynamic characteristics of the Tangier wind potential. This study used the measurements carried out by the DMN at the Tangier synoptic station, located at Tangier airport, considering a period of 12 years of HAWS (1978–1989), that is to say 105,000 brut data (12 × 365 days) and 2 years of hourly data for wind direction (1988–1989). The data are archived on cards at DMN in Casablanca. Nfaoui seized and stored them on the computer support. Thus, it has set up a wind speed database for Tangier site. Part of the results of this research will be presented in Chap. 4.

In 1991, CDER launched a campaign of measures of short duration ranging from 3 months to 2 years, in order to locate the windiest sites that could be economically profitable for installing wind farms. In this context, CDER installed more than 17 measuring stations in all the windy regions of Morocco. Table 13.9 shows nine stations of them. The data are published in a document in recapitulatory table forms [14]. In 1995, nine anemometers would be also installed by the CDER to improve and complete the assessment of the Moroccan wind potential in the northeast and the south regions [14].

It emerge from the second campaign of wind speed measurements, carried out by the CDER, that Morocco has a very important wind potential. For some sites, the annual average of wind speed can reach 10.9 m/s, the northwest region of Morocco, Blanco1 site, for example (Table 13.10). Table 13.11 summarizes the main results for the wind distributions for the considered sites. In general, it can be seen that the more windy sites, such as the Haouma one, have wide distributions, whereas the less windy sites have relatively narrow distributions and are concentrated around a lower wind speed, in general, 5 m/s. In addition, the frequency of calm periods decreases in windy locations, and the percentage of wind whose speed is greater than or equal to 5 m/s increases in these same areas, the case of the Bouznika site, for example (Table 13.11).

13.3.5 Conclusions and Perspectives

The data from 17 stations operated by the National Meteorology Department permitted CDER to assess the wind potential in Morocco. The windiest regions of Morocco are the extreme north, the extreme south, and to a lesser extent, a narrow strip of the central coast of the country. Some mountainous areas could also be windy.

The use of wind machines will be economically competitive with other energy sources only for geographical areas where the wind is sufficiently strong and regular. In general, wind turbines are considered economical for average wind speeds greater than 5 m/s. This is the case for several regions across Morocco.

Table 13.9 Characteristics of CDER measuring stations [14]

Stations	Altitude (m)	Period of measurements	Average wind speed (m/s)[a]	Maximal wind speed (m/s)[a]
Mouly Boussalham (Kenitra)	Sea level	1991–1992 (2 years)	3.78	20.9
Tamagrout-Zagora (Ouarzazate	800 m	1/1991–6/1992 (18 months)	4.04	26.6
El Gaada (Tiznit)	150 m	7/1991–9/1992 (15 months)	4.56	21.7
Tiniguir (Dakhla)	Sea level	1994 (1 year)	8.21	18.7
Tan-Tan	Sea level	10/1992–12/1994 2 years 3 month	5.12	23.8
Bouznika (Benslimane)	Sea level	7/1992–6/1993 (1 years)	3.85	16.2
Blanco 1 (Tetouan)	400 m	2/1993–8/1994 (19 months)	10.94	36.5
Sendouk 1 (Ksar Sguir)	150 m	2/1993–8/1994 (19 months)	8.59	35.1
Haouma (Tétouan)	250 m	2/1993–4/1994 (15 months)	9.43	34.4

[a]Anemometer heights: 9 m

It should be noted that there are undoubtedly other places, where the topography allows a better exposure, or even an accentuation of the winds, where the winds are stronger than indicated by the results of the 17 meteorological stations. Airports, where most of these stations are located, are generally not located in the windiest area. Therefore, the conclusions presented in the document, published by CDER on the wind potential in Morocco, are incomplete. Further work is needed to improve the "wind speed database" presented in this document. This is what CDER had done in 1991.

CDER started a program combining a further study of the existing data, and a selective and specific project of installation of new anemometers, to improve the assessment of the wind potential in Morocco. Some of the objectives to be carried on include:

- the acquisition of additional and more precise data of wind speed, in windy regions, such as the extreme north (Tangier region), mountainous regions, the Taza corridor, the interior of the Sahara provinces in the south of Morocco,
- influence of the period of measurements on wind potential assessment for a given site,
- the study of wind penetration from the Atlantic coast to the interior,
- the study of the vertical profile of the wind, by installing anemometers at different heights,

Table 13.10 Wind speed monthly data for the CDER measuring stations (1991–1994) [14]

Stations (Month)	Mouly Boussalham (Kenitra)	Tamagrout-Zagora (Ouarzazate)	El Gaada (Tiznit)	Tiniguir (Dakhla)	Tan-Tan	Bouznika (Benslimane)	Blanco1 (Tetouan)	Sendouk1 (Ksar Sguir)	Haouma (Tétouan)
1/91	2.54	2.47							
2	3.74	3.64							
3	4.91	4.61							
4	4.06	4.33							
5	4.34	5.53							
6	4.16	5.12							
7	4.38	4.83	4.42						
8	3.99	4.18	3.78						
9	3.74	4.38	4.08						
10	3.99	3.72	3.86						
11	2.96	2.60	4.04	6.88					
12/91	2.72	2.26	4.72	5.53					
1/92	2.64	3.00	4.40	7.31					
2	3.70	4.52	4.92	7.78					
3	4.38	4.52	4.64	8.54					
4	5.05	4.70	5.08	8.47					
5	4.30	4.60	5.58	9.88					
6	4.20	4.00	5.22	10.17					
7	4.44		4.86	9.73		3.41			
8	4.28		4.51	9.84		3.59			
9	4.28		4.40	7.96		3.41			
10	4.22			5.86	4.18	4.19			
11	2.74			6.88	4.59	3.19			
12/92	3.40			7.53	3.94	3.78			

1/93					5.70	3.81			
2					5.12	3.89	13.26	10.03	11.42
3					6.16	4.19	10.32	8.81	9.00
4					6.59	4.48	10.96	8.95	9.38
5					5.61	4.33	9.11	7.30	7.77
6					5.63	3.85	11.97	8.52	9.35
7					4.67		13.36	9.31	10.65
8					4.42		13.17	9.35	10.73
9					4.45		11.14	7.91	8.92
10					4.47		9.03	8.77	8.96
11					5.40		9.49	10.53	10.38
12/93					5.54		10.32	6.90	8.00
1/94					6.47		11.34	8.45	9.31
2					5.08		9.86	8.23	8.31
3					5.81		12.34	9.85	10.19
4					5.56		12.16	9.06	9.31
5					5.58		9.39	7.76	
6					5.15		12.71	9.71	
7					4.67		7.64	6.33	
8					4.62		9.67	6.72	
9					4.00				
10					4.21				
11					5.31				
12/94					5.01				

Table 13.11 Annual frequency distribution of wind speed for the CDER measuring stations (1991–1994) [14]

Stations (Month)	Mouly Boussalham (Kenitra)	Tamagrout-Zagora (Ouarzazate)	El Gaada (Tiznit)	Tiniguir (Dakhla)	Tan-Tan	Bouznika (Benslimane)	Blanco1 (Tetouan)	Sendouk1 (Ksar Sguir)	Haouma (Tétouan)
0	6.14	5.45	1.56	1.19	2.23	1.36	0.24	0.97	0.66
1	20.03	21.57	9.24	1.47	6.12	11.04	1.74	3.51	2.86
2	16.34	19.58	14.33	2.87	9.79	20.72	3.05	5.01	3.67
3	13.62	14.43	16.65	4.55	14.22	24.12	4.05	5.87	4.09
4	11.83	10.38	17.06	7.49	17.82	19.78	4.47	7.28	5.40
5	10.74	7.73	13.90	12.38	17.09	12.05	5.03	8.61	6.91
6	8.95	5.74	10.23	15.60	13.02	5.87	5.77	9.29	8.70
7	5.79	4.34	6.28	15.46	8.07	2.68	6.41	9.79	10.40
8	3.14	3.39	3.88	12.52	4.64	1.26	7.24	9.29	10.88
9	1.63	2.43	2.71	9.37	2.74	0.60	7.79	8.43	10.77
10	0.84	1.77	1.80	4.69	1.65	0.30	7.84	7.37	8.53
11	0.38	1.25	1.14	2.87	0.73	0.14	7.75	6.33	6.60
12	0.20	0.81	0.63	1.61	0.58	0.07	7.61	4.95	5.09
13	0.08	0.59	0.30	0.77	0.37	0.03	6.81	3.89	4.17
14	0.05	0.37	0.15	0.28	0.27	0.01	5.89	2.94	3.28
15	0.02	0.29	0.08	0.07	0.18	0.00	4.88	2.20	2.47
16	0.01	0.18	0.04	0.00	0.10	0.00	3.92	1.53	1.81
17	0.01	0.11	0.02	0.00	0.05	0.00	2.95	0.99	1.31
18	0.00	0.07	0.00	0.00	0.02	0.00	2.12	0.62	0.89
≥19	0.00	0.07	0.00	0.00	0.02	0.00	4.44	1.13	2.43

According to the wind measurements carried out by the CDER in the North of Morocco between 1993 and 1994, as part of a techno-economic feasibility study for the realization of wind farm projects, the most favorable site was that of Koudia Al Baida (Province of Tétouan). The average and maximum speeds recorded at the height of 9 m are, respectively, 10.94 and 36.5 m/s.

In addition, wind energy over open oceans has wide potential for electricity generation. Wind speeds can be as much as 70% higher than on land. With 3500 km of the coast, Morocco has a promising potential for offshore wind energy. It needs to be harnessed by conducting sites assessment through installing and operating meteorological towers and buoys, which are economic and valid for offshore wind farms installation.

13.4 Statistical Characteristics of Wind Potential of the Windiest Sites in the Sahara Provinces of Morocco

13.4.1 Introduction

The research on wind potential assessment and its exploitation for the electricity production on a large scale in the windiest sites in the Sahara provinces in south of Morocco remain limited until the early twenty-first century, despite the existence of wind speed data dating back to 1978 which are available at Directorate of National Meteorology (DNM). A statistical analysis of these data is necessary and important, to assess quantitatively and qualitatively the wind potential available in this region.

In addition to its location in an arid zone, south of Morocco enjoys a significant wind potential. This allows this region to be a favorable area and economic for the installation of wind farms to produce electricity on a large scale. Thereafter, we present the characteristics and the assessment of wind potential in this region, mainly in Tan-Tan, Laâyoune, and Dakhla sites.

13.4.2 Wind Characteristics and Resources in the South

13.4.2.1 Annual Averages of Wind Speed

The annual averages of wind speed for Tan-Tan, Laâyoune, and Dakhla, are in order of 5.12 m/s, (5.90 ± 0.89 m/s) and (7.64 ± 1.15 m/s) respectively (Table 13.12). Comparing these averages show that Dakhla is the windiest site.

Figure 13.11 shows that the months of August and July are the windiest for Laâyoune and Dakhla, respectively, with monthly averages of 7.67 and 10.12 m/s. November is the least windy month for both sites, with monthly average of 4.57 and 5.35 m/s respectively. However, for Tan-Tan site, the monthly averages are

Table 13.12 Annual Weibull Hybrid distribution parameters and the available annual averages of wind potential [2, 7, 14, 17]

Sites	Period of measurements	Average wind speed (m/s)	Maxima (m/s)	Average wind potential (W/m^2)	Annual parameters of Weibull hybrid *K* *C* (m/s)		Frequency (%) for $V = 0$ m/s (non-wind)
Tan-Tan	2 years, 11/93–12/94	5.12	24	122	2.07	5.43	4
Laâyoune	13 years, 1/78–12/90	5.9	27	204	2.47	6.79	2
Dakhla	10 years, 1/80–12/90	7.6	34	462	2.32	8.78	2

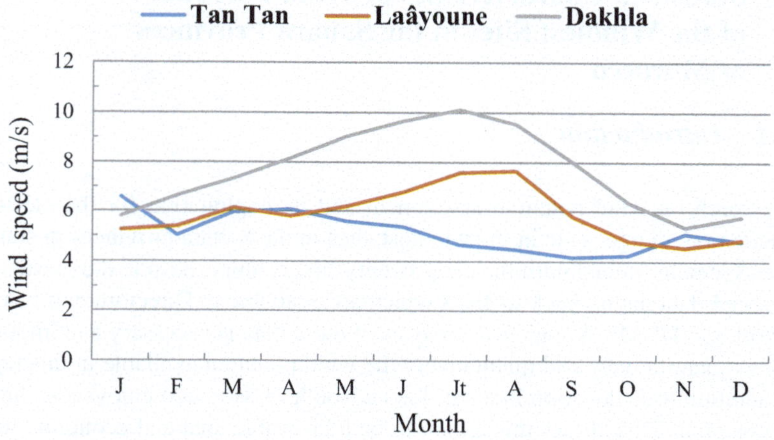

Fig. 13.11 Seasonal variations of wind speed [2, 6, 13, 16]

contained between 4 and 6 m/s. The seasonal variation is not regular compared to the other sites. For Laâyoune, the maxima speed is 34 m/s, August 1978, while it is 27 m/s for Dakhla, obtained for July 1983.

13.4.2.2 Diurnal Variations

Figure 13.12 shows, in general, for the three considered sites that the wind is strong in the afternoon and reaches its maximum at 16 h local time, and weak at night. This can be explained by the influence of the breeze sea/land, a localized circulation particularly characteristic of coastal areas. Knowing this type of variation permits to harmonize the power recuperated from the wind and the energy needs.

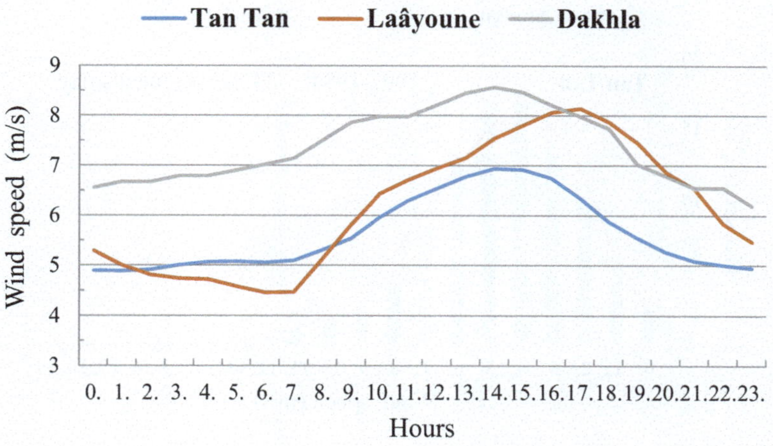

Fig. 13.12 Diurnal variations of wind speed [2, 7, 14, 17]

13.4.2.3 Frequency Distributions

The annual frequencies for class 0 m/s corresponding to $V = 0$ m/s (no wind) are small for the three considered sites. They are about 2% for Laâyoune and Dakhla and almost double for Tan-Tan. These figures need to be known, when it is a question, for example, of storage sizing for some wind power applications (Fig. 13.13).

We notice also that the classes that correspond to wind speeds between 3 and 5 m/s (medium wind) are more important for Tan-Tan compared to those of Laâyoune and Dakhla. Their frequencies are higher than 10. On the other hand, their classes corresponding to high values of wind speed (strong wind) are not empty compared to those of Tan-Tan. This shows that Laâyoune and Dakhla are windier; the latter two sites are windier compared to Tan-Tan (Fig. 13.11).

13.4.2.4 Weibull Hybrid Distributions

Figure 13.13 represents the observed distributions and those estimated using the Weibull Hybrid function on an annual scale. We remark that the observed frequencies are well modeled by Weibull Hybrid. For Laâyoune and Dakhla, the observed frequencies and the estimated ones are almost the same compared to those corresponding to Tan-Tan.

13.4.3 Annual Averages of Wind Potential Available in the South of Morocco

The annual averages of wind potential are calculated for the three considered sites with $\rho = 1.18$ kg m^{-3}. Dakhla has the highest annual average of wind potential. It is

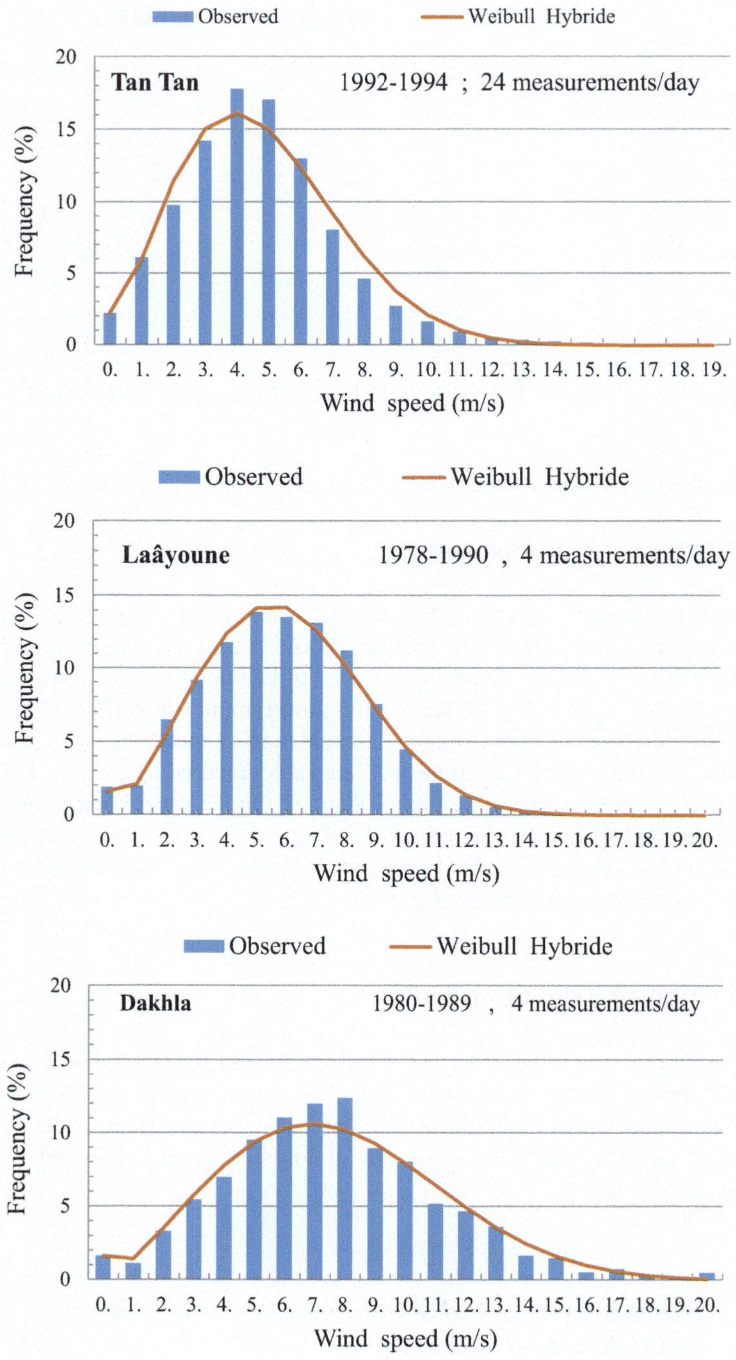

Fig. 13.13 Annual distributions of wind speed [2, 7, 14, 17]

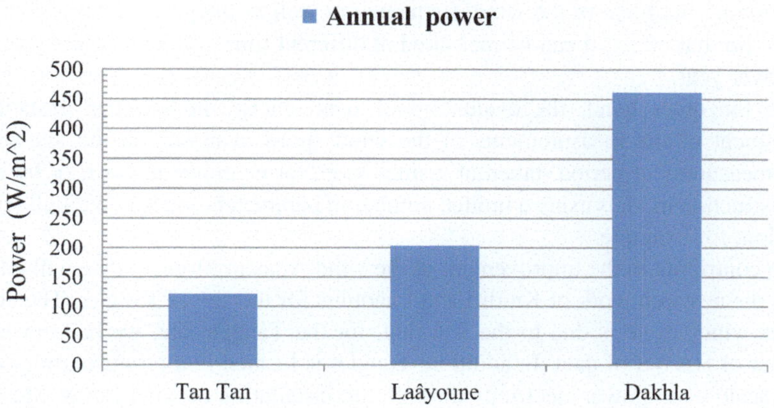

Fig. 13.14 Annual averages of wind potential [2, 7, 14, 17]

in order of 462 W/m^2 almost double that of Laâyoune and four times that of Tan-Tan (Table 13.12, Fig. 13.14).

13.4.4 Conclusion

The statistical characteristics of hourly averages of wind speed for the three considered windiest sites and the most promising ones for wind farms installation are studied on the basis of long term measurements especially for Laâyoune (13 years) and Dakhla (10 years). For the three sites, the long-term annual average of wind speed is greater than 5 m/s, a value from which the wind farms produce electricity competitive with that generated by thermal or nuclear power plants.

Overall, for the three sites, the wind is strong in the afternoon, especially in summer, and lower at night. This is due to the influence of the breeze sea/land. The study of the frequency distribution of wind speed shows that the observed frequencies are well modeled by Weibull Hybrid function. The southern provinces, such as Dakhla site, have a very significant wind potential. At a height of ten meters, wind power can reach 462 W/m^2. So it is a very promising area for large-scale electricity generation.

13.5 Influence of the Period of Measurements on Wind Potential Assessment for a Given Site

13.5.1 Introduction

In order to evaluate and exploit wind energy, statistical and dynamic analysis of wind, related to the climatic data of the concerned region, consists essentially of determining the force of the wind, frequency as well as the periods during which it blows. So that effect, it can be measured at different times of the day, the year, and year over year.

On the other hand, the evaluation of wind energy necessitates a statistical assessment of the measurements of the wind speed available on the site over a long measurement period, taken at a small step, for example an hour, or by using reconstruction models using a limited number of parameters, such as Weibull Hybrid function, for example.

To contribute to the improvement of the wind Atlas in Morocco, we will take up again the research work of Knidiri and Laâouina for the site of Tangier. The choice of this windy zone is due to the fact that, for the Tangier site, we have available 12 year of HAWS of data. In addition, Tangier is located in a promising region for large scale wind power electricity through the installation of wind farms. Measurements are taken by DMN at 10 m height.

Measuring wind speed every hour needs the use of measuring instruments (anemometer, acquisition chain, etc.). They operate all the time in addition to their permanent control and maintenance, in order to minimize errors due to breakdowns, as well as their calibration in order to increase the reliability of the measurements. This requires an investment and a skilled staff working continuously, which is not always easy to achieve. To facilitate taking measurements in an isolated site in order to evaluate its wind potential, it is more practical and economic when we optimized taking measurement by selecting only one to four measurements daily over a short period so as to evaluate the wind potential with a good accuracy for a given site.

13.5.2 Statistical Characteristic and Resource of Wind Potential in Tangier

The choice of this windy zone is due to the fact that, for the Tangier site, we have available 12 year of HAWS of data. In addition, Tangier is located in a promising region for large scale wind power electricity through the installation of wind farms.

The use of monthly and annual averages as well as the maximum wind speed is often satisfied. This presentation provides interesting information to wind energy users on wind speed fluctuations: the annual average wind speed gives an order of magnitude of the wind importance at the site under consideration, the average and maximum monthly speeds inform on seasonal and interannual variations. They are also served to estimate usable annual and monthly wind power. The maxima permit to know the critical speed of the wind turbine and then to predict the resistance of its rotor.

During the considered period (1978–1988), Table 13.13 shows July is the most windy month with a monthly average of 6.5 m/s and June is the least windy month, $V = 5.5$ m/s. The maximum monthly average reached during July 1986 is of the order of 10.5 m/s, but the minimum value of the order of 2.41 m/s is recorded for December 1978. The windiest year is 1986, while the least windy year is 1978. The maximum speed that can be reached is of the order of 30 m/s [2].

Table 13.13 Monthly, annually, and standard deviation of wind speed, *V* (m/s) (24 measurements/24 h, 1978–1989) for Tangier [2]

Ann Mont	78	79	80	81	82	83	84	85	86	87	88	89	Aver	σ
J	5.02	4.05	3.95	6.07	4.48	6.69	5.07	6.29	6.78	7.73	4.44	5.97	5.54	1.16
F	3.67	5.57	3.62	5.70	7.10	6.07	5.70	7.27	8.07	5.84	6.11	7.77	6.04	1.35
M	4.05	4.48	5.09	4.49	6.48	6.68	6.93	7.04	5.72	5.80	4.28	5.51	5.55	1.03
A	4.50	4.43	6.00	5.21	5.95	6.81	5.70	6.50	7.15	7.45	4.77	5.95	5.87	0.96
M	4.42	6.16	5.01	5.11	6.34	4.78	6.14	5.67	7.72	5.81	5.22	6.07	5.70	0.85
J	2.97	5.92	4.44	6.28	5.56	4.95	6.98	4.23	6.33	7.47	5.68	5.54	5.53	1.19
Jt	4.27	6.39	5.55	8.74	4.60	3.38	7.57	6.08	10.5	7.24	7.50	6.30	6.5	1.9
A	4.59	5.31	5.36	5.85	6.87	5.18	7.78	6.15	5.71	5.20	5.38	4.50	5.66	0.89
S	5.68	5.89	7.24	5.48	6.61	6.56	4.83	6.51	7.28	6.32	10.2	4.37	6.41	1.41
0	4.99	5.44	6.65	5.21	5.91	6.18	8.91	8.37	5.35	4.44	5.81	7.02	6.19	1.2
N	5.21	6.05	5.56	8.46	6.14	5.24	6.96	6.44	6.23	6.44	5.88	6.24	6.24	0.8
D	2.41	5.48	6.13	6.88	5.36	6.91	6.12	7.02	5.15	5.17	5.76	5.44	5.65	1.18
Aver	4.31	5.43	5.38	6.12	5.95	5.79	6.56	6.46	6.82	6.24	5.92	5.89	5.90	0.6
σ	0.90	0.71	1.01	1.26	0.79	1.04	1.14	0.95	1.39	1.01	1.52	0.91	0.3	

13.5.3 Influence of the Frequency of Measurements and Number of Years on Wind Speed

The study of the influence of the frequency of measurements and the sample size (length of the series of measurements) is a good test of representativeness of this sample from the standpoint of climatological norms.

13.5.3.1 Frequency of Measurements

Measuring wind speed every hour needs the use of measuring instruments (anemometer, integrator or acquisition chain, etc.). They operate all the time in addition to their permanent control and maintenance, in order to minimize errors due to breakdowns, as well as their calibration in order to increase the reliability of the measurements. This requires an investment and a skilled staff working continuously, which is not always easy to achieve. To facilitate taking measurements in an isolated site in order to evaluate its wind potential, it is practical and economic to optimize the taking of measurement by selecting only one to four measurements daily over a short period so as to evaluate the wind potential with a good accuracy.

Figure 13.15 shows that there is a more pronounced approach between the curves corresponding to 4 and 24 measurements per day. So it is possible to limit to 4 measurements per day (0, 6, 12, and 18 h) to estimate the monthly and annual averages with a good precision for Tangier. For realizing their book on wind potential in Morocco, Knidiri and Laâouina used, also, only four measurements per day [7].

13.5.3.2 Number of Years

The visual examination of Figs. 13.16 and 13.17 shows that as the number of years of data increases, the curves tend to approximate to each other, those corresponding to 9 and 10 years being virtually coincided. We conclude therefore that we can limit to 9 years of measurements (1978–1986) with four measurements per day to perform a statistical analysis HAWS for Tangier to assess its wind potential.

13.5.4 Validation of the Results Obtained

To confirm that only four measurements per day at 0, 6, 12 and 18 h are required over a period of at least 9 years, and in order to characterize the wind potential for Tangier and to reduce the enormous wind speed data, the results of the statistical assessment of wind speed data are presented in different forms by comparing the

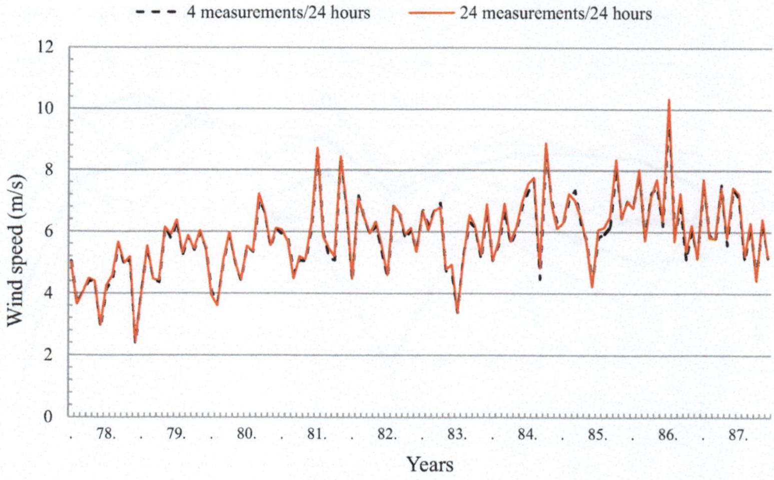

Fig. 13.15 Interannual variations of wind speed for Tangier, V (m/s) [2]

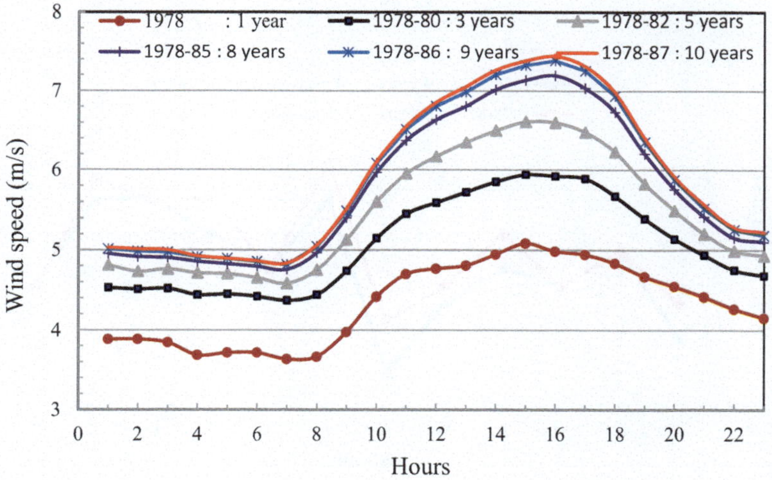

Fig. 13.16 Influence of number of years on daily variation of wind speed [2]

original series (10 years, 24 measurements/day) and 9 years of measurements, using only four measurements per day.

13.5.4.1 Seasonal Variations of Wind Speed

The monthly variation of wind speed for 9 years series with four measurements per day is comparable to that of the original series (1978–1987) (Fig. 13.18). This confirms the possibility for Tangier site to use only four measurements per

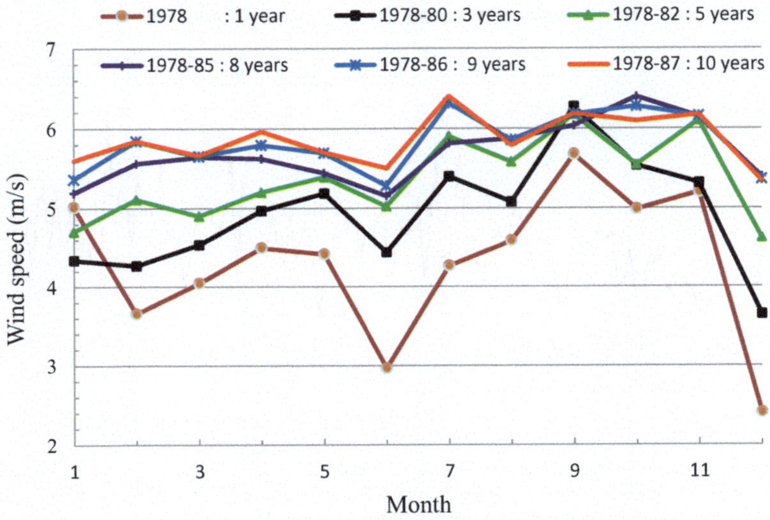

Fig. 13.17 Influence of number of years on monthly averages of wind speed [2]

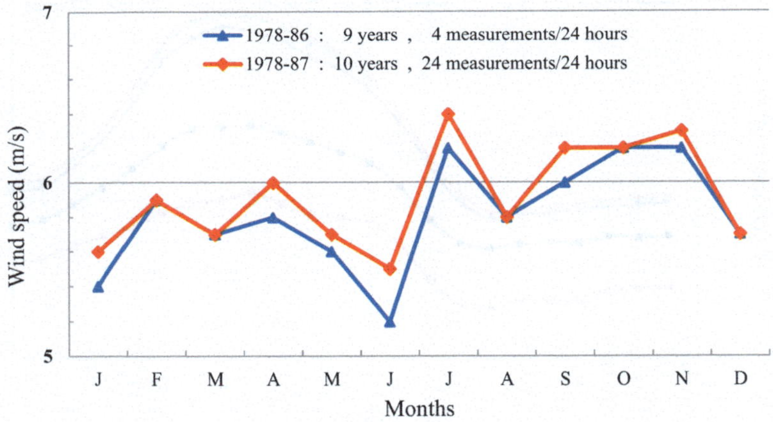

Fig. 13.18 Monthly variations of wind speed [2]

day over the period 1978–1986 (9 years) to estimate the monthly averages of wind speed.

13.5.4.2 Modeling of Wind Speed Frequencies by Weibull Hybrid

Figure 13.19 shows that Weibull hybrid distributions for the two considered series 9 years with 4 measurements/day and 10 years with 24 measurements/day coincide and adapt well to the observed data. As for estimating monthly averages of wind

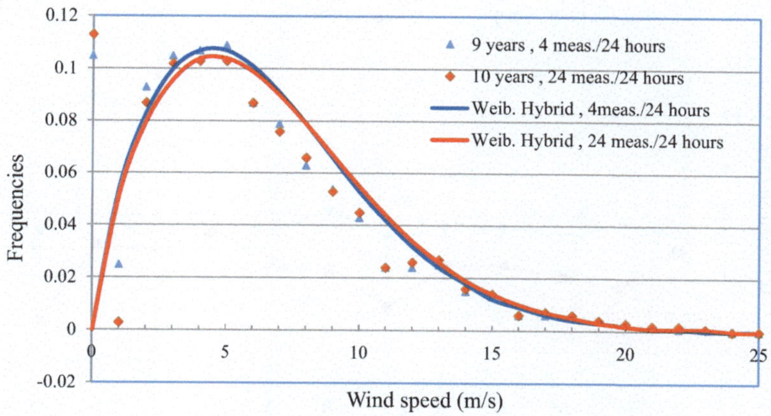

Fig. 13.19 Yearly frequencies of wind speed [2]

speed, only four measurements per day over a 9 year period can be used to calculate *K* and *C* (Fig. 13.19).

13.5.5 *Influence of the Choice of the Most and the Least Windy Year of the Considered Series on the Estimation of Wind Potential*

13.5.5.1 Seasonal Variations of Wind Speed

In general, wind speed varies with the seasons. This seasonal variation is usually represented by a series of average velocities (Fig. 13.20). The monthly averages for the considered period (12 years, 24 measurements/day) are close to the long-term average (5.90 m/s). On the other hand, the seasonal fluctuations corresponding to the windiest year and the least windy are more important. For the first, they vary between 5 and 10 m/s and generally, are above the annual average, while for the other it is the opposite and vary between 2 and 5 m/s.

13.5.5.2 Wind Speed Histograms

The histogram (Fig. 13.21) represents a number of hours for which the wind speed is equal to 0, 1, 2 m/s, etc. (on the abscissa). It informs us about the distribution of hours in different classes. For the windy year, however, there is a maximum of 1565 h per year for class 0 m/s corresponding to $V = 0$ m/s (no wind), or about 18%, a figure that needs to be known, for example when it is about sizing wind energy storage. For the less windy year, we notice that classes 3 and 4 (low wind) are higher

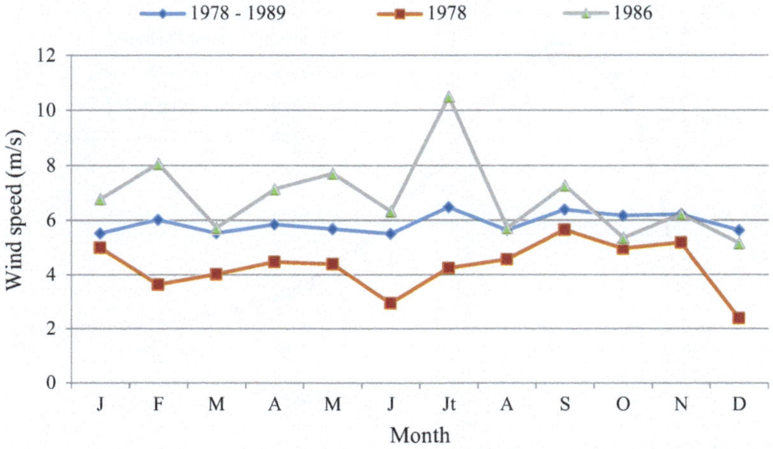

Fig. 13.20 Seasonal variations of wind speed

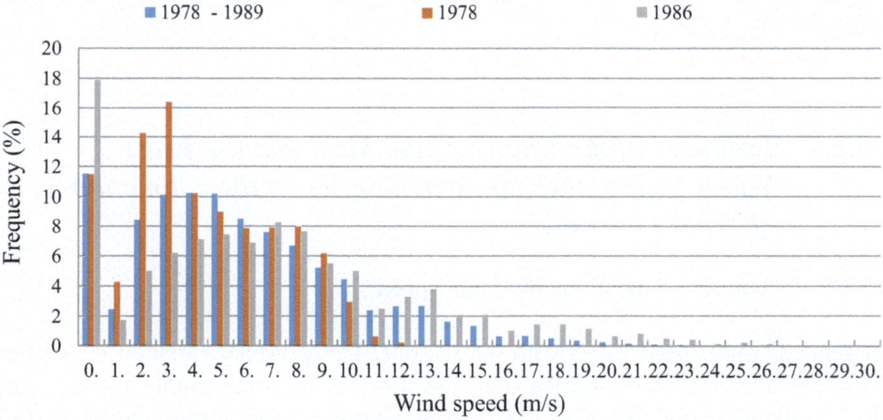

Fig. 13.21 Wind speed histograms

corresponding 1253 and 1441 h per year, respectively, that is to say the frequencies 14% and 16% on average. The classes with high values are not empty, for the original series and the windy year. They correspond to the strong wind that can reach 30 m/s for Tangier.

13.5.5.3 Modeling of Wind Speed Frequencies by Weibull Hybrid

Parameters *K* and *C* define the shape and the scale of Weibull distribution. A high value of *K* implies a narrow distribution, with wind speed concentrated around a value, while a low *K* value leads to widely dispersed wind speed. The scaling factor,

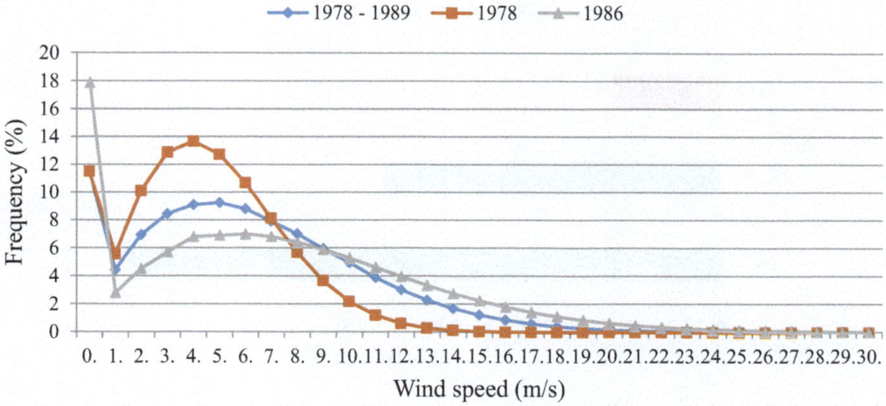

Fig. 13.22 Modeling of wind speed frequencies by Weibull hybrid

C (m/s), adjusts the curve to the frequency histogram. Generally, its value is high for windy sites.

Figure 13.21 shows the annual adjustment of the observed distributions for the three datasets under consideration using Weibull hybrid distribution. For the most windy year, Weibull hybrid distribution is more flattened and shifted towards high wind speed values, the scale factor ($C = 9.35$ m/s) is high compared to that of the least windy year ($C = 7.53$ m/s), but the form factors are comparable (Fig. 13.22).

13.5.5.4 Estimation of Wind Potential

Figure 13.23 shows that the long-term wind potential (366 W/m^2) for the period considered (12 years, 24 measurements/day) is three times that corresponding to the least windy year, but is only half that of the most windy year, which confirm the results obtained previously, that is to say the importance of taking a long series of measurements to estimate the wind potential. By limiting to 1 year, the overestimation and underestimation of the wind potential can reach 75% (more windy year) and 68% (less windy one), respectively, for Tangiers, which is enormous and hence the importance of the period, the step, and the reliability of measurements to study the characteristics of wind speed and consequently to assess the wind potential for a given site.

13.5.6 Conclusion

The series of wind speed data used for the establishment of the Moroccan Wind Atlas, published by Knidiri and Laâouina, are insufficient despite the interest of the subject being treated. The map of the average wind speed of the winds in Morocco is

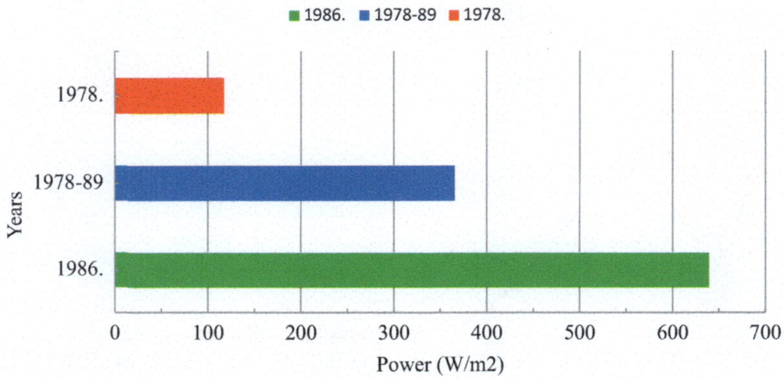

Fig. 13.23 Annual power (W/m^2) (24 measurements/day)

interesting, but does not take into account corrections of the local environment (topography, obstacles, etc.) as described, for example, in the ""European Wind Atlas".

The results of the statistical analysis of the data obtained for Tangier show that the minimum period of data that should be considered is 9 years with 4 measurements per day (6, 9, 12, and 18 h) in order to study adequately the statistical characteristics of the wind speed in Tangier, and consequently evaluate its wind potential. The influence of the choice of the most and the least windy year of the considered series on the estimation of wind potential confirm those obtained previously, that is to say the importance of taking a long series of measurements to estimate the long-term wind potential for Tangier. By limiting itself to 1 year, the overestimation and underestimation of wind potential can reach 75% for a windy year (1986) and 68% for a less windy year (1978), which is enormous. From this, one may conclude the importance of the period, the step and the reliability of the measurements to study the statistical characteristics of wind speed, and consequently to assess the wind potential for a given site.

13.6 Wind Energy Progress in Morocco

13.6.1 Introduction

Morocco imports more than 90% of its fossil fuels. This energy dependence has a negative impact on state budget in terms of foreign currencies. The electricity consumption increases annually by 6%. In 2016, the contribution of wind electricity to net national electricity consumption reached 9%.

Fortunately, Morocco has a strategic location in the North West of Africa and is only 14 km from Europe, across the Strait of Gibraltar. In addition, it is characterized by a varied hilly relief (the Atlas and Rif Mountains, the Sahara, etc.) and the length of its coasts (the Atlantic Ocean and the Mediterranean Sea) constitute 3500 km of coastline. This allows Morocco to have a significant potential of wind resources that can be exploited for the production of large electricity power connection to the national grid. There are several pathways for the generation of electricity power from renewable energy sources, among the most developed techniques are wind farms. So it has good solar and wind potential.

The launch of wind projects in Morocco allows local resources to contribute, in addition to its energy independence, to the economic and social sustainable development efforts. Thus, the high energy demand, the need to create jobs and reduce oil imports, and the environmental protection of the country support the installation of wind farms. For these reasons, in 2009, a large-scale solar energy use strategy was launched by the creation of the Moroccan Agency for Solar Energy (MASEN) to develop renewable energy projects. This policy permits Morocco to promote regional integration between sub-Saharan African and European countries and to strengthen the links with the main players in this sector on a worldwide basis. Thanks to its model of sustainable development, Morocco will be able to identify the latest innovations, energy efficiency, and clean technologies. Morocco has undertaken an oath to ensure its energy independence and become a leader in renewable energy in Africa. Thus, plans for development of renewable energy have been launched. The Moroccan project of solar energy, controlled by MASEN and its wind counterpart, led by the National Electricity Office (ONE), comes within this framework. The aim is to bring to 42% the share of electricity produced from renewable energy by 2020, 14%, respectively, from hydraulic, solar, and wind power each.

In this perspective, a full renewable energy option will be fully discussed based on what is feasible and appropriate for Morocco. The new strategy adopted is based on increasing the contribution of renewable energies to the national electric power installed at 52% by 2030 and on the development of a clean energy economy. This will enable Morocco to reduce its greenhouse gas (GHG) emissions and to fight against climate change (CC) and thus contribute to the preservation of the environment. The twenty first century will be a century of less and less water, sustainable development and environmental emigration. Morocco is known for its know-how of dam construction and its renewable energy models. The sharing of these skills with African countries will contribute to its development and consequently to the reduction of emigration to Europe.

13.6.2 Background of Wind Energy

Up until 1986, the only application of wind energy in Morocco has been to pump water. 96% of bulk generation of electricity is done by means of a number of imported oil-fired power stations, the rest is generated by hydro power stations. In 1982, the CDER installed a 1 kW wind turbine in Essaouira region, and in 1986 10 kW wind turbine in Eljadida region. Later in 1989 two 10 kW turbine were

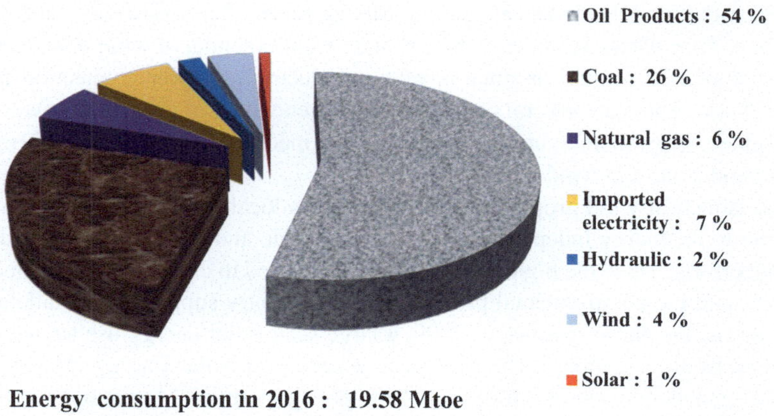

Fig. 13.24 Distribution of national net energy consumption in 2016 [17]

installed in Oujda region. This makes a total of four wind turbines in the country at that time. The first wind farm (50 MW) installed in 2000 by ONE. It produces approximately 225 GWh per year at a competitive cost. Financing and operation is provided by a private company [15].

13.6.3 Energy Sector Characteristics

13.6.3.1 Energy Consumption

The Moroccan energy balance remains characterized by the predominance of petroleum products which represents, in 2016, 54% of the national net energy consumption, evaluated at 19.58 M ton. Coal ranks second with 26%, the most polluting source of energy (Fig. 13.24).

13.6.3.2 Electricity

Until 1996, of thermal and hydraulic origin, almost all electricity production is provided by the ONE [16, 18]. The gradual liberalization of the electricity sector began in 1997, either by deprivation of refining companies or by concessional electricity production. The electrical power installed by the ONE and the autoproducers is 10,938 MW in 2018 (Table 13.14, Fig. 13.25). Figure 13.25 shows that the hydraulic power has not changed much since 2005. Thermal and wind power have increased significantly since 2008, especially for thermal power.

Figure 13.26 shows the evolution of the net electrical energy demand has increased significantly, from 19.52 TWh in 2005 to 35.87 TWh in 2016, an increase of 84%. The contribution of hydroelectricity depends on rainfall. For the same

Table 13.14 Comparison of the installed electric power in Morocco in 2018 to that of 2017 [17–23]

	Installed electrical power (MW)		
	2018	2017	Var: 18/17 (%)
Hydraulic	***1770***	***1770***	***0***
Thermal	**7237**	**5851**	**23.7**
Steam coal	4281	2895	47.9
Fuel vapor	600	600	0
Gas turbine	1230	1230	0
Diesel	263.7	263.7	0
Combined cycle	834	834	0
Gas oil	28.3	28.3	0
Wind	***1220***	***1018.4***	***19.8***
Wind ONEE	204.5	204.5	0
Wind private	37.3	37.3	0
Wind IPP	351.7	351.7	0
Wind law 13-09	626.5	424.9	47.4
Solar	***710.8***	***180.8***	***293.1***
Solar ONEE	20.8	20.8	0
Solar MASAN	690	160	331.3
Total	***10,937.8***	***8820.2***	***24***

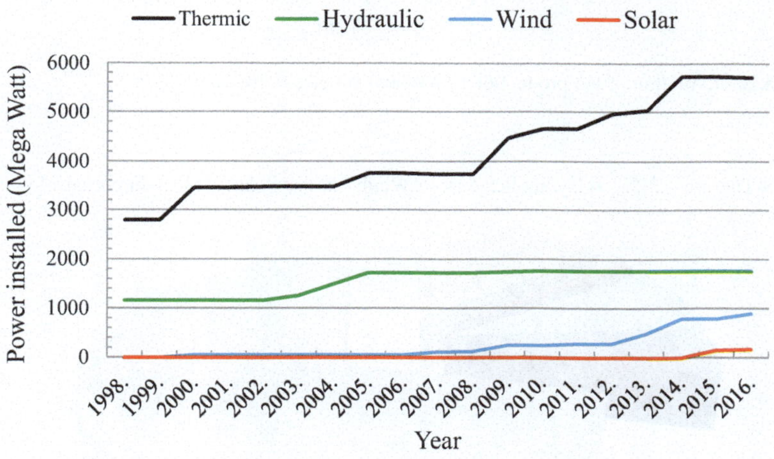

Fig. 13.25 Evolution of the national electric power [17]

installed capacity, 1800 MW, hydroelectric power generation was 3631 TWh in 2010 and it dropped to 1816 TWh in 2012 [17, 24, 25]. The contribution of wind energy to electricity consumption became significant since 2012 (Fig. 13.26). On the other hand, the first kWh of electricity from solar energy is injected into the national electricity grid only in 2015 (Figs. 13.26 and 13.27). Since 1999, the consumption of primary energy and electricity has increased annually on average by 4.5 and 6% respectively (Fig. 13.28).

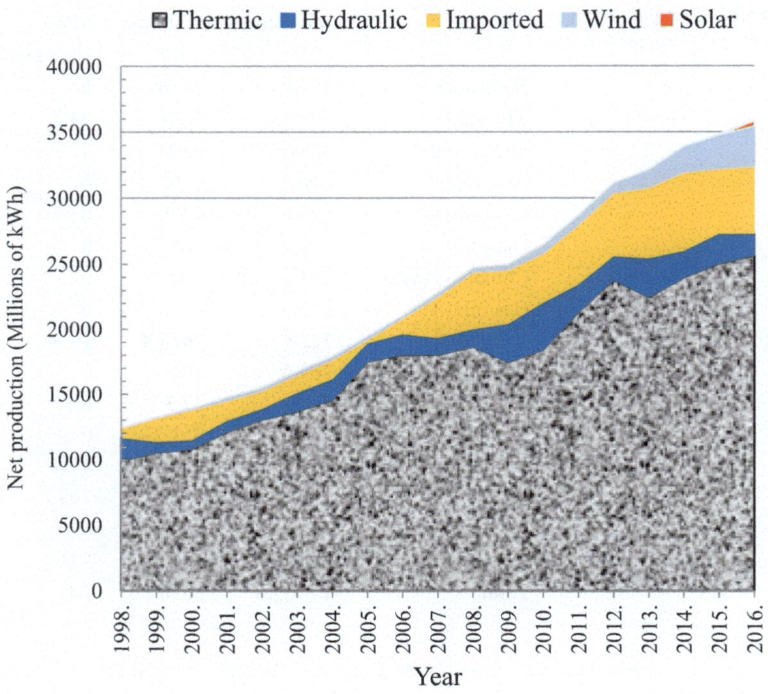

Fig. 13.26 Evolution of the production of national net electricity called [17]

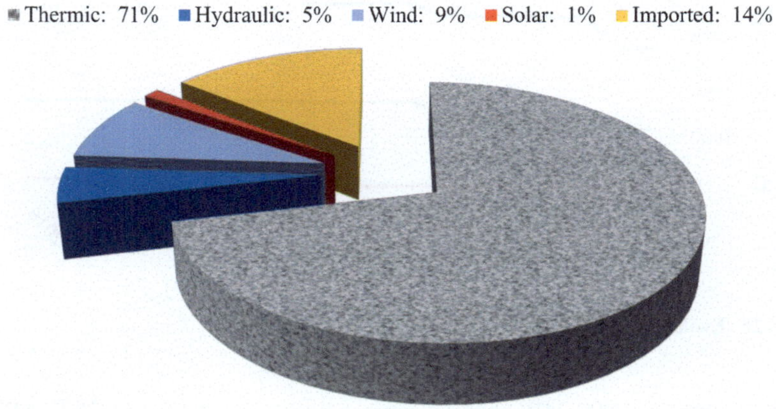

Fig. 13.27 Distribution of national net electricity consumption in 2016 [17]

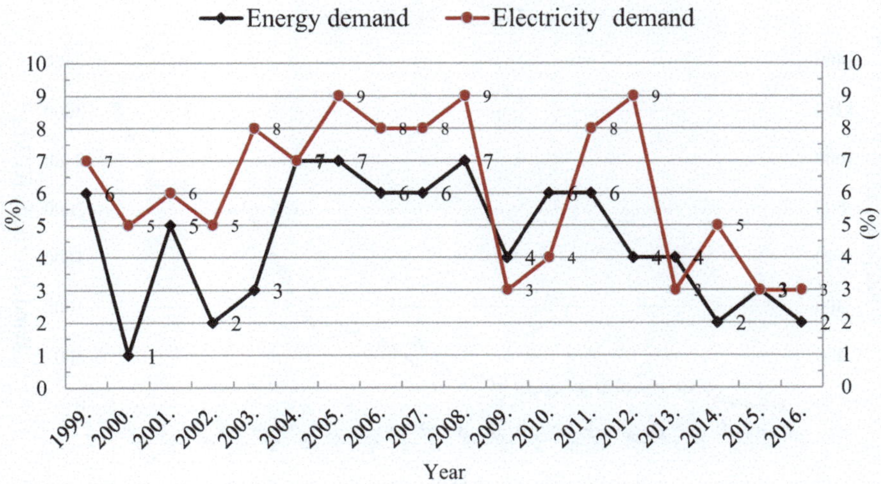

Fig. 13.28 Evolution of electricity demand growth [17]

13.6.4 Wind Energy

The objective of the solar plan is to increase Morocco's energy independence and to make alternative and renewable energies the cornerstone of the Moroccan national energy policy. Morocco made the choice of an energy transition towards sustainable, by integrating renewable energy sources into its energy mix.

In 2015, MASEN became the Moroccan Agency for Sustainable Energy. It will be tasked to do the overall strategic planning of renewable energy projects. It will take charge of all projects for electricity generation from renewable sources, including feasibility studies, planning, financing, development, and construction as well as operation and maintenance. Among the concerns and objectives of MASEN, we quote also the following [18]:

- The implementation and the evaluation of the national energy strategy, in particular that related to renewable energies,
- The cost-effectiveness appropriation for energy storage technologies,
- The effects of the reduction of storage costs on the expectations of electricity demand and the supply strategy of the national market.

In 2018, wind power installed in Morocco reached 1220 MW, or 19.8% of its national power (Table 13.14, Fig. 13.29). For wind energy, in Africa, Morocco ranks second after South Africa (2.2947 GW) [26]. However, the largest wind farm is Tarfaya in southern Morocco (300.1 MW), putting into service in 2014. It is the largest in Africa, outclassing that of "Gulf of EL-Zay" (200 MW) in Egypt. The most recent wind farm, putting into service in 2018, is "Khalladi" (Tangier), with a capacity of 120 MW (Table 13.15) [26].

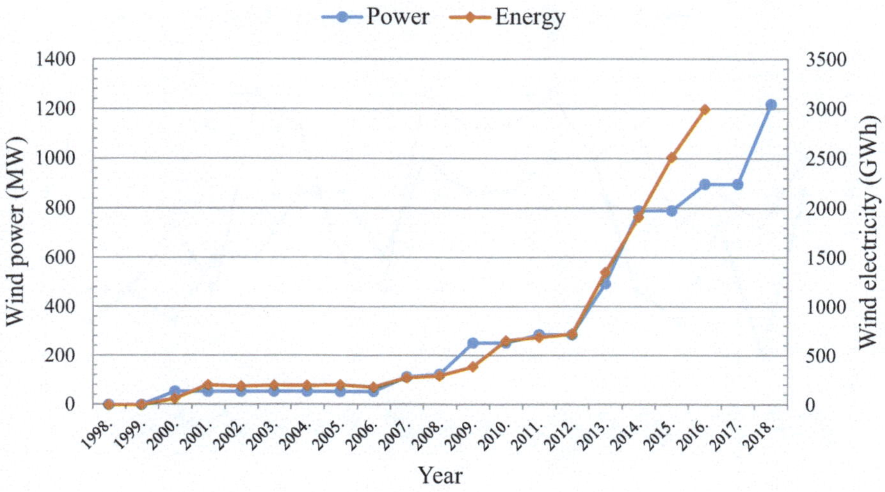

Fig. 13.29 Evolution of the wind power and energy [17–23]

Under the stated objective by Morocco to reach 42% of its electric power installed from renewable resources (mainly hydro, solar, and wind) by 2020. Table 13.16 shows four other wind farms in progress with a total capacity of 830 MW. Three of them will be installed in south of Morocco (Fig. 13.30). The concerned sites were selected in different windy regions across Morocco, in order to provide better production and stability of the electricity grid. In addition, other projects are being considered for electricity production or for supplying sea water desalination [18–23].

Major cities in south of Morocco such as Tarfaya, Tan-Tan, and Laâyoune are connected to the national grid. On the other hand, Dakhla city is electrified by a diesel power plant (37.5 MW). To cope with the significant increase in electricity consumption for Dakhla region, a new diesel generator of 16.5 MW capacity is planned to be installed by ONE [27–30].

13.6.5 Wind Energy Perspectives

Increasing the renewable energy share, especially wind power in the Moroccan electricity production will contribute to the reduction of the share of coal and petroleum products imported, the two most polluting energy products. The realization of wind farms will also participate, in the economic and social development of Morocco through the involvement of Moroccan firms (feasibility studies, electrical works, civil works, etc.), industrial integration (parts manufacturing of wind turbines, etc.), creation of indirect jobs during the construction and direct jobs during the operational phase. Furthermore, the wind power generation will contribute to the Moroccan energy independence and reduce the purchase of fossil fuels, and

Table 13.15 Wind farms installed [17–23]

Sites	Number of turbines and manufacturer	Capacity (MW)	Commissioning	Annual production (GWh/an)	Operator/ owner
Koudia Al Baida (Tétouan)	84 Vestas	50.4	2000	226	Detroit Energy Company
Koudia Al Baida (Tétouan)	7 Enercon	3.5	2001	12	National Office of Electricity, ONE
Tétouan	12	10.2	2005		Lafarge Ciments
	5	10	2008	38	
	6	12	2009		
Amogdoul (Essaouira)	71	60.35	2007	210	ONE
Tanger	165	140.25	2009 (107.1 MW)	526.5	ONE & Gamesa Ealica
Laâyoune	6 Gamesa	5.1	2011	16	Ciments du Maroc
Akhfenir (Tan-Tan)	61 Ecotecnia	101.87	2013		Nareva Holding
Foum ElOued (Laâyoune)	22 Siemens	50.6	2013		Nareva Holding
Haouma (Tétouan)	22 Siemens	50.6	2013		Nareva Holding
Tarfaya	131	300.1	2014 (50 MW)	1084	Tarfaya Energy Company (TAREC)
Khalladi (Tanger)	40 Vestas	120	2017		ACWA Power
Total		***914.97***			

Table 13.16 Wind farms under construction [18–23]

Sites	Capacity (MW)	Wind speed (m/s) at 10 m	Annual production (GWh/an)	Operator/owner
Taza	150	**		ONE
Midelt	180	**		Nareva/Enel Green Power Company
Akhfenir (Tan-Tan)	100	5.12		Nareva
Tiskrad (Laâyoune)	300	8.45	1000	Nareva
Boujdour	100	8.40	385	Nareva
Total	***830***			

** In addition, other projects are being considered for electricity production or for supplying sea water desalination

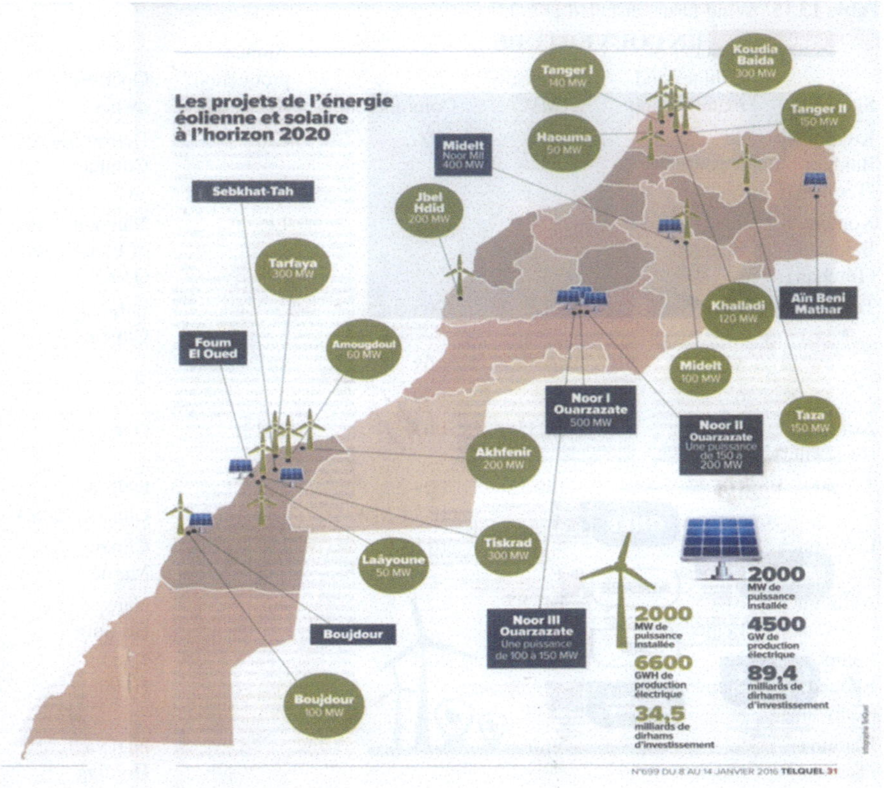

Fig. 13.30 Projects of solar power plants and wind farms in 2030 [18]

consequently to the currency economies. In addition, the wind power consumption contributes to the environmental protection by reducing CO_2 emissions, greenhouse gases which Limits climate change such as floods and droughts.

The Dakhla region, that has not been connected to the national grid until now is supplied with electric power generated by a power Diesel plant. So wind power can contribute to the electrification of this region by coupling it the Diesel plant.

Compared to solar energy, the electricity from wind and hydro power are twice less expensive. Installing more wind farms, the objective set by the Moroccan national energy strategy, namely, the wind power share will be established at 14% of the national electric power capacity (14,580 MW), is likely to be realized in 2020.

On the top of that, in 2018, Blockchain Saluma firm and DMG Blockchain Solutions have signed a new partnership to support Saluma's large scale wind farm in Morocco. The new agreement will facilitate Saluma's ambitious project to develop 900 MW wind farm in Dakhla, in Morocco's southern provinces [31].

13.6.6 Morocco: Wind Energy, An Exemplary Model in Africa

Since the beginning of the twenty-first century, the contribution of renewable energies to the national electricity consumption has become significant and in constant growth, reflected by the completion in 2018 of the installation of the wind farm "Jbel Khaladi" in Tangier (120 MW). The share of solar and wind power will be 40% in 2030 compared to 18% in 2018 and only 2% in 2009 [17]. Morocco managed to break through the 1000 MW in wind power installed in 2017, overtaking Egypt and approaching the level of South Africa. The acceleration of the deployment of the national strategy in this sector will consolidate its leading position on African scale.

Morocco, with this energy policy, opens up new horizons for the future of clean energy development in Africa. Its strategy includes, in addition to the development of solar, thermal, photovoltaic, and wind energy, the introduction of a series of legislative, regulatory, and institutional reforms, while improving tax benefits to encourage investments in this sector. Moreover, starting from 2015, for a rapid and successful energy transition, it will have increased the share of renewable energies to 52% of its installed electricity capacity by the year 2030.

It also considers itself to be one of the most active African countries for investing in this sector. This makes it one of the pioneering African countries, and it is the first in Africa in the solar energy sector. Morocco has made renewable energies a real lever for south–south cooperation and a vector of development for sub-Saharan African countries, having proven renewable potential. It also plays a leading role in African development through training and experience provided for a range of African countries.

Moroccan renewable energy projects are arousing an increasing interest from international and local investors, allowing Morocco to attract more investments in this sector, such as the installation, in Morocco, of an industrial unit of the group German Siemens, with an investment of nearly 100 million Euros for the manufacture of wind turbine blades.

13.6.7 Morocco, Driving Force of South–South Cooperation in Africa

13.6.7.1 New Foreign Policy in Africa

Regaining its membership in the African Union (AU) in 2017, Morocco set up a new foreign policy in Africa by mobilizing internal cooperation between African countries rather than remain at the mercy of foreign powers. It based on two priorities: sustainable development and economic integration. Moreover, Morocco is an active member of the African Climate Committee. It contributes to the African Monetary Fund for the Congo Basin, with different agendas [26]:

1. Supporting the African countries, in order to face the negative impacts of climate change on the African agricultural sector by taking advantage of Moroccan experience in this sector,
2. Activating the African Monetary Fund for the Congo Basin, in accordance with the recommendations adopted at the African Summit held on the sidelines of the conference of the of the United Nations (COP22) on climate change in Marrakech, in 2016, as well as the two regional commissions, namely, Sahel Committee and Island States Committee.

13.6.7.2 Transfer of Its Know-How to African Countries

Morocco's national renewable energy model is consolidating its leadership position at African level. It has made this sector a real lever for South–South cooperation in addition to its know-how and recognized expertise in the field of assistance services. Here are examples of cooperation carried out by Morocco with African countries [26]:

- Improving the energy services provided to African populations by facilitating access to training for technicians and strengthening their capacities,
- Management and supervision of energy production, transmission and distribution infrastructure projects,
- Completion of specific training courses for African national electricity companies.

In addition, cooperation in the water sector through the construction of dams, there is cooperation in the electrical energy sector (organization of training, exchange of experiences, technical assistance by the study feasibility of rural electrification projects, incentives for private investors to carry out electricity generation projects using renewable energies) [26].

13.6.7.3 Electricity

The European electricity grid is the largest interconnected network in the world in terms of power transmitted. Since 1997, the Moroccan power grid is the only African network connected to the Europe one. Morocco's planning is to be a regional center for the export of competitive clean electricity to its neighbors, Spain and Algeria (Fig. 13.31). In addition to the project of electrical connection with Mauritania through the cities of Laâyoune and Dakhla, in 2018, Morocco and Portugal have agreed on the construction of a submarine electrical cable connecting the two countries of 1000 MW capacity. Thus, technico-economic feasibility studies are undertaken in the field. In 2019, Spain and Morocco decided to increase the capacity exchanged between them.

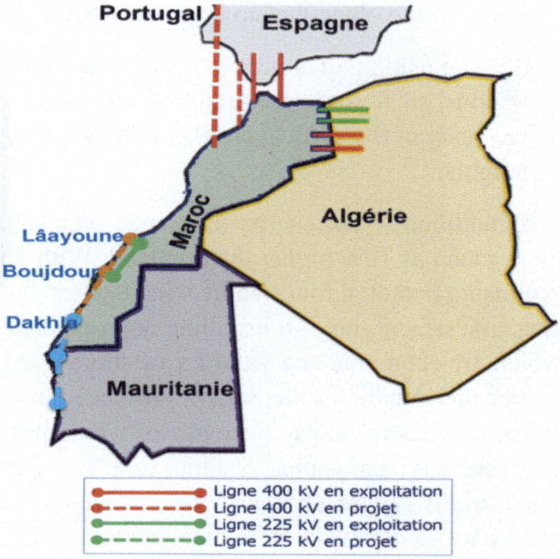

Fig. 13.31 Electrical interconnection [32]

13.7 Conclusion

The data from seventeen stations operated by the National Meteorology Department permitted CDER to assess the wind potential in Morocco. The windiest regions of Morocco are the extreme north, the extreme south, and to a lesser extent, a narrow strip of the central coast of the country. Some mountainous areas could also be windy.

The use of wind machines will be economically competitive with other energy sources only for geographical areas where the wind is sufficiently strong and regular. In general, wind turbines are considered economical for average wind speeds greater than 5 m/s. This is the case for several regions across Morocco.

It should be noted that there are undoubtedly other places, where the topography allows a better exposure, or even an accentuation of the winds, where the winds are stronger than indicated by the results of the seventeen meteorological stations. Airports, where most of these stations are located, are generally not located in the windiest area. Therefore, the conclusions presented in the document, published by CDER on the wind potential in Morocco, are incomplete. Further work is needed to improve the "wind speed database" presented in this document. This is what CDER had done in 1991.

CDER started a program combining a further study of the existing data, and a selective and specific project of installation of new anemometers, to improve the assessment of the wind potential in Morocco. According to the wind measurements carried out by the CDER in the North of Morocco between 1993 and 1994, as part of a techno-economic feasibility study for the realization of wind farm projects, the most favorable site was that of Koudia Al Baida (Province of Tetouan). The average and maximum speeds recorded at the height of 9 m are respectively 10.94 and 36.5 m/s.

Some of the objectives to be carried out include the following:

- the acquisition of additional and more precise data of wind speed, in windy regions, such as the mountainous regions,
- the study of the vertical profile of the wind, by installing anemometers at different heights,

In addition, Wind energy over open oceans has wide potential. Wind speeds can be as much as 70% higher than on land. With 3500 km of the coast, Morocco has a promising potential for offshore wind energy. It needs to be harnessed by conducting sites assessment through installing and operating meteorological towers and buoys, which are economic and valid for offshore wind farms installation.

For the interior of the Sahara provinces in the south of Morocco, the statistical characteristics of hourly averages of wind speed for the three considered windiest sites, namely, Laâyoune, Dakhla, and Tan-Tan, the long-term annual average of wind speed is greater than 5 m/s, a value from which the wind farms produce electricity competitive with that generated by thermal or nuclear power plants. Overall, for the three sites, the wind is strong in the afternoon, especially in summer, and lower at night. The study of the frequency distribution of wind speed shows that the observed frequencies are well modeled by Weibull Hybrid function. The southern provinces, such as Dakhla site, have a very significant wind potential. At a height of ten meters, wind power can reach 462 W/m^2. So it is a very promising area for large-scale electricity generation.

The series of wind speed data used for the establishment of the Moroccan Wind Atlas, published by Knidiri and Laâouina, are insufficient despite the interest of the subject being treated. The map of the average wind speed of the winds in Morocco is interesting, but does not take into account corrections of the local environment (topography, obstacles, etc.) as described, for example, in the "European Wind Atlas".

The results of the statistical analysis of the data obtained for Tangier show that the minimum period of data that should be considered is 9 years with 4 measurements per day in order to study adequately the statistical characteristics of the wind speed in Tangier, and consequently evaluate its wind potential. The influence of the choice of the most and the least windy year of the considered series on the estimation of wind potential confirm those obtained previously, that is to say the importance of taking a long series of measurements to estimate the long-term wind potential for Tangier. By limiting itself to 1 year, the overestimation and underestimation of wind potential can reach 75% for a windy year (1986) and 68% for a less windy year (1978), which is enormous. From which one may conclude the importance of the period, the step and the reliability of the measurements to study the statistical characteristics of wind speed, and consequently to assess the wind potential for a given site.

In 2018, the wind power installed in Morocco has reached 1220 MW, and over 60% is located in the south. In addition, the largest wind farm in Morocco is located in the south. Its capacity is 301 MW and generates 1.084 GWh/year in average, the equivalent of the electricity consumption of Marrakech city. Five other wind farms totalizing 530 MW are in progress in the south of Morocco. Also contributing to

Morocco's energy independence, the electrical energy obtained from the use of renewable energy, particularly wind energy is a clean energy which contributes to the environmental protection from the emission of greenhouse gases.

On the other hand, wind energy contribution in national net electricity consumption increased from 1.1% in 2003 to 9% in 2016. Though, wind energy varies on the wind force, generally, the electrical energy produced increases with the installed capacity. This shows that wind energy is one of the most promising renewable energy sources for Morocco. Its contribution to the energy independence of Morocco depends on the installed power capacity. This is possible as Morocco has good wind potential and land for wind farms installation. On the other hand, during the last decade, the weight and the price of wind turbines have been significantly reduced. Thiers ratings and efficiency have improved significantly.

Morocco has good wind potential. On top of building more than 11 wind farms, it developed a great expertise in the field of studies and construction of wind farms, recognized in Africa. Morocco's reintegration into African Unity provides African countries with knowledge, skills and experience in the use of solar and wind energy. It offers endogenous development and strong economic integration in Africa comparable to that of the European Union.

Morocco became one of the pioneering African countries investing in renewable energies. Concerning wind energy, the largest wind farm in Africa is the Tarfaya one, installed in 2004. Morocco ranks second in Africa, after South Africa (2295 MW). Wind energy generated 2858 GWh in 2018.

Morocco's Renewable Energy model consolidated its leadership position at African level and opened up new horizons for the future of clean energy development in Africa. In 2017, Moroccan adopted a new foreign policy in Africa, based on two priorities: Sustainable development and Economic integration. Since 1997, the Moroccan power grid is the only African network connected to the Europe one. Morocco's planning is to be a regional center for the export of competitive clean electricity to its neighbors, Spain and Algeria.

References

1. J. Manwell, J. McGown, A.L. Rogers, *Wind Energy Explained: Theory, Design and Application* (Wiley, Hoboken, 2010)
2. H. Nfaoui, Caractéristiques du gisement éolien marocain et optimisation d'un système aérogénérateur/groupe électrogène pour l'électrification des villages isolés. Thèse de Doctorat d'Etat, Faculté des Sciences de Rabat, 2004
3. E.H. Lysen, *Introduction to Wind Energy* (CWD, Amersfoort, 1983)
4. G.L. Johnson, *Wind Energy Systems* (Prentice Hall, Englewood Cliffs, 1985)
5. P.D. Dunn, *Renewable Energies: Sources, Conversion and Application* (Peter Peregrinus, London, 1986)
6. Z. Bahraoui, F. Bahraoui, M.A. Bahraoui, Modeling wind energy using Copula. Open Access Library J. **5**, e4984 (2018)
7. F. Knidiri, A. Laâouina, *L'énergie éolienne au Maro* (CDER, Marrakech, 1986)
8. C.G. Justus, *Winds and Wind System Performance* (Franklin Institute Press, Philadelphia, 1978)

9. A. El-Mallah, A.M. Soltan, A nomogram for estimating capacity factors of wind turbines using site and machine characteristics. Solar Wind Technol. **6**, 633–635 (1989)
10. N.L. Jonson, S. Kotz, *Continuous Univariate Distributions*, vol 2 (Houghton, Mifflin, 1970)
11. A.S.K. Darwish, A. Sayigh, Wind energy potential in Irak. Solar Wind Technol. **5**, 215–222 (1988)
12. J. Buret-Bahraoui, M.N. Bargach, M.L. Benkadour, Le gisement solaire marocain. Editions SMER (1983)
13. J. Buret-Bahraoui, Caractéristiques du gisement solaire marocain, Etude spécifique du site de Rabat. Thèse de Doctorat d'Etat, Faculté des Sciences de Rabat, 1986
14. M. Enzili, A. Nayssa, F. Affani, *Le gisement éolien marocain* (le Centre Développement des Energies Renouvelables, Marrakech, 1995)
15. I. Troen, E. Lundtang, *European Wind Atlas* (Riso National Laboratory, Roskilde, 1988)
16. H. Nfaoui, J. Buret, A. Sayigh, Wind characteristics and wind energy potential in Morocco. Solar Energy **63**(1), 51–60 (1995)
17. Annuaires statistiques du Maroc: 1988–2017. Direction de la statistique.
18. www.masen.gov.ma
19. www.mem.gov.ma
20. www.one.gov.ma
21. www.aderee.ma
22. Morocco, Energy policies. Beyond IEA countries (2014), https://www.iea.org/publications/freepublications/publication/Morocco2014.pdf
23. M. Enzili, Renewable energy, energy efficiency in Morocco, situation and prospects. Cairo, Egypt (2008), www.needs-project.org/
24. https://www.theguardian.com/environment/2018/apr/11/day-zero-water-crises-spain-morocco-india-and-irak-at-risk-as-dams-shrink.
25. https://docplayer.fr/51193702-La-politique-et-les-bienfaits-des-barrages-au-maroc-presentee-par-mr-khalid-el-ghomari-directeur-des-amenagements-hydrauliques.html
26. H. Nfaoui, A. Sayigh, New horizons for renewable energies in Morocco and Africa, in *Proceedings of World Renewable Energy Congress-18*, London, UK, 2018
27. H. Nfaoui, E. Essiarab, A.A.M. Sayigh, Wind energy and its economy in south of Morocco, in *Proceedings of the 8th WREC*, Denver, USA, 2004
28. H. Nfaoui, A.A.M. Sayigh, Renewable energy, large project in Morocco 500MW–solar and wind potential in Ouarzazate Region, Morocco, in *Proceedings of WREF-12*, Denver, CO, USA, 2012
29. H. Nfaoui, A.A.M. Sayigh, Contribution of renewable energy to energy independence in Morocco: wind energy case, in *Proceedings of World Renewable Energy Congress-14*, London, UK, 2014
30. H. Nfaoui, A.A.M. Sayigh, Evaluation of renewable energy in the south of Morocco and perspectives, in *Proceedings of Med Green Forum*, Florence, Italy (2015)
31. A. Eljechtimi, Blockchain firm Soluna to build 900MW wind farm in Morocco: CEO (2018), https://www.reuters.com/article/us-cryptocurrency-morocco-soluna-energy/blockchain-firm-soluna-to-build-900mw-wind-farm-in-morocco-ceo-idUSKBN1KU2BA
32. https://au.int/sites/default/files/documents/32251-doc-lome150317vfhaj_maroc.pdf

Chapter 14
Wind Energy Program in Republic of Korea: Present and Future

Soogab Lee

14.1 Introduction

The importance of renewable energy is fully expressed by world energy conditions. The energy consumption is increasing steadily, but the sources are decreasing gradually in the world. According to the BP Statistical Review of World Energy 2018, the years remaining for fossil energy source (petroleum, coal, and natural gas) mining are 50.2, 134, and 52.6, respectively [1]. The energy consumption has increased annually with average of 2.3% by 2017. As shown below, Fig. 14.1 indicates that the world oil production and consumption are increasing every year. Furthermore, the international energy market is unstable, which is demonstrated by the increment of oil consumption and price. In Fig. 14.2, the oil price is increasing consistently since January, 2016. In a view of environment, the global warming can be main issue for using renewable energy. For three decades, the average temperature of earth has increased by 0.51 °C. Depending on the increment of temperature, the arctic sea ice is decreasing every year. According to NSIDC (National Snow & Ice Data Center), the arctic sea ice extent from 1979 to 2012 has a decline of 10.2% per decade.

The alternative energy should be needed for those issues above, so the South Korea has developed a renewable energy with various methods and policy. The representative of the renewable energy which prevents those issues is wind energy. In this chapter, the present and future of wind energy technology are handled.

S. Lee (✉)
Department of Mechanical and Aerospace Engineering, College of Engineering, Seoul National University, Seoul, South Korea
e-mail: solee@snu.ac.kr

A. Sayigh, D. Milborrow (eds.), *The Age of Wind Energy*, Innovative Renewable Energy, https://doi.org/10.1007/978-3-030-26446-8_14

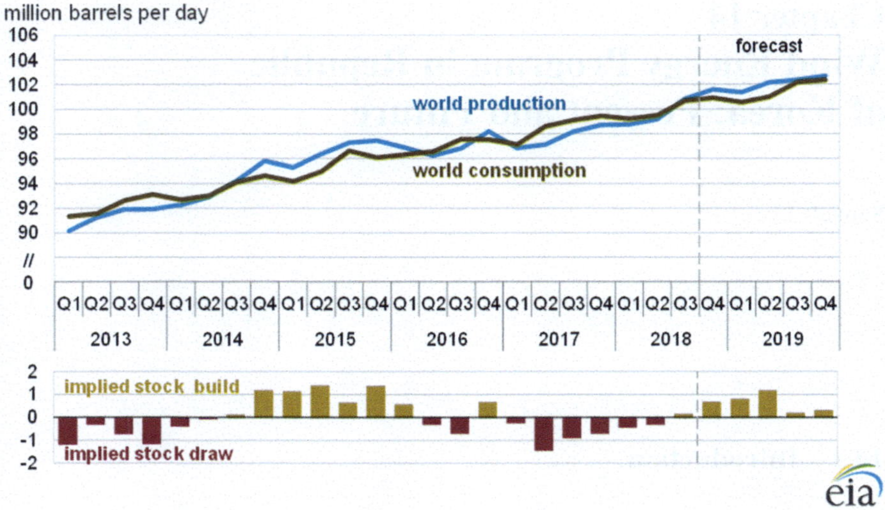

Fig. 14.1 World liquid fuels production and consumption [2] (Source: Short-Term Energy Outlook, November 2018)

Fig. 14.2 Oil price chart for last 3 years (source: EIA) [3]

14.2 The Present

14.2.1 The Total Energy Status in South Korea

Figure 14.3 shows the energy source consumption in South Korea. As mentioned above, the overseas dependence is 94.7%, and the Middle East dependence accounts

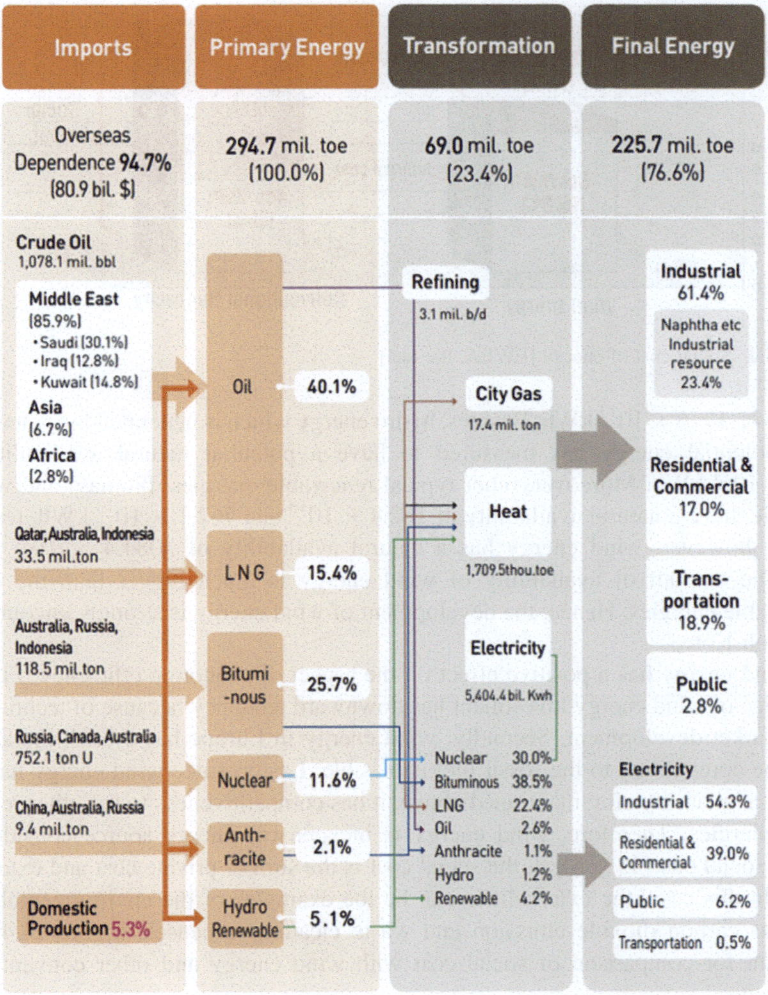

Fig. 14.3 The energy balance flow in South Korea [4]

for 85% in crude oil. The domestic production is only 5.3%. The primary energy source is mostly oil, LNG, bituminous, and nuclear. The renewable energy only accounts for 5.1%.

14.2.2 *Wind Energy*

As the importance of renewable energy has been known, a demand of wind energy also rises. Wind energy has a massive efficiency and availability among other renewable energies. For example, the amount of world energy consumption in

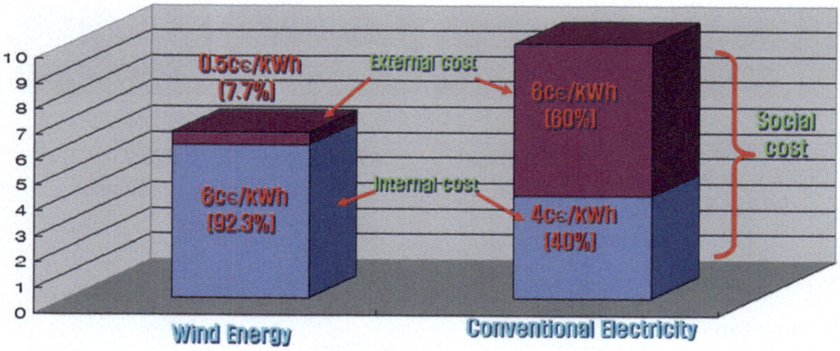

Fig. 14.4 Social cost of energy [EWEA, the fact]

2017 is 2.1776×10^{13} kWh. Besides, hydro energy which is generated by water flow and potential energy has measured to have a potential natural availability of 4.6×10^{13} kWh. Moreover, other typical renewable energies, biomass and wave-oceanic, have a natural availability of 152.4×10^{13} and 762.1×10^{13} kWh, respectively; however, wind energy has a natural availability of 3084.4×10^{13} kWh [5]. The amount of availability of wind energy is much bigger than the other renewable energies. Hence, the development of wind energy is strongly encouraged in South Korea.

Wind energy has a positive effect on the aspect of economic efficiency. Firstly, the price of wind energy investment has downward tendency because of technology and market development. Secondly, wind energy in Europe has a brisk market so that the competition to the fossil fuel is feasible. Lastly, since wind energy has the strong availability aforementioned above, it has competitiveness to the other renewable energies. Therefore, wind energy is inexpensive energy source considering entire social cost. In general, the social cost is the sum of private cost and externalities. In this case, the externalities can be the examples of the environmental cost such as carbon dioxide emission and waste cleaning expenses. Figure 14.4 is a diagram for comparison of social cost with wind energy and other conventional electricity.

14.2.2.1 Onshore/Offshore Wind Energy

In general, the wind turbine can be divided into two by location of installation: Onshore and Offshore wind turbine. The onshore wind turbine is typically installed on the land. Hence, the installation is not difficult, and the management can be simply worked. Shortly, the advantages and disadvantages of onshore wind turbine are listed below:

1. Advantages
   - Uncomplicated installation and management.
   - Facilitation of system connection.

2. Disadvantages
   - The noise which can affect to human.
   - Disturbing the landscape.
   - Significantly large lands occupation.

The offshore wind turbine is installed on the sea. There are two different methods to install: By mounting on land or floating substructure. The offshore wind turbine is located on the sea, so it has a great quality of wind. Furthermore, it can solve the problem of noise, landscape, and the need of large land. Therefore, the magnificent scale of wind farms can be modeled on the sea. However, it also has several problems as well. It is difficult to make a system connection, installation, management. Moreover, it takes more costs to install and development than the onshore wind turbine. In addition, the vibration of offshore wind turbine affects to the marine lives. The vibration occurs from the turbine, and it transfers along the tower. The vibration of tower radiates in the sea. Hence, it affects to the marine lives under the sea. Below lists show the summery of the advantages and disadvantages of offshore wind turbine.

1. Advantages
   - Good quality of wind which has more intense and velocity.
   - Fewer problems of noise and landscape.
   - Large wind farms availability.
2. Disadvantages
   - The increment of development and installation costs.
   - Difficult system connection.
   - Environmental issue which affects to marine lives.

There are more onshore wind turbines than the offshore wind turbines. The ratio of onshore and offshore wind turbine is now 9:1. In recent, however, the demand of offshore wind turbine is increasing for resolving the problems of the land limitations and environmental issues such as landscape and noise. According to the EWEA, the number of offshore wind turbine installations will increase continuously, and the onshore to offshore ratio will be changed to 6:4 in 2030 as shown in Fig. 14.5.

Since the onshore wind turbine has crucial environmental issues, the South Korea has shifted the policy for focusing on the offshore wind turbine. The object of this policy is mainly minimizing the environmental impacts, facilitating establishment of large-scale wind farm, and synergy effects with strongly related industries such as shipbuilding, plant, etc. A similar tendency of wind farm installation is shown in Korea as well. According to the KWEIA, Korea wind energy industry association, the plan for accumulative installation will be expected to the amount of 17.7 GW. Likewise, a large amount of the offshore wind turbine will be accounted for this wind energy electricity generation, which is about 13 GW or more. Figure 14.6 shows the development plan of wind farms distribution in South Korea. In the regional distribution, Seonam and Jeonnam which are located on the left side of the sea

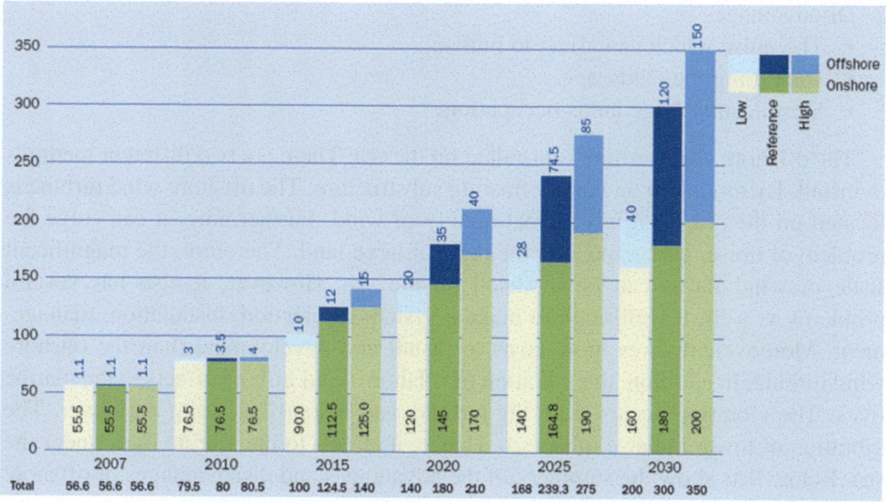

Fig. 14.5 The expectation of onshore/offshore installation [EWEA]

marking on red color, show the biggest capacity in South Korea. The planned wind power capacities are 665.6 and 8056 MW for onshore and offshore, respectively.

14.2.2.2 Noise Reduction Technology in Korea

Aforementioned above, the noise problem is one of the major environmental issues in wind energy program. The research of airfoil noise with numerical method has been conducted by some of universities in Korea, but they are not validated with experiments. However, some researches for investigation of noise source by using microphone array has been conducted by several universities and Korea Aerospace Research Institute. Moreover, the Seoul National University in Korea has developed "low noise airfoil and serrated blade for broadband noise reduction technology development" to reduce the noise from wind turbine. The broadband noise accounts for most of the aerodynamic noise which comes out from wind turbine blade. It is a procedure for design a new shape of serrated blade and low noise airfoil and rotor for maximize noise reduction. The validation has conducted by a wind tunnel test with a small size model.

The research has been conducted with the procedure as follow. First of all, the numerical method for design of low-noise airfoil has conducted. In detail, the research on the broadband noise of airfoil has studied. Afterward, the design for optimized shape of low-noise airfoil has accompanied. Secondly, the wind tunnel test for design of low-noise airfoil shape has examined. The test is based on the comparisons of base model and low-noise airfoil for aerodynamic performance and noise. In third, another wind tunnel test has conducted for the serrated trailing edge applying on the two dimensional airfoil. Subsequently, the numerical methods for

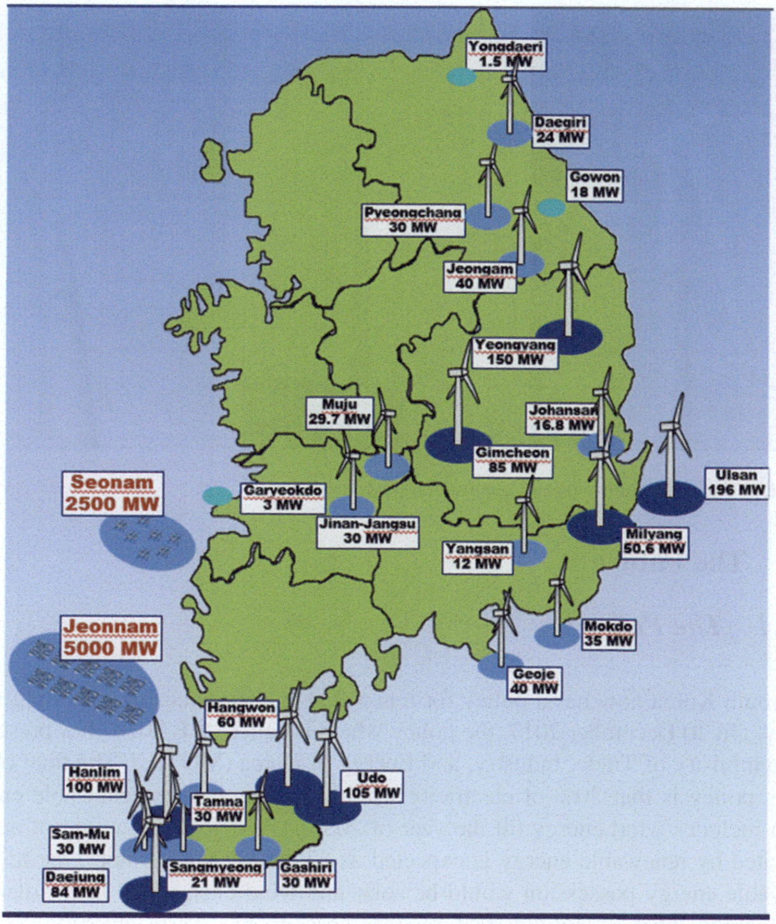

Fig. 14.6 Wind farm development plan in South Korea (2014-) [SNU]

low-noise rotor platform shape design has conducted. All the wind tunnel tests and numerical methods are for the aerodynamic performance and noise reduction. Finally, the low-noise serrated blade has been designed and manufactured for small size wind turbine. On-the-spot inspection for investigation of low-noise serrated blade performance and noise has conducted and analyzed. Figure 14.7 shows three pictures below show the three different serration installation on the blade of small sized wind turbine.

After the research, the Seoul National University expected that the low-noise blade can be successfully developed with the research procedure above. Furthermore, the technology which has been used in this research can be applied to all rotors including helicopter, propeller, and turbo machine.

Fig. 14.7 Three different types of serration on blades [SNU]

14.3 The Future

14.3.1 The Policy

The South Korea now has a policy for renewable energy extension including wind energy. On 20 December 2017, the policy which is called "RE 3020" has presented by the ministry of Trade, Industry, and Energy in Korea (MOTIE). The final object of this policy is that 20% of electricity would be generated by renewable energy which includes wind energy till the year of 2030. The objective amount of energy generated by renewable energy is expected as 63.8 GW. Furthermore, the 85% of renewable energy possession would be solar and wind energy [6]. Both solar and wind energy have a common feature that these do not waste anything. It only generates the electricity, unlike the other renewable energy such as biomass. Therefore, the core of this policy is using a "clean energy supply". In 2016, the offshore wind farm construction-industry had reduced and delayed, so lots of companies had drawn off including Samsung and Hyundai; however, this policy, RE 3020, encouraged several companies to invest in the wind energy industry. Moreover, the employments for renewable energy can be generated by the policy which takes more renewable energy utilization [7].

14.3.2 The Projects

Since South Korea has a plan for development of wind energy, the huge scale of project is on the way. Here is an example of 2.5 GW offshore in the southwest sea which is called SEONAM by 2020. Table 14.1 below shows a description of the

Table 14.1 The description of 2.5 GW offshore project in South Korea

	First phase (demonstration)	Second phase (standardization)	Third phase (deployment)
Objective	• Test beds • Core tech. development	• Track records • Project feasibility	• Commercial operation. • Large-scale wind farm
Targets	100 MW	400 MW	2000 MW
Periods	2011–2015 (4 years)	2016–2017 (2 years)	2018–2020 (3 years)
Participants	KEPCO/Doosan/Hyundai	KEPCO	Free competition
Budgets	$350M	$1400M	$7125M

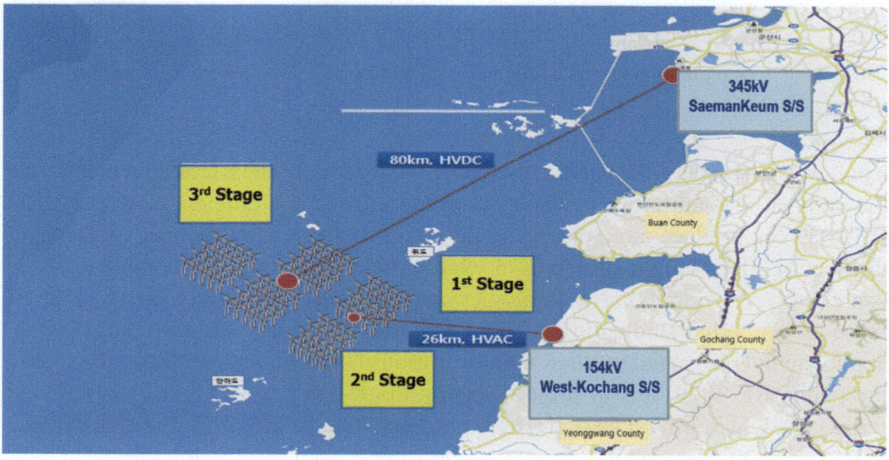

Fig. 14.8 The location of 2.5 GW offshore wind farms installation

offshore wind farm development project. The target energy generation by 2020 is 2.5 GW, and the total budget would be $8875M. The location of installation is described in Fig. 14.8 below.

In addition, five research and development projects have been selected for more than 100 MW offshore wind energy foundation recently. These consortiums consist of the areas Jeonbuk, Jeonnam, Gyeongbuk, Gyeongnam, and Ulsan. Table 14.2 below shows the detailed information of the projects [8].

14.4 Conclusion

The renewable energy development is essential for financial and environmental issues. Especially, wind energy can solve these problems, although the development and technology is not fully enough comparing to Europe. The South Korea now has

Table 14.2 Five projects of 100 MW or more offshore wind farm development

Subjective organization	Jeonbuk Technopark	Jeonnam Technopark	Gyeongbuk Technopark	Gyeongnam Technopark	Ulsan Technopark
Participants	Jeonbuk, Gunsan, GNU, Jeonbuk research center, western generation, Doosan	Green energy research center, Hyundai engineering, KWEIA, KHNP	Gyeongbuk, Yeongduk, KHNP, KEPCO, KIER	Gyeongnam research center, Doosan, Unison, IAE	Ulsan, Dongseo, KRS, SNU, KMOU, Ulsan University, Changwon University
Objects (offshore)	110 MW	220 MW	100 MW	100 MW	200 MW
Plans	1 GW extension	1 GW extension	Completion	400–500 MW extension	1 GW extension
Budgets (KRW)	4.6 billion	4.2 billion	4 billion	3.1 billion	4 billion

a policy for the renewable energy, including wind energy. The RE 3020 will increase the number of wind turbines which can produce clean and efficient electricity. The RE 3020 can produce more electricity by installing more wind turbines offshore till the year of 2030, and the problems of wind turbines including onshore and offshore can be solved by applying technologies, such as noise reduction. The research of noise reduction has studied in the Seoul National University, and the results can be applied to not only wind turbines but also various fields which use blades. To satisfy the policy RE 3020, lots of projects are in progress for installation of huge wind farms. The wind farms at least 100 MW are planned to be installed offshore, so those wind farms are expected to generate magnificent energy for South Korea.

References

1. https://www.bp.com/content/dam/bp/en/corporate/pdf/energy-economics/statistical-review/bp-stats-review-2018-full-report.pdf
2. https://www.eia.gov/outlooks/steo/
3. https://www.txfnews.com/News/Article/6360/Understanding-oil-prices-A-tale-of-two-charts
4. http://www.keei.re.kr/keei/download/FFS2017.pdf
5. Agency/Eurosolar, WIP: power for the world—a common concept (energy consumption stats)
6. https://www.eiic.or.kr/info/sub3_9_1
7. http://interestingenergyfacts.blogspot.com/2010/10/wind-power-jobs-facts.html
8. Wind Power Journal 2018 summer vol. 32 by Ministry of Trade, Industry & Energy (MOTIE) of the Republic of Korea

Chapter 15
Energy Storage on a Power System

Donald Swift-Hook

15.1 Pumped Water Storage

Large-scale storage of electrical energy on power systems is nearly as old as power systems themselves. The first public electricity system was built to light the streets of Godalming in Surrey, England in 1881. It was run-of-the-river hydroelectric but the next one, which was opened by Thomas Edison in January1882, was coal-fired and driven by steam. It was at Holborn Viaduct in London and, later that year, Edison opened a similar one at Pearl Street in Manhattan. Within another 5 years, there were 121 Edison power stations in the United States alone delivering electricity to customers—Edison famously favoured direct current—but there were many hydroelectric schemes around the world as well.

There is less demand for electricity in the middle of the night and on average it costs less to generate then because plant with the lowest running costs can be used: water, for instance, instead of coal. When more plant is needed, to meet demand during the day, it will be more expensive to run because the cheapest plant will have been used first.

Many of the early hydroelectric schemes used to run out of water by the end of the day and by the 1890s it was convenient to pump water back up to reuse the dam that was otherwise running less than full. The electrical energy to pump the water up above the dam is then effectively stored until the water runs down again to regenerate electricity. If the difference in cost between the night-time pumping and the day-time generation is enough to pay for the cost of installing the water pumping equipment plus the loss of energy in the storage and regeneration round-trip, such a storage system can be economic. Because such pumped water systems made use of spare capacity that was left behind existing dams, they could be much cheaper than installing new generation, needing a new dam. The economics of any storage system relies on this arbitrage.

D. Swift-Hook (✉)
Kingston University, London, UK

WREN, London, UK
e-mail: donald@swift-hook.com

A. Sayigh, D. Milborrow (eds.), *The Age of Wind Energy*, Innovative Renewable Energy, https://doi.org/10.1007/978-3-030-26446-8_15

15.2 Solar Power Cannot Be Stored (Economically on a Power System)

An immediate consequence is that solar power cannot be stored economically on a well-run power system. A store needs to buy power at night when it is cheap and (by definition) there is no sunshine and therefore no solar power available at night.

This conclusion could be invalidated if ever there was so much solar power generation installed that daytime generating costs fell below might time generating costs but such a scenario is very far off.

It is sometimes claimed that, if a consumer installs batteries alongside the solar panels on the roof of their house, "demand and peak charges can be reduced if not eliminated, and independence from the utility, to some degree can be achieved". This is wholly misleading.

Independence from a utility can be achieved easily enough: it is a simple matter of disconnection. Just switch off! But few people, once they are used to modern living, are prepared to accept the unavailability and loss of a continuous power supply which that involves. No individual can possibly provide a power system for himself to match the redundancy and diversity which enable a large power system to provide high levels of reliability. And you cannot be *almost* disconnected. The overhead costs of providing a connection are considerable and must be paid, however long the connection is used for.

By the same token, peak demand and peak power costs are not reduced if they are covered privately; they are simply transferred, probably becoming part of the capital payment and installation cost. There are almost certainly increased pro rata in the process, because of the small scale involved. If not—and this is perhaps most likely—they will just be dropped, if the peak power available is simply reduced and the consumer has to go without more often, as will then be necessary. We can expect a less reliable system to cost less.

15.3 Solar Power with No Mains Connection

For many applications of solar power, a mains connection is simply not possible, for instance, mobile applications. Figure 15.1 shows Vanguard 1, one of the first satellites to fly—it was actually the fourth, in 1958. It was powered by solar panels, as have most satellites been since then.

For many earthbound static applications that are remote, a power supply, although not impossible, would be prohibitively expensive; for instance, a microwave repeater link at the top of a tall tower on a remote hill-top or an automatic buoy out at sea.

Other applications are not at all remote but the cost of laying a mains connection in an urban environment when only a few milliamps of current is required would be prohibitive; for instance, digging up the road or pavement in a High Street just to power a parking meter or the cost of cabling a flowerbed to power a few garden lights, see Fig. 15.2.

Fig. 15.1 Vanguard 1 satellite

Fig. 15.2 Solar powered garden lantern

Most of these applications require power at night as well as during the day, so storage is essential if there is no mains supply and, for such small systems, rechargeable batteries are economic.

15.4 Providing Spinning Reserve

Storage systems can, of course, bring other benefits apart from cheap power. If, for instance, they have a rapid response, they can be brought online rapidly to provide generation immediately while other plant is being started up (which can be a time-consuming process for coal, gas or oil combustion plant). It is quite normal to keep some plant permanently fired up but turning over as "spinning reserve", so that it is ready to be ramped up if and when needed.

This is wasteful and many power systems around the world have some large-scale pumped water storage to avoid this need for wasteful spinning reserve. In Europe by the year 2000, more than 5% of all electrical generation involved pumped water storage (see Fig 15.3).

The UK is not untypical. Table 15.1 lists the four pumped water storage stations currently operational. The largest by far is Dinorwig (see Fig. 15.7) in Snowdonia National Park.

The energy storage capacity of Dinorwig is around 11 GWh (so it can store the output of a 1 GW power station for 11 h, say). For quick start capacity to act as spinning reserve, the six 450 tonne generators can take up their full loads (around 300 MW each) from stand-still in 75 s but if they are already spinning in anticipation,

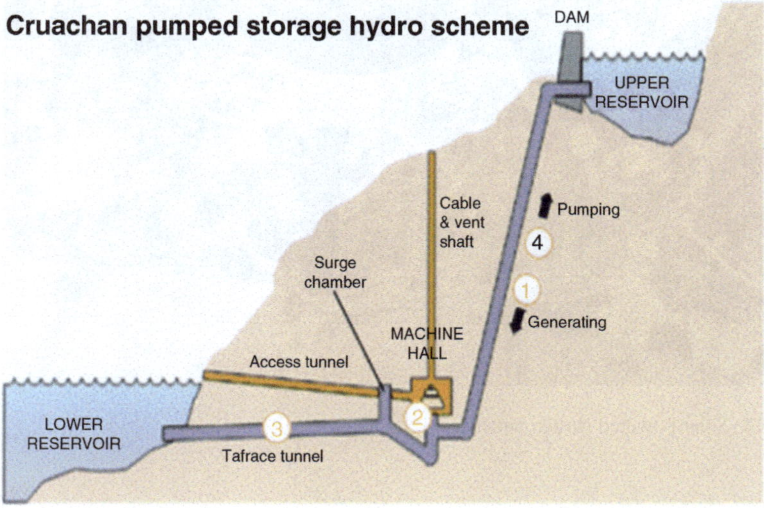

Fig. 15.3 Typical layout of a pumped water storage scheme

Table 15.1 Pumped water storage power stations in the UK

Power station	Location	Capacity (MW)	Date	Illustration
Blaenau Ffestiniog	Wales	360	1963	Fig. 15.4
Cruachan	Scotland	400	1965	Figs. 15.3 and 15.5
Foyers	Scotland	305	1969	Fig. 15.6
Dinorwig	Wales	1728	1984	Fig. 15.7

Fig. 15.4 Rheidol Ffestiniog (1963) 360 MW upper dam

Fig. 15.5 Cruachan (1965) 400 MW upper dam

Fig. 15.6 Falls of Foyers (1969) 305 MW

for example, of the end of a popular television programme or sporting event, they can go from 0 to 1800 MW in 16 s. This power station is built in tunnels and caverns deep inside the mountain Elidir Fawr (see Fig. 15.7c, d).

15.5 Other Types of Storage

Pumped water is the only system of bulk energy storage in practical use on power systems today.

In the last century (in 1978), a 219 MW demonstration project using pumped air instead of pumped water was put into operation at Huntorf in Germany. Thirteen years later a smaller 110 MW plant working on. Compressed air was built at McIntosh, Alabama. Neither of these schemes proved economic and they have not been replicated since then. Studies on compressed air systems do continue, including recent experimental units at Gaines in Texas in 2012 generating 2 MW and at Seabrook, New Hampshire in 2013 generating 1.65 MW, but so far nothing on a commercial scale has materialised.

Fig. 15.7 (**a**, **b**) Dinorwig (1984) 1728 MW. Marchlyn Mawr upper reservoir and dam and Llyn Peris lower reservoir. (**c**, **d**) Outside view of Elidir Fawr Mountain between the two reservoirs and a view of the generating plant inside the mountain

Ter-Gazarian [1] has described many other methods of storage based on many different physical principles including various types of thermal energy storage, flywheels, compressed air, various other electrochemical systems, capacitor banks and superconducting coils. Although his book is now into its Third Edition, few methods have been added to his original list, despite continuing research activities and development projects. It bears repeating, pumped water is the only viable method of bulk electrical energy storage on power systems that can be installed economically today.

15.6 Electrochemical Systems

15.6.1 Batteries

Most road vehicles carry a modest size lead-acid battery for powering lights and all the electrical systems when the engine is not running but, on the large scale, banks of lead-acid batteries have been used to power submarines silently for nearly 100 years. Although batteries can handle substantial power levels they cannot keep going long enough to store serious amounts of energy in power system terms—one power station's output for one night, say—because battery costs are far greater than for pumped water storage.

Today, huge efforts are going into the development of lithium-ion batteries for electronics applications such as mobile phones. Lithium-ion batteries have much higher energy density, much lower loss-of-charge and no memory effects so, although they cost more at present, they are making some inroads into vehicle markets, too, and prices are coming down quite rapidly.

Batteries are used on power systems for some power system control and balancing purposes. For instance, the U.K.'s National Grid plc recently tendered for Enhanced Frequency Response equipment and of the 64 sites that bid, 61 were for batteries, most of them lithium-ion. Successful companies included EDF Energy Renewables Ltd., Vattenfall AB, EON SE, Low Carbon, Element Power Limited, RES, and Belectric Solar Limited. This was the first time batteries have been introduced into the UK power system on such a scale but the amount of energy stored is not significant in bulk energy terms—these equipment will not be called upon to supply electricity for more than 15 min in the event of a fault. A dam for pumped water storage would not be economic on such a small scale.

15.6.2 Fuel Cells

One of the oldest electricity storage systems is the fuel cell which was first invented in 1838/1839. Many demonstration units have been built over the years, most famously by NASA for the Apollo space missions, see Fig. 15.8a, b and in the Space Shuttles, but as yet there are no widespread commercial applications on power systems. Nevertheless, the development of large fuel cells is often included in the research portfolio when the large-scale use of renewable energy is being considered. It will be shown that this connection between fuel cells—or indeed any other type of storage—and renewable energy is inappropriate.

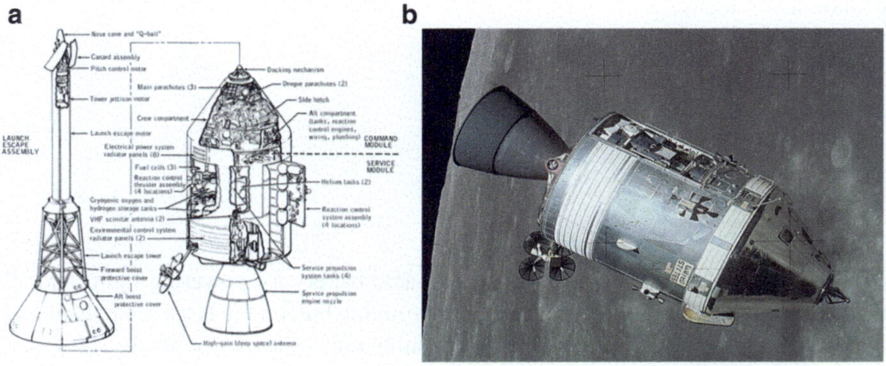

Fig. 15.8 (**a**, **b**) The Apollo mission relied upon fuel cells

15.6.3 Electric Cars

Electric vehicles provide storage if plugged into a power system but they cannot easily be recharged rapidly (i.e. much faster than they are discharged) without the provision of huge power levels. With normal amounts of power available, a vehicle has to be recharged all night. For many commercial applications, such as milk floats in urban areas, this is no problem but for more general use it may well raise difficulties.

15.7 Wind Is Not for Base Load or Peak Load

It is often suggested that storage would be a useful way of making intermittent wind into baseload power. A little thought will show that it is during periods of base load that most spare capacity is standing idle and is readily available for any generation that may be needed. The value of yet more capacity at such a time is clearly negligible, so the installation of storage or, indeed, any other type of power plant, cannot be justified economically at such a time.

The time when capacity would be valuable would be at times of peak demand and indeed power systems often pay a "capacity credit" for generating plant that can be available at that time. Swift Hook [2] showed analytically, back in 1987, that to first order (penetrations up to 30%) wind's firm power contribution is equal to its average power, which is also true for any other power source. Assertions to the contrary simply show lack of understanding about how power systems work; Millborrow [3] gives a fuller discussion. Swift Hook's analysis [2] assumed that all plant failures are statistically independent and independent of demand at all times. In fact, the capacity available only really matters during periods of peak demand. Technically, to be definite and to avoid statistical distortions due to exceptional, one-off occurrences, power system operators (in the UK, at least) look at a plant's availability during the "triad" [3] of separate half-hour periods of peak demand, which is when capacity is most needed.

Wind in the UK has about 30% availability year-round but much higher than that, 38% or more, during triad peaks, because the wind blows harder in the winter, when it is cold and demand is at a maximum. So wind has a capacity credit. It is not zero as many people wrongly assume. The public are unduly concerned that wind and solar plant are intermittent but they fail to realise that, in fact, all plant is always intermittent, including "base load" nuclear. All their concerns are already well taken care of by the spare capacity that is needed for all plant, not just for wind.

15.8 Intermittency

Solar power is intermittent but fairly predictably so, as night follows day and day follows night. Wind intermittency is far more irregular, see Fig. 15.9a.

The classic work of van der Hoven [4] on the wind frequency spectrum shows three peaks (see Fig. 15.9b), one due to local turbulence (with a period of 1–2 min), a small diurnal (day to night) peak and a broad synoptic peak with a period of 2–10 days when the wind typically blows for up to 10 days, as a weather pattern passes, and ceases for up to 10 days.

An energy store coping with 10 days of wind power must be up to 20 times larger and more costly than a typical night-to-day store, making effective storage for wind power prohibitively expensive.

Swift Hook [5] points out that storage and wind power are contractually incompatible. The wind only blows for a fraction of the time, typically around one third on average in the UK but even less in other countries. Any associated storage would therefore find wind taking three times as long and therefore three times the cost to store as nuclear or any other continuously available type of power would.

It has been pointed out that a store cannot afford to buy power round-the-clock but only when it is cheap—at night—and it can normally only take power for a single

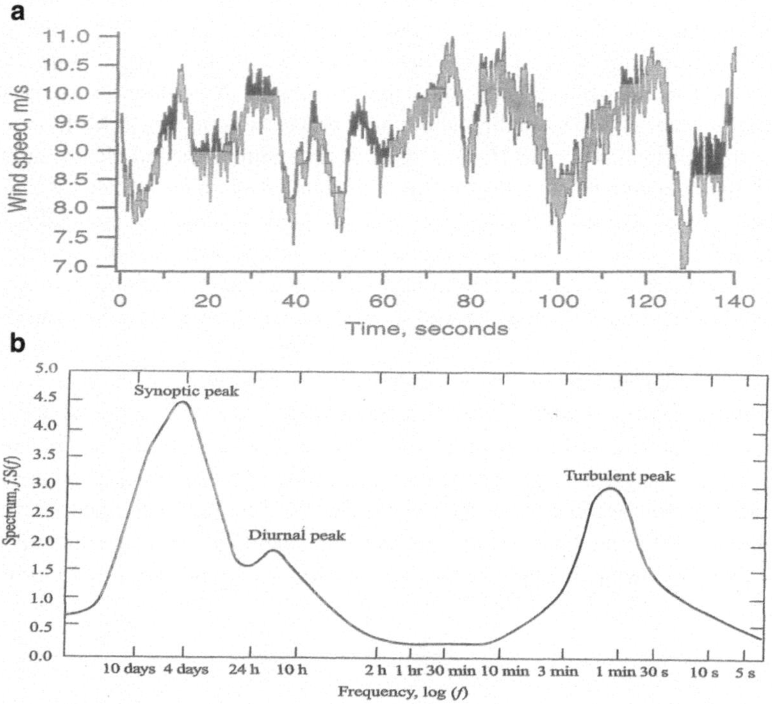

Fig. 15.9 (**a**) Wind time series data. (**b**) Wind frequency spectrum

night before it is full. A wind generator, on the other hand, needs to be able to sell its power to a store or elsewhere whenever the wind blows, day or night, and often for many days on end. As already noted, a store with a much larger capacity (say 10 days) would be prohibitively (20 times more) expensive. Contrary to frequent misunderstandings, the installation of storage cannot be justified on a power system to store wind energy.

15.9 Sources of Confusion

It is frequently asserted that the variability of wind makes power system energy storage essential and often batteries and other systems, such as capacitor banks, are proposed for this purpose. This is a misuse of terminology. What can be needed with high penetrations of wind power is not bulk energy storage but short term balancing and control for some additional power system fluctuations which only become significant for wind capacity penetrations well above 30%.

There are several sources of confusion here. These other types of plant (batteries, etc.) can perform their control tasks without having to cope with large-scale energy storage. Much shorter timescales are involved: just while other plant is brought up to speed or equipment is reconnected. Such control functions are quite distinct from that of bulk energy storage and this misuse of terminology can be misleading.

The other way round, pumped water systems that are installed for bulk energy storage purposes are also widely used to provide power system balancing and control services. Their rapid response times are very suitable for providing spinning reserve and other essential power system control features. Dinorwig can go from full pumping (1.6 GW) to full generation (1.8 GW) in about 2 min, a swing of 3.4 GW which is well able to cope with the loss of two large nuclear plants on the UK power system almost simultaneously—as has happened!

Like most other electrical engineering components and systems (such as capacitors, inductors and transmission lines), the physical processes underlying system balancing and control involve the physical process of energy storage. It is misleading however to call such control equipment "power system energy storage equipment", just as it would be misleading to describe a school bus as a "fuel transportation system". Fuel is certainly transported as an essential part of the operation of a school bus but fuel transportation is not the aim or intention and to describe it so would be wholly misleading in an engineering context.

Batteries and other control devices can indeed be used for energy storage applications but not on a power system, only on a much smaller scale, notably for transport and electronic applications. They are far too expensive to be used for bulk power system energy storage.

15.10 Wind Is the Last to Be Stored

Although the installation of storage cannot be justified in conjunction with wind, storage can have a useful role to play on any power system and it will often be in place regardless of any wind power that may also be place. Pumped water storage systems have been in place for more than a century whereas it is only relatively recently that significant amounts of wind power generation have been installed. It might be thought that, when electricity is being put into a store, typically during base load periods, and the wind happens to be blowing at the same time, wind is as likely as any other source of power to be stored or perhaps even more likely. More careful consideration shows that this is not the case.

When many different types of power plant are online, to discover which source of energy is being stored at any particular time, we need to look at the state of the power system with and without storage. The difference between the two tells us which power is being stored. The power plant which would need to be switched off if storage ceased (because the store failed or became full) would be the power plant that was filling the store at that particular time. In a well-run power system, which always uses the cheapest power available, the plant to be switched off would be the dearest generation on the system at the time. But "the wind blows free" and has zero marginal cost so wind will always be amongst the cheapest plant on line and never the dearest (with one or two provisos), unless the cheapest generation is also the dearest, that is to say, wind is the only generation left on the system. In that case, it can be said that wind is the last to be stored.

There are two provisos under which some plant may have less than zero marginal cost and be prepared to pay to stay online. Firstly a Combined Heat and Power system (CHP) in the small hours of a cold winter night may have periods when heat is needed but no electric lights and it can't produce one without the other. Then it will be prepared to pay from the sale of heat to be allowed to continue operating, generating electricity.

Another situation is when high performance thermodynamic cycles have complex welds to withstand high temperatures and these welds can suffer from thermal fatigue, with cracks developing when they are subjected to temperature cycling. Such plant will not want to be switched on and off and may be prepared to pay to stay online. Nuclear plant is not particularly high performance as far as temperatures go but thermal fatigue of its welds can be substantially increased by radiation damage, so it will often need to stay online and be prepared to pay to do so to avoid thermal cycling. With these provisos it can be said that wind is the last to be stored.

15.11 Conclusions

1. Despite continuing research and development activities, **pumped water** is still the only system of bulk energy storage being installed on power systems today.

2. Energy storage is economically justified by **arbitrage** between cheap (night-time) and more expensive (daytime) electricity.
3. Pumped water storage plant can also provide many **control and balancing services such as spinning reserve** but these are quite distinct from storage. Batteries, capacitor banks and other plant can also offer such control and balancing functions but it is misleading and confusing to claim that they are "for power system energy storage".
4. With no sunshine at night, **pv solar power cannot be stored economically on a power system**.
5. Contrary to frequent misunderstandings, the installation of storage cannot be justified on a power system to store wind energy either. **Storage and wind power are contractually incompatible**.
6. **Wind power is to save fuel** and to store wind power would waste fuel on round-trip losses. There is **no arbitrage between day and night for fuel** as there is for electricity, so **storage cannot be economic for a fuel saver**.
7. Wind only blows for about one third of the time on average, so **it will take three times as long and cost three times as much to store only wind** compared with any continuously available type of power, such as nuclear.
8. The wind frequently blows for up to 10 days at a time (and frequently ceases for up to 10 days). To provide such storage capacity **for wind, a store would cost 20 times as much** as current night/day energy storage costs.
9. If storage is already installed with various other types of generation, normally **wind is the last to be stored** on a well-run power system.

References

1. A. Ter-Gazarian, Energy storage for power systems, in *IEE Energy Series*, ed. by D. T. Swift-Hook, vol. 6, (Peter Peregrinus, London, 1994)
2. D.T. Swift-Hook, Firm power from the wind, in *Wind Energy Conversion*, ed. by J. M. Galt, (MEP, London, 1987), p. 33
3. D.J. Milborrow, Is wind power reliable? UK: New Power (1), 40 (2009)
4. L. van der Hoven, Power spectrum of horizontal wind speed in the frequency range from 0.0007 to 900 cycles per hour. J. Meteorol. **14**, 160 (1957)
5. D.T. Swift-Hook, Wind energy really is the last to be stored and solar energy cannot be stored economically. Renew. Energy **50**, 971–976 (2013)

Chapter 16
Wind Energy Economics

David Milborrow

16.1 Introduction

There is no single answer to the question, "What is the price of wind energy," nor to the question, "Is wind economic?" The answer depends on the local wind regime, the price of competing fuels, and institutional factors. The price of wind-generated electricity needs to be set in context and compared with the prices from other renewable energy sources, and from the thermal sources of electricity generation.

With these caveats, 2018 was a turning point for wind energy. Several respected authorities concluded that the price of wind energy was comparable with those of the thermal sources and so wind energy could be viewed as a "mainstream" generation source. This view was underscored by the fact that a number of projects secured contracts in Germany and elsewhere that did not require any subsidies. So, although most wind energy developments still attract financial support—mostly in the form premium payments for the energy, but occasionally in the form of capital subsidies—the gap between the generation costs of wind energy and of electricity from conventional thermal plant has narrowed or disappeared completely. The situation, however, depends on the geographical location. In areas where good wind speeds can be found, wind is likely to be competitive—unless coal or gas is cheap. Exactly what is meant by "good wind speeds" and "cheap coal and gas" is explored later.

16.2 The Cost of Electricity from Wind Energy

The majority of energy cost figures in this chapter are quoted in \$/MWh and, strictly speaking, are energy prices, inasmuch as they invariably include an element of profit. They are calculated using the "levelised cost concept" which is the "cost of

D. Milborrow (✉)
Consultant, Lewes, East Sussex, UK

A. Sayigh, D. Milborrow (eds.), *The Age of Wind Energy*, Innovative Renewable Energy, https://doi.org/10.1007/978-3-030-26446-8_16

production of one unit (kWh) levelised over the wind power station's entire lifetime." [1]. The concept is a standard one and is used for assessments of other electricity-generation technologies. With the thermal sources of energy, the only difference is that the cost of the fuel must be taken into account in the calculation.

There are three parameters associated with wind turbine projects that are key to the calculation of electricity generating costs:

- The installed (capital) cost of the project, usually quoted in terms of the cost per installed kilowatt, in $/kW, €/kW or £/kW. The installed cost includes charges for any preparatory activities, such as Environmental Impact Statements and other costs associated with securing planning permission. The installed cost, broadly speaking, is primarily a function of the size of the installation (due to economies of scale). Large wind turbines tend to be cheaper than small wind turbines and large wind farms cheaper than small wind farms.
- Operating costs: These are sometimes expressed in $/kW/year, sometimes in $/MWh. The full spectrum of operating costs includes not just the operation and maintenance costs of the wind farm, but also such factors as local taxes and grid connection and usage charges. As it is not always clear whether the latter are included, this can invalidate comparisons between wind farms in different locations.
- Site wind speed: The higher the wind speed, the greater the energy productivity—and so the lower the generation cost. The usual measure of productivity is the capacity factor (average power/rated power). This is not a totally reliable parameter, as different wind turbines yield different capacity factors, even if the wind speed is the same. This point is dealt with in Chap. 2.

16.2.1 Institutional Factors

The funding regime in the state where the wind energy project is located largely dictates the basis of the financial calculations. What is loosely termed the "interest rate" varies considerably. Strictly speaking, it is the "Weighted Average Cost of Capital" (WACC), as it depends on the interest charged on any loan and the return on the equity contribution. For state-funded projects that draw funding directly from governments, the WACC is the rate of return demanded by the Exchequer, sometimes termed the Test Discount Rate.

Values of WACC, in a recent analysis carried out by Grant Thornton [2], vary between 5% and 9.75%, depending on the technology and whether projects are financed wholly by the developer (unlevered), or include equity contributions (levered). An earlier assessment (2011) carried out for the International Energy Agency, found rates for levered projects varied from 4.7 and 7.5% [3]. The highest interest rate applied in the USA and the Grant Thornton report—details of which are shown in Table 16.1—also found that interest rates there were amongst the highest.

Table 16.1 Interest rates for renewable technologies from the Grant Thornton report for levered and unlevered projects

	Hydro		PV		Onshore wind		Offshore wind	
	Levered	Unlevered	Levered	Unlevered	Levered	Unlevered	Levered	Unlevered
Australia	11	8.5	9.5	6.75	10	7.5		
Canada	6	4.75	6.25	5.25	7	6	8.5	7.5
France	6.25	5.5	6.5	5.5	7.25	5.75	9.75	8.25
Germany			5.5	4.25	6.5	5	8	6.25
Ireland	8.25	7	7.75	6.25	8.5	6.5	9.5	8.25
Italy	6.25	5.25	8	6.25	9	7.25		
Nordics	5.75	5	7.25	6.25	7.25	5.75	8.25	7
Spain			7.75	7	8.5	7.25		
UK	7.75	6.5	7.25	6	8.25	6.75	9	7.75
USA	7.5	5.75	8.5	6.75	8.5	6.75	9.25	7.5
Average	7.34	6.03	7.43	6.03	8.08	6.45	8.89	7.50

The data in Table 16.1 reflect the perceptions of risk that attach to each of the technologies. Hydro, being long established, attracts the lowest interest rates, although there is very little difference between Hydro and PV. Onshore wind is seen as slightly more risky than PV and offshore wind as more risky again. It is possible, of course, to make generation cost comparisons using a uniform interest rate, but it is considered preferable to reflect the "real world" economics, with interest rates dependent on the particular technology.

Amortisation periods, like discount rates, vary widely, and are not necessarily as long as the expected life of the plant. The latter represents an upper limit, and is rarely used outside public sector utilities. The amortisation period is frequently governed by the nature of the support mechanism and varies between 12 and 20 years.

To illustrate how the various factors (technical and institutional) combine to yield a wide range of generation cost estimates for onshore wind, Ref. [3] gives figures for typical installations that vary from $95/MWh (Denmark) to $167/MWh (Switzerland). The latter figure is high because the assumption about installed costs was high (€1790/kW) and the productivity was low (2000 kWh/kW).

16.3 Installed Costs

16.3.1 Historical Trends: Large Turbines, Onshore

In a study for the International Energy Agency [4], the authors suggested that the levelised cost (LCOE) of onshore wind energy in Denmark fell from $250/MWh in 1981 to about $60/MWh in 1999. In the USA, the LCOE fell from around $170/MWh in 1991 to $60/MWh in 2003. After the turn of the century, however, partly due to increased commodity prices (particularly steel and copper) wind turbine prices rose, followed by installed costs. That trend reversed from around 2008, and costs are now on a downward trend. Recent trends for wind turbine and project costs are shown in Fig. 16.1.

The latest cost estimates for wind turbines and wind projects in the USA average around $850/kW and $1610/kW, respectively [5]. Data from a European manufacturer suggests turbines prices there are almost identical.

16.3.2 Regional and Size Variations

The data shown in Fig. 16.1 are averages and most projects are utility-scale wind farms in the size range 20 MW and upwards. There are wide variations in installed costs and one of the key factors that influence these is project size. Figure 16.2 shows how these fall from over $3000/kW, for small projects less than 5 MW in size, down to $1549/kW for large projects of 200 MW installed capacity and upwards.

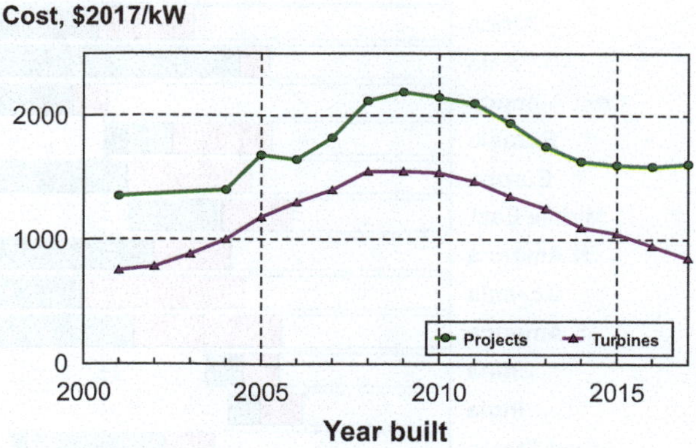

Fig. 16.1 Installed costs of onshore wind projects, and wind turbine prices in the USA, 2001–2017

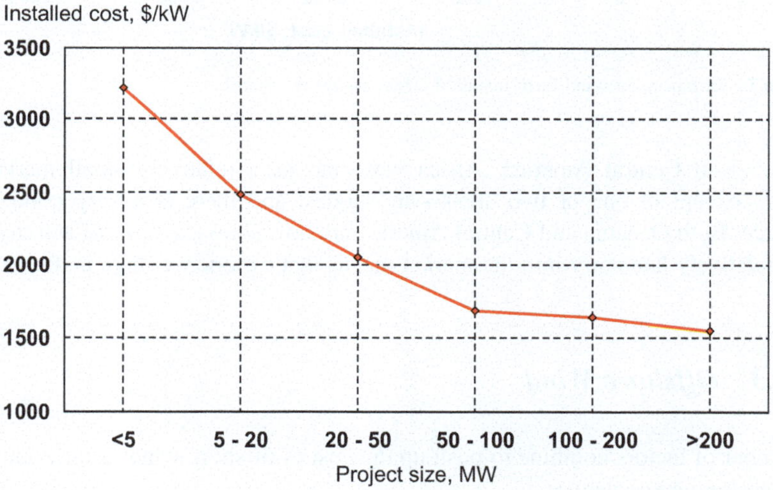

Fig. 16.2 Variations in project costs as a function of the size of the project. Data drawn from the Berkeley Laboratories report, as used for Fig. 16.1, covering 91 projects with a total capacity of 12,410 MW

There are also variations about the mean, depending on the degree of difficulty in building the wind farm, with remote hilltop sites costing more to develop than lowland sites. In addition, national factors influence prices. Although prices are broadly similar in Europe and the USA, wind turbine prices are significantly cheaper in China and India, with major Chinese manufacturers reporting selling prices for their turbines around $570/kW. This is about two-thirds of the selling price in the USA, and installed costs of wind farms are also about two-thirds of the American level. Figure 16.3 shows how installed costs vary around the world [6]. The cheapest figure comes from India, at $850/kW, below the Chinese price of $989/kW. India and China reported the lowest average figures, at $1097/kW and $1197/kW, respectively. The highest averages-all over $2000/kW—were reported from Africa,

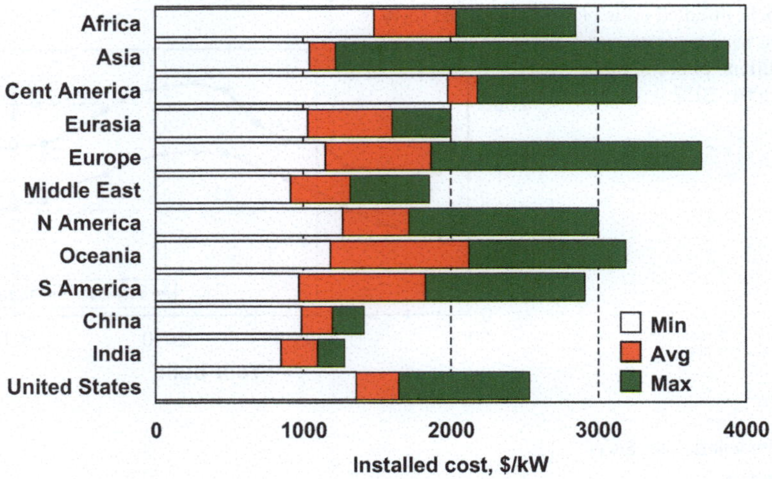

Fig. 16.3 Variations in wind farm installed costs around the world

Oceania and Central America. Africa was, and is, a relatively small market as winds—except in one or two areas—are modest and there is a very good solar resource. Both Oceania and Central America are now showing signs of activity, but their relatively late entry into the market means that experience there is limited.

16.3.3 Offshore Wind

A number of factors combine to push up the cost of offshore wind farms relative to their onshore equivalents:

- The extra time and cost associated with construction works.
- The cost of the cable connection from the wind farm to the shore.
- The need for more expensive foundations, where a number of options have been examined.
  - Gravity-based structures, simple, but heavy.
  - Piled structures.
  - Tethered, floating structures, which support individual turbines or groups.
- Operation and maintenance costs are increased with the risk of lower availability due to difficulties in obtaining access to the wind turbines during bad weather.
- The need to "marinise" the wind turbines, to protect them from the corrosive influence of salt spray.

As a consequence of these extra costs (in comparison with onshore) wind turbines typically account for just under half the installed cost and a typical breakdown is shown in Fig. 16.4. Onshore, the turbines account for 60–70% of the total installed cost and a typical breakdown is shown in Fig. 16.5.

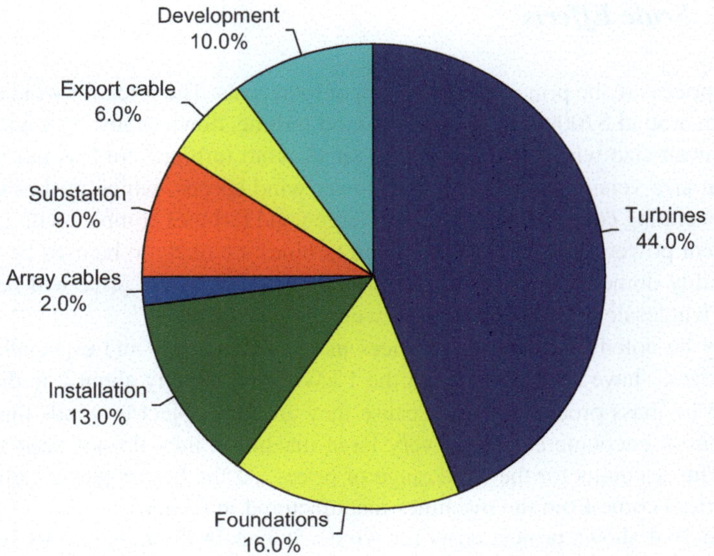

Fig. 16.4 Cost components of a typical offshore wind farm (Source: Danish Energy Agency)

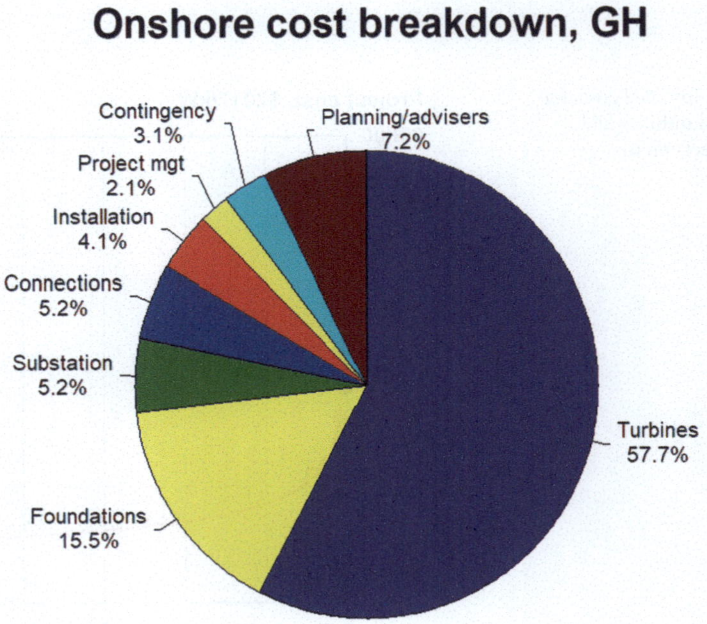

Fig. 16.5 Cost components of a typical onshore wind farm [7]

16.3.4 Scale Effects

Turbine prices are the principal component of total costs. The prices of wind turbines vary from around $7000/kW for a 2 kW wind turbine, down to just over $1000/kW for megawatt-size wind turbines. As the small wind turbines are less tall than the megawatt size versions, they encounter lower wind speeds, which pushes up electricity generating costs further. However, large wind turbines compete with gas, coal and nuclear power stations, but small wind turbines are likely to be used by farmers and possibly domestic consumers, and so the target electricity price will be higher than the wholesale price on the grid network.

It may be noted that the quoted prices in all size ranges—and especially at the smaller sizes—have a wide range. At the 1 kW size (typically about 2 m diameter) units can be mass produced and, because they are not subject to winds that are as high as those encountered by the very large machines, they do not need to be as robust. This accounts for the wide range of prices. At the higher power ratings, the lowest prices come from the machines manufactured in China.

Figure 16.4 shows project costs for wind schemes in the range up to 1000 kW [8]. These mostly comprise single wind turbines and it may be noted that the costs fall towards the figure of $3500/kW shown in Fig. 16.2 for small wind farms. There is a dearth of information for projects in the gap between 100 kW and 5 MW; even if it were available it is likely that there would be considerable scatter (Fig. 16.6).

Fig. 16.6 Installed costs for small wind turbines and wind projects up to 1000 kW

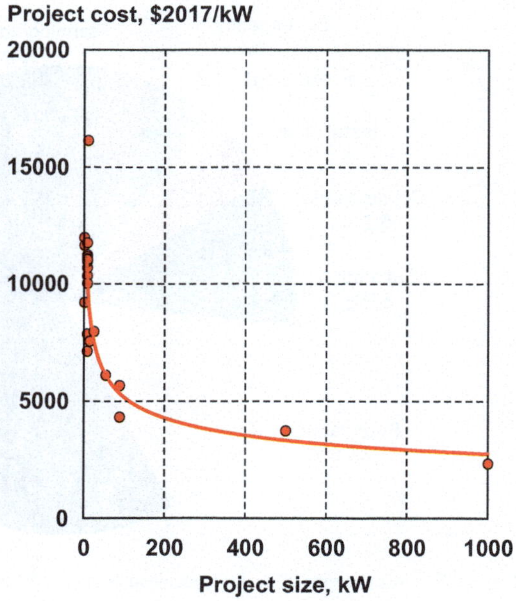

16.4 Operation and Maintenance Costs

Operation and maintenance costs vary with wind farm size and location. There is an additional complication, inasmuch as the exact scope is not always clarified. Routine maintenance and the operational costs are only one part of the total cost of running a wind farm. Unless the operator owns the land, rents will be payable, together with business rates and "use of system" charges to the utility network to which the wind farm is connected. The total costs are therefore very site-specific.

O & M costs for onshore wind vary from \$40/kW/year for small machines (<10 kW) to \$31/kW/year for machines in the range 100–1000 kW [9] but a later estimate from the US Department of Energy quotes \$47.5/kW/year for utility-scale projects [10]. The Fraunhoer Institute [11] quotes €30 (\$35)/kW/year plus €5 (\$6)/MWh, which is very similar. The precise magnitude and the form of the operation and maintenance costs depends very much on the institutional framework, particularly in the case of business rates and taxes.

Operation and maintenance costs for offshore wind are quoted as \$115/kW/year plus \$6/MWh by the Fraunhofer Institute and \$79/kW/year by the US Department of Energy.

In practice, operation and maintenance costs vary from year to year, depending on the amount of unscheduled maintenance that is required. Unforeseen costs can be mitigated by warranties—but at higher cost. Operation and maintenance costs also tend to increase with age. An analysis by IHSMarkit [12] found that average costs for 5-to 10-year-old wind turbines were in the range \$45-50/kW/year but for 10–15-year-old wind turbines the figure rose to \$50–60/kW/year. A similar increase is reflected in an analysis by Fischer [13] that suggested the operation and maintenance cost for old onshore wind turbines was virtually the same as that for offshore turbines. These data are shown in Fig. 16.7. This showed the maintenance costs for new onshore wind turbines as around \$13/MWh, whereas the figure for old wind turbines and offshore machines was around \$28/MWh. However, it must be emphasised that other authorities give different estimates—some higher, some lower.

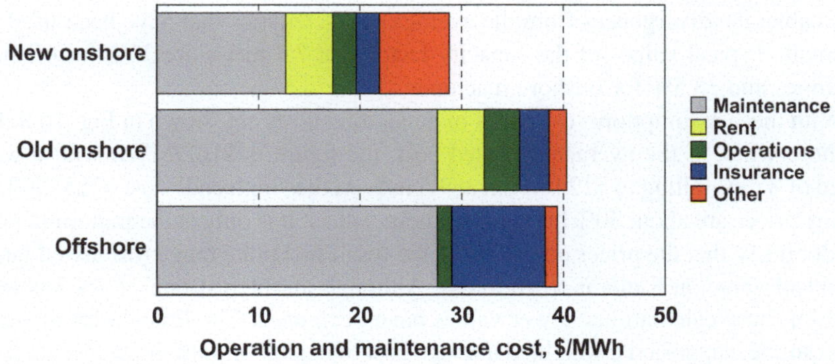

Fig. 16.7 Operation and maintenance costs for onshore and offshore wind turbines

16.5 Generation Costs

16.5.1 Key Figures for Calculating Levelised Costs of Energy

Capital costs are the most important parameters in the case of the renewable energy sources. For onshore wind, the "average of the averages" plotted in Fig. 16.5 is \$1650/kW and one standard deviation either side, using round figures, suggests an upper bound of \$2000/kW and a lower bound of \$1300/kW. The weighted average cost of capital for onshore wind onshore wind has been set at 6% and the amortisation period at 20 years. This is the duration of several "contracts for difference" awards, although a few run for shorter periods (15 years in the UK), and one recently announced contract will run for 29 years. This is believed to be the world's longest contract for wind power. In July, Norsk Hydro, a Norwegian aluminium company signed the contract with Green Investment Group and will build a 235 MW wind farm in Sweden. Fewer data are available for offshore wind, where costs are evolving rapidly, but, following a similar procedure, the overall average can be set at 3900/kW. The minimum figure is 2200/kW, based on a recent paper published by the Yale School of Forestry and Environmental Studies [14]. The upper bound can be set at 5600/kW. These figures are difficult to quantify, partly because of the shortage of data and partly because averages are skewed by the high prices that came, until recently, from the relatively immature American market. The National Renewable Energy Laboratory's Annual Technology Baseline quotes installed costs for floating wind turbines which range up to \$8200/kW, but these have not been taken into account. The weighted average cost of capital for offshore wind has been set at 7%, and the amortisation period, as for onshore wind, at 20 years. In order to establish a link between sites wind speeds and costs of electricity, it was necessary to make assumptions about the performance of typical wind turbines. The energy productivity of a number of commercial designs was therefore examined so as to establish a link between wind speed and capacity factor. As offshore wind turbines tend to have higher ratings, per unit of swept area, than onshore wind turbines, different relationships were used for offshore and onshore. Although these links between capacity factor and wind speed will not be universally applicable, the divergences from the average characteristics that have been used are be small. Typical values of the capacity factor – at 7.5 m/s – are 30% for offshore machines and 38.5% for onshore machines.

With these assumptions, estimates of generating costs are shown in Fig. 16.8. For offshore wind, at the average installed cost, the figure is \$162/MWh with a wind speed of 8 m/s, falling to \$124/MWh at 9.5 m/s. At the "high end" cost of \$5150/kW, energy prices are about 30% higher than these values. It is only at the minimum price of \$2650/kW that the prices come within the fossil fuel price range (discussed later). A typical value, at 8 m/s is \$116/MWh. Although an interest rate of 7% has been used for these calculations, lower values have been used. The Fraunhofer Institute, for example, suggested a weighted average cost of capital of 4.8% in its recent report [11] and that would bring the electricity cost, at \$2650/kW and 8 m/s, down to around \$83/MWh.

16.6 Comparative Generation Costs

The information on wind energy generation costs plotted in Fig. 16.8 is of little value in isolation and needs to be compared with the corresponding information on the generation costs of other renewables and of the fossil fuels. As with wind, there is no single value of electricity generation cost that applies everywhere. Construction costs of all plant vary from region to region and, in the case of the fossil fuel sources, there are wide variations in the costs of gas and coal. Broadly speaking, fuel prices tend to be low in the USA, but higher in Europe, although there are significant variations. Nuclear, like wind, is capital intensive and there are wide variations in estimates of installed costs around the world.

Appropriate levels of wind speed also need to be taken into account when making comparisons. In the case of onshore wind, the highest wind speeds are to be found in most coastal zones and hilltop sites.

As noted earlier, the other factor that has a strong influence on the quoted generation costs of renewable energy technologies is the Weighted Average Cost of Capital. In the USA and the UK, the value is typically around 9–10%, but in mainland Europe, it can be significantly lower. It may be noted that the relevant levels may not always be quoted when estimates of electricity generation costs are specified. In the case of coal and gas—especially the latter—the absence of this information is not crucial as capital cost repayments tend to account for a relatively modest proportion of the total cost.

Figure 16.9 shows data from recent analyses. The primary source of data comes from the USA [15], which accounts for the minimum values, but account has been

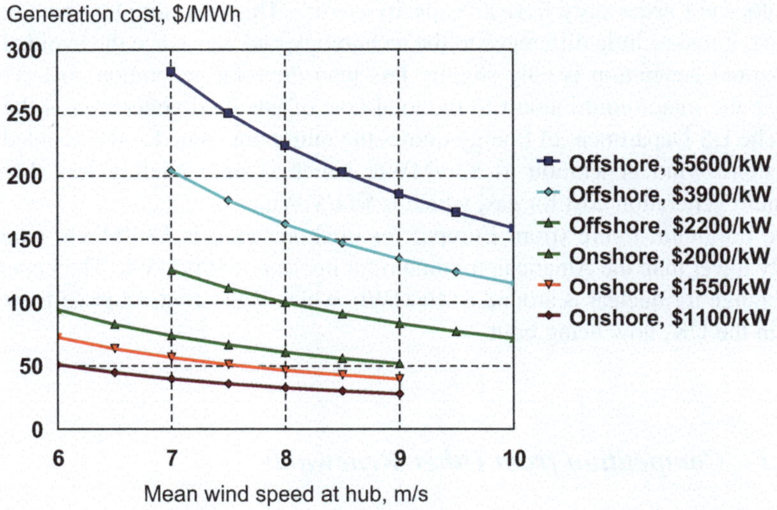

Fig. 16.8 Estimates of electricity generating costs for offshore and onshore wind, for a range of wind speeds. It is unlikely that any offshore projects would be built in areas with wind speeds below 7 m/s and there are only a few onshore sites with wind speeds greater than 9 m/s

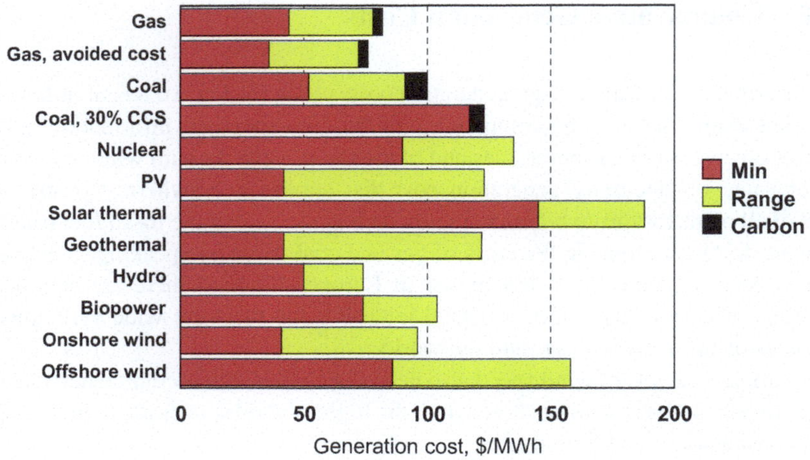

Fig. 16.9 Electricity generating costs for fossil fuel sources and renewables in the USA and Europe

taken of European data [11], where fuel costs are generally higher. Both these authorities quote ranges for wind prices, but a more comprehensive picture is given in Fig. 16.8. The range of generation costs for onshore wind is around $41–96/MWh, making it the cheapest renewable technology; gas is sometimes cheaper in the USA, although the minimum figure quoted is $44/MWh; coal is somewhat more expensive. It is sometimes argued that generation costs for wind should be compared with the avoided cost of the electricity that is displaced, because wind does not necessarily have a "capacity credit". This is a complex issue but, in practice, it makes little difference to the technology rankings since the avoided cost of gas-fired generation is only slightly less than the total generation cost. This is because the major contribution to the total cost of gas-fired generation is the fuel cost. The US Department of Energy quotes the minimum value for the avoided cost of onshore wind generation as $36/MWh, which is only slightly less than the minimum generation cost for gas, which is $44/MWh.

The minimum figure (from Europe) for offshore wind is $86/MWh, which is slightly lower than the American minimum for nuclear of $90/MWh. The upper end of the range for nuclear is around $146/MWh, which is the contract price for a new plant in the UK, now being built.

16.6.1 Competition from Other Renewables

Although there are places, particularly in Asia and China, where hydro projects can be developed to deliver electricity at low prices, Fig. 16.9 suggests that wind is now, for the most part, the cheapest of the renewable energy sources. The American

Department of Energy suggests the minimum price for wind ($41/MWh) is now very slightly lower than the corresponding price for utility-scale solar photovoltaics. The close similarity between the minimum PV and wind prices suggests that solar is now becoming increasingly competitive with wind. It is an oversimplification to suggest that PV or wind is cheaper as the location is crucial in each case. Some of the best wind resources are to be found in the centre of the USA, such as the Rocky Mountains, the northeast coast, and the southern tip of South America. Across the Atlantic, the best resources are in Northern Europe, the West coast of north Africa, the Horn of Africa, the southern tip of Africa, and South West Australia. Solar resources are not, as might be expected, evenly distributed either side of the equator and the best resources are in the southern USA and Mexico, most of the coastal regions of Chile, most of Africa with the exception of a region around Zaire and the Congo, and northern Australia. The only regions where there are significant overlaps where there are good wind and PV resources are the West coast of Australia and the Horn of Africa.

16.6.2 Levelling the Playing Field

It is sometimes suggested that generation cost comparisons, of the type discussed in the previous paragraph, is not an equitable way of assessing the competitiveness of variable renewable energy sources, such as wind and photovoltaics. Account should be taken of the extra costs incurred by system operators in providing resources to deal with uncertainties in the balance between demand and generation. This is a fair point, but it opens up a wider discussion on the so-called External Costs of electricity generation sources.

Although there is general agreement as to the broad definition of external costs—costs attributable to an activity that are not borne by the party involved in that activity—there are widespread variations in defining the boundaries. There is an argument, for example, that a substantial proportion of Western defence budgets should be regarded as an "external cost" of oil—for fairly obvious reasons.

External costs—or at least some—may be difficult to quantify, but they are real. If the enormous costs of the clean-up operation after the Chernobyl nuclear disaster had been taken into account when the plant was constructed, it is unlikely it would have been built. The task facing energy policy makers is how best to go about the job of reducing pollution in electricity generation when, in most countries, external costs are not reflected in the market price of electricity. This has led to the concept of "carbon prices" which are now widely used.

External costs may be positive or negative and in the early days of wind energy development, one bonus for wind energy (and other small-scale renewables) was that many were connected into distribution networks. They therefore delivered electricity closer to the consumer than centralised generation and thereby avoided the costs of the grid network. That argument still applies, although, with the increasing size of wind energy developments it is less cogent and therefore will not be pursued. The principal components of external costs are the following:

- Subsidies paid to electricity generation from all sources.
- The damage caused to buildings, ecosystems and human health from harmful emissions from the fossil fuel sources.
- The costs of global warming, due to the emissions of carbon dioxide from gas and coal-fired power stations.
- Integration costs for the variable renewable energy sources.

Each of these issues has been investigated in some detail by various authorities. They are all location-dependent and the second and third items are difficult to quantify.

16.6.2.1 Subsidies

Subsidies have been paid to many renewable energy sources for some time, and it is argued that it is cheaper for the electricity consumer if modest subsidies are provided to the renewables, rather than attempt to apply the very significant external costs of damage and climate change to the fossil fuel sources. There is a wide range of schemes, although in recent years, rather than give premium payments to renewables, they have been invited to bid in competitive auctions, with successful bidders being awarded fixed prices for periods between 10 and 20 years. More recently still, some auctions for wind (onshore and offshore) have attracted bids that requested no premium, demonstrating the fact that wind is now commercially viable, without subsidy.

16.6.2.2 Costs of Damage and Global Warming

The costs of damage from the pollutants emitted from fossil-fuel power stations and of damage due to carbon dioxide emissions are a complex subject, and the results are inevitably subject to some uncertainty. Nevertheless a number of studies have been carried out and work continues in this field. An authoritative report commissioned by the UK Treasury [16] on the consequences of climate change concluded "Preliminary calculations . . . suggest that the social cost of carbon today is of the order of $85 per tonne of CO_2" . . ."This number is well above marginal abatement costs in many sectors". It may be noted that $85/t$CO_2$ adds $77/MWh to coal-fired gencosts, ~$34/MWh to gas.

A report that looked at subsidies and external costs of electricity generation within the European Union was published in 2014 [17] and the results are shown in Fig. 16.10.

It may be noted that the generation cost estimates for wind and solar PV are now out of date, but the other figures are still reasonably accurate. The apparent absence of subsidies for nuclear is due to the fact that not all the European states support a

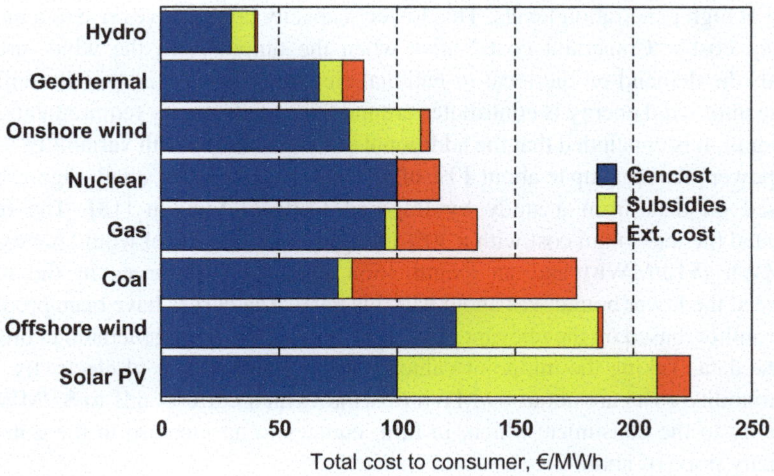

Fig. 16.10 Subsidies and costs of electricity generation within the European Union

nuclear programme and so the subsidies on a €/MWh basis—averaged across all 28 states—are small. However, external costs related to negative environmental impacts of nuclear power are estimated to be in the range €18–22(2012)/MWh.

16.6.2.3 System Integration Costs

As the cost of wind energy falls increasing amounts are likely to be installed and the variability of this power needs to be managed. Although aspects of the management of wind variability can be controversial, utilities the world over generally agree that there is no fundamental technical reason why high proportions of wind energy cannot be assimilated into the system. An understanding of the impacts of the variable sources of renewable energy must take into account the wider issues associated with managing electricity systems. Modern integrated networks are designed to cope with 'shocks' such as the sudden loss of large thermal power stations and with uncertainties in consumer demand. As the tools to deal with these are already available the key question is the extent to which the introduction of large amounts of wind energy will increase the overall uncertainty in matching supply and demand. This extra uncertainty means that additional short-term reserves are needed to guarantee the security of the system.

The costs of additional reserves are one component of "the costs of wind variability". A second is the backup cost and the third is "constraint costs". No special backup provisions need to be made for wind energy. All generating plants make use of a common pool of backup plant that is typically around 20% of the peak demand on the electricity network. When wind is introduced, system operators do not rely on the rated power of all the installed wind farms being available at the times of peak demand, but a lower amount—roughly 10–30% of the rated capacity at low penetration levels, falling at high penetration levels. This lower "capacity credit" gives rise to a modest "backup cost". 'Constraint costs' arise when the output from the wind turbines exceeds the demand on the local or national electricity network. They are unlikely to arise until wind energy is contributes around 25% of electricity requirements.

Overall, it is concluded that the additional costs associated with variability—with wind power providing up to about 40% of all electricity, are quite small. Figure 16.11 is based on data from a study by Imperial College, London [18]. The report suggested the maximum cost with a 30% energy penetration level would not exceed £10/MWh ($13/MWh) and an examination of the costs curves in the report suggested the lower bound was about half this level. The curves have been produced by the author, based on the relevant mathematics and have been generated in order to fit these data. Taking the midpoint value, if wind provides 30% of electricity, then the variability costs are about $9/MWh of wind, which corresponds to $3/MWh of electricity to the consumer, which, in turn, equates to an increase in the domestic electricity price of about 2%.

It is sometimes argued that the introduction of variable renewables into an electricity network makes the provision of storage essential. Whilst this is not necessarily the case, reductions in the cost of storage in recent years may make the proposition more attractive. A recent report from Australia [19] focused on the use of storage, since new electricity market regulations put emphasis on the provision of "firm power". However, this report notes that, *"little or no storage would be required up to 50% variable renewable energy share"* Even with 50–75% variable renewables, only 2–3 h storage is required. These estimates come from a new method of calculation which produces lower estimates than several other studies and suggests the "variability cost" is around $10/MWh with 50% variable renewables.

Further increases in the level of wind penetration beyond 40% are feasible and there are numerous technical innovations at various stages of development that can

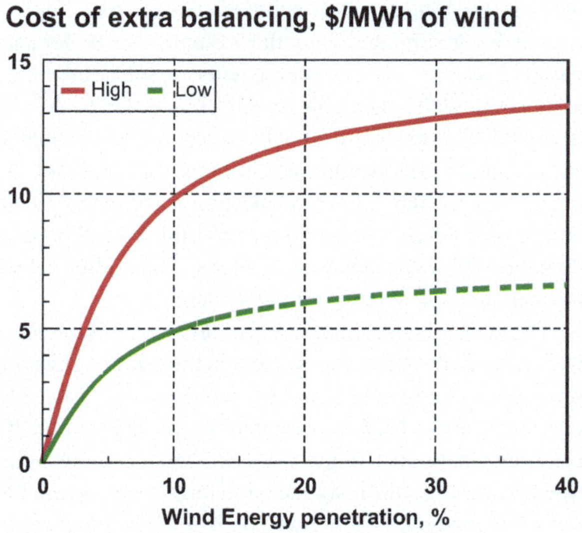

Fig. 16.11 The cost of extra balancing for wind energy

mitigate the costs associated with variability. Improved methods of wind prediction are under development worldwide and could potentially reduce the costs of additional reserve by around 30%. Most other mitigation measures reduce the costs of managing the electricity network as a whole. "Smart grids", for example, cover a range of technologies that may reduce the costs of short-term reserves; additional interconnections with neighbouring utilities would also be beneficial. Electric cars hold out the prospect of reduced emissions for the transport network as a whole and could act as a form of storage for the electricity network—for which the electricity generator would not have to pay.

16.7 Future Costs

There is an general expectation that the cost of wind energy will continue to fall. Although lower capital costs are likely to be the principal determinant, improved productivity and further reductions in interest rates are also expected to play a part. An important study, published in 2016, reported on the views of 163 of the world's foremost wind energy experts, aimed at better understanding future wind energy costs and potential technology advancement [20]. The study sought to gain insight on the possible magnitude of future cost reductions, the sources of those reductions, and the enabling conditions needed to realize continued innovation and lower costs. Wind applications covered by the survey include onshore, fixed-bottom offshore, and floating offshore wind. Key results from this study are shown in Fig. 16.12.

Between 2014 and 2020 the median cost of energy from onshore wind is expected to come down by 10% and then come down by another 16% by 2030. Offshore wind with fixed structures is expected to come down by 10% between 2014 and 2020 and

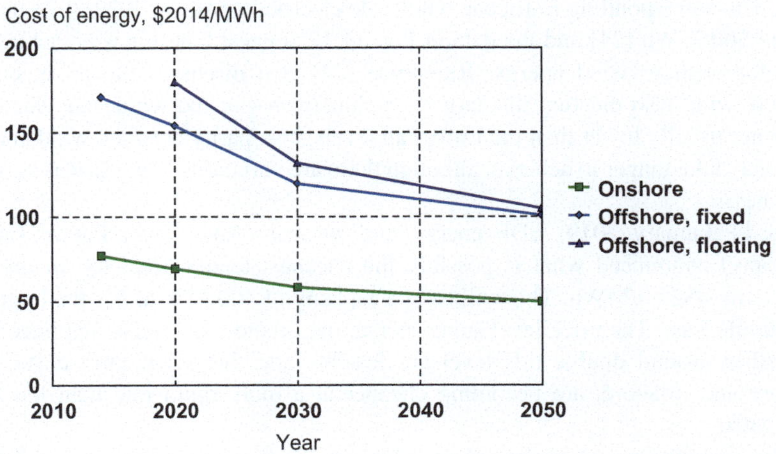

Fig. 16.12 Results from a study that sought the views of wind experts on future cost trends for onshore and offshore wind

to fall to $100/MWh by 2050. These cost reductions are similar to those projected by the Danish Energy Agency (DEA) [21], which expects installed costs of onshore wind to come down by 8% between 2015 and 2020 and then by another 16% by 2050. Given that the DEA also anticipates increases in productivity, the similarity is very close. The DEA expects installed costs for offshore wind to come down by 20% between 2015 and 2020 and to continue falling to 2050, by which time the anticipated installed costs are around $1970/kW. Somewhat lower costs of energy are anticipated by Ram et al. [22], who suggests that the minimum value for both wind and photovoltaics by 2030 is likely to be around $18/MWh. They suggest a minimum value for offshore wind of $74/MWh, which is considerably lower than the median value from the expert study group.

16.7.1 Recent Tender Results

From around 2017, a number of wind farm projects—in Germany, The Netherlands and the UK submitted bids into auctions for fixed prices that would require no subsidy. In other words, they anticipated that they would be profitable, based on revenues from the prices in the wholesale market, alone. These bids came from onshore and offshore projects.

Zero subsidy bids make sense if it is anticipated that the increase in wholesale electricity prices will combine with lower costs for wind projects and result in electricity costs from wind that are lower than wholesale prices. Most commentators expect gas prices to rise in the near future, with the wholesale price of gas in Europe possibly set to rise from around €23/MWh in 2018 to around €36/MWh in 2030 [23]. As the costs of gas-fired generation tend to have a strong influence on wholesale electricity prices, this suggests the latter may be around $75/MWh by 2030. The corresponding European wholesale electricity price in 2022/2023 may be around $60/MWh [24] and the data in Fig. 16.12 suggest that this is an achievable price for onshore wind energy. Reference [24] also discusses the prospects for offshore wind also meeting this target. As American gas and wholesale electricity prices are usually lower than the European levels, cost parity between wind and gas there may take longer to achieve, although there are variations in gas prices between the states.

On 11 January 2019, EDF energy and Masdar (Abu Dhabi Future Energy Company) announced what is possibly the cheapest tender price for an onshore wind farm—$21.3/MWh. This will have a capacity of 400 MW and is the largest in the Middle East. The price level suggests that the offshore contracts that have been secured at around double this level are feasible and that wind farm costs, both onshore and offshore, are becoming cheaper at a more rapid rate than has been anticipated.

This assumption was confirmed on 9 January 2019, when details of an 800 MW American offshore wind farm were announced [25]. The contract will run for 20 years and the price is $85/MWh. This suggests that offshore generation costs below $100/MWh are feasible and that the American offshore market had closed the gap with the European market.

16.8 Conclusions

The combination of rising fuel prices and falling wind prices has created a "virtuous circle" and means that the rate of development of wind energy may accelerate. Another factor that may lead to further cost of energy reductions is the likelihood that the financial terms may become more favourable, with lower interest rates both for the debt and equity contributions. Further small reductions in the cost of wind energy may come about from the adoption of 25-year contract periods, rather than the 20 years that is now common. Photovoltaics are likely to provide increasing competition to wind energy, but for the most part these systems are likely to be complementary.

Although the contribution of variable renewable energy sources to electricity networks is limited, that ceiling is at least 50% (on an energy basis) and possibly more. The modest extra costs of providing additional reserves to deal with the variability do not negate the competitive position of wind, relative to the fossil fuels, and, freed from subsidies, the growth of wind is likely to be strong.

References

1. O. Tande, Estimation of cost of energy from wind energy conversion systems. Expert group study on recommended practices for wind turbine testing and evaluation. Submitted to Executive Committee of the International Energy Agency Program for Research and Development on Wind Energy Conversion Systems, 1994
2. Grant Thornton and Clean Energy Pipeline Initiative, 2018. Renewable energy discount rate survey results—2017
3. P. Schwabe, S. Lensink, *IEA Wind Task 26: Multi-national Case Study of the Financial Cost of Wind Energy* (National Renewable Energy Laboratory, Lakewood, 2011). NREL/TP-6A2-48155
4. E. Lanz, R. Wiser, M. Hand, IEA Wind Task 26, The Past and Future Cost of Wind Energy (2012). NREL/TP-6A20-53510
5. R. Wiser, M. Bolinger, *2017 Wind Technologies Market Report* (US Department of Energy, Lawrence Berkeley National Laboratory, Berkeley, 2018)
6. Renewable Energy Policy Network, (REN21), *Renewables 2018 Global Status Report* (REN21, Paris, 2018)
7. J. Holt, D. Eaton, *Assessment of the Energy Potential and Estimated Costs of Wind Energy in British Columbia* (Garrad Hassan Canada Inc., Ottawa, 2008)
8. Pacific Northwest National Laboratory for the United States Department of Energy, 2018. 2017 Distributed Wind Market Report
9. National Renewable Energy Laboratory, Distributed Generation Renewable Energy Estimate of Costs (2016)

10. US Energy Information Administration, Cost and Performance Characteristics of New Generating Technologies, Annual Energy Outlook 2018 (2018)
11. Fraunhofer Institute for Solar Energy Systems, Levelised cost of electricity renewable energy technologies (2018)
12. North American Wind Power O&M Costs to Increase Nearly 40 Percent in Coming Decade, https://news.ihsmarkit.com/press-release/energy/north-american-wind-power-om-costs-increase-nearly-40-percent-coming-decade-ihs. Accessed 1 Feb 19
13. K. Fischer, Utilising wind-turbine failure and operating data for route-cause analysis, in *Analysis of Operating Wind Farms. Wind Europe Technology Workshop*, Bilbao, 14–15 April 2016
14. R. Drouin, *After an Uncertain Start, U.S. Offshore Wind Is Powering Up* (Yale School of Forestry & Environmental Studies, New Haven, 2018)
15. United States Department of Energy, Energy Information Administration, 2018 Annual Energy Outlook (2018)
16. The Economics of Climate Change, *UK Treasury* (Cambridge University Press, England, 2006)
17. Ecofys, Subsidies and costs of EU energy, Report prepared for the European Commission (2014)
18. R. Gross et al., *The Costs and Impacts of Intermittency—2016 Update* (UK Energy Research Centre, London, 2017)
19. P.W. Graham et al., *GenCost 2018. Updated Projections of Electricity Generation Technology Costs* (Commonwealth Scientific and Industrial Research Organisation, Canberra)
20. R. Wiser et al., *Forecasting Wind Energy Costs and Cost Drivers: The Views of the World's Leading Experts* (National Renewable Energy Laboratory, Lakewood, 2016). LBNL-1005717
21. Danish Energy Agency, (with revisions to Nov 2018). Technology data for energy plants for electricity and district heating generation (2016)
22. R. Ram et al., A comparative analysis of electricity generation costs from renewable, fossil fuel and nuclear sources in G20 countries for the period 2015-2030. J. Clean. Prod. **199**, 687–704 (2018)
23. European Commission, Brussels, *Energy Prices and Costs in Europe. Report from the Commission to the European Parliament, the Council* (The European Economic And Social Committee and the Committee of the Regions, 2019)
24. Institute for Energy Economics and Financial Analysis, Technology gains help drive a rush from capital markets into offshore wind (2017)
25. Institute for Energy Economics and Financial Analysis, A sea change in American offshore wind (2019)

Chapter 17
Conclusions

Ali Sayigh

The overwhelming case for renewable energy, and specifically wind energy, is more apparent than ever before—the planet is on the cusp of irreversible climate change.

Over the last few decades the case has been made by scientists and economists for the replacement of fossil fuels by renewable sources of energy. Technology has brought all renewables into a position whereby they are a viable alternative to fossil fuels. Now we find ourselves in the position where it makes not only economic and technical sense to adopt renewables, but also it is vitally crucial to the future of the planet that all non' renewable sources of energy be phased out completely in order to halt the global temperature rise caused by carbon emissions whether from transport systems or natural emissions caused by the melting of the permafrost and glaciers.

It is evident from all the chapters in this book that wind energy and wind turbines are a major contributor in the reduction of CO_2 emission in the atmosphere. It is reliable, cost-effective, can be applied onshore or offshore and require a relatively small space to operate. Wind energy technology is well matured and has advanced to an industrial level so that a single turbine can represent a power station by itself (10 MW). As David Milborrow, has pointed out in his chapter "Earlier projections from various sources suggested that 500 GW would be reached by 2020, (in fact, it was achieved during 2019). The moderate projection from the Global Wind Energy Council suggested that 1000 GW will be reached by 2025, or shortly after. As the industry is continuing to innovate, and generation costs are still falling, its low carbon credentials should ensure that it remains an attractive choice for electricity generation.

Now that the industry has moved closer to being able to operate free of subsidies, growth is likely to be strong. The basic design concepts are likely to remain unchanged, with three-blade upwind machines being the norm. There are likely to

A. Sayigh (✉)
World Renewable Energy Congress, Brighton, UK
e-mail: asayigh@wrenuk.co.uk

A. Sayigh, D. Milborrow (eds.), *The Age of Wind Energy*, Innovative Renewable Energy, https://doi.org/10.1007/978-3-030-26446-8_17

be further developments in drive train technology and direct-drive machines may become more common. With sizes steadily increasing, 20 MW machines are likely to become established within a few years, but whether sizes will continue to increase beyond that point is more difficult to forecast.

In a review of the various countries mentioned in this book, most agree one of the best sources of electricity generation is the wind turbine. For example in Greece where there are many isolated islands with small communities, wind energy application is ideal. Turbines can also be used as hybrid source of electricity with storage system. Several examples were mentioned by Prof. John Kaldelis in his two chapters.

In Japan following the tragedy of Fukushima both Ministry of Economy, Trade and Industry (METI) and Ministry of Environment (MOE), have utilized wind energy so that by 2017 more than 3400 MW has been installed and produces 5000 GWh. Japan also has the largest floating wind turbine in the World with a capacity of 7 MW, and started operation in 2015. Another floating turbine is Hitachi with 5 MW capacity. This clearly explained by Izumi Yushiyama in his chapter.

Ali Hamzeh and his co-authors describe the importance of using wind energy in Jordan since presently Jordan imports around 97% of its energy and fuel requirements, nearly 10% of the country's Gross Domestic Product (GDP), bearing in mind that some locations in Jordan have a wind speed 7 m/s. The government is planning to have 1000 MW from wind projects by 2020. As of 2018, there are six operational wind farms with a total capacity of 369.5 MW already built and connected to the grid. For more details see their chapter.

In Australia at the end of January 2019, there are 3296.5 MW of wind power installations. The authors dedicate their chapter to the memory of Geoff Hill, who founded Westwind Turbines and was a pioneer and innovator in the age of wind energy in Australia.

The risk analysis of using wind turbines is fully discussed in a chapter which highlights some of the investigations needed to be carried out before wind turbines are installed. Such risk assessments are necessary to counter safety objectives to wind farms. In fact wind energy is a comparably risk free energy source.

Although there is no legislation for advancing the use of wind energy in Argentina, the potential is as large as 200 GW. However, a few wind turbines of 2.5 MW capacity with 80 m hub height have been installed recently. Most wind turbines in Argentina are of low power 10–20 kW, and erected on flat roofs. This is similar to the Gulf region in the Middle East. Both regions, Argentina and the Gulf countries, are yet to start tapping in a meaningful way into this important source of energy.

In North Africa, Morocco has made reasonable progress. In 2018, wind power installation reached 1220 MW, or 19.8% of its national renewable energy power supply. However, the Moroccan national energy strategy aims at renewable energy for 42% of Moroccan electricity by 2020. Wind power share will be supplying 14% of the national electric power capacity (14,580 MW), and this is likely to be realized in 2020.

South Korea has a policy plan, RE 3020, for using renewable energy. An offshore wind farm of 100 MW was recently approved. By 2020 the plan is to have more than

2.5 GW offshore wind power. Also according to the KWEIA (Korea Wind Energy Industry Association), the plan for accumulative installation will be expected to reach 17.7 GW. Likewise, the number of the offshore wind turbines will be increased to the extent of supplying power to about 13 GW or more.

The topic of power storage has been addressed by Swift-Hook and the conclusion has been reached that in the case of wind energy it is better to load the generated electricity into the grid directly or use it in hybrid manner with other sources of renewable energy such as photovoltaic installations in order to avoid unnecessarily costly and complex systems.

Finally the crucial topic of the economics of wind electricity in comparison with other renewable energy sources has been presented by David Milborrow.

Index

A. Sayigh, D. Milborrow (eds.), *The Age of Wind Energy*, Innovative Renewable Energy, https://doi.org/10.1007/978-3-030-26446-8

The manufacturer's authorised representative in the EU is Springer Nature Customer Service Centre GmbH, Europaplatz 3, 69115 Heidelberg, Germany. If you have any concerns regarding our products, please contact ProductSafety@springernature.com

Printed and bound by CPI Group (UK) Ltd, Croydon, CR0 4YY
30/09/2026
02225657-0005